Y0-BCK-962

Anthony Brabazon and Michael O'Neill (Eds.)

Natural Computing in Computational Finance

Studies in Computational Intelligence, Volume 100

Editor-in-chief
Prof. Janusz Kacprzyk
Systems Research Institute
Polish Academy of Sciences
ul. Newelska 6
01-447 Warsaw
Poland
E-mail: kacprzyk@ibspan.waw.pl

Further volumes of this series can be found on our homepage: springer.com

Vol. 78. Costin Badica and Marcin Paprzycki (Eds.)
Intelligent and Distributed Computing, 2008
ISBN 978-3-540-74929-5

Vol. 79. Xing Cai and T.-C. Jim Yeh (Eds.)
Quantitative Information Fusion for Hydrological Sciences, 2008
ISBN 978-3-540-75383-4

Vol. 80. Joachim Diederich
Rule Extraction from Support Vector Machines, 2008
ISBN 978-3-540-75389-6

Vol. 81. K. Sridharan
Robotic Exploration and Landmark Determination, 2008
ISBN 978-3-540-75393-3

Vol. 82. Ajith Abraham, Crina Grosan and Witold Pedrycz (Eds.)
Engineering Evolutionary Intelligent Systems, 2008
ISBN 978-3-540-75395-7

Vol. 83. Bhanu Prasad and S.R.M. Prasanna (Eds.)
Speech, Audio, Image and Biomedical Signal Processing using Neural Networks, 2008
ISBN 978-3-540-75397-1

Vol. 84. Marek R. Ogiela and Ryszard Tadeusiewicz
Modern Computational Intelligence Methods for the Interpretation of Medical Images, 2008
ISBN 978-3-540-75399-5

Vol. 85. Arpad Kelemen, Ajith Abraham and Yulan Liang (Eds.)
Computational Intelligence in Medical Informatics, 2008
ISBN 978-3-540-75766-5

Vol. 86. Zbigniew Les and Mogdalena Les
Shape Understanding Systems, 2008
ISBN 978-3-540-75768-9

Vol. 87. Yuri Avramenko and Andrzej Kraslawski
Case Based Design, 2008
ISBN 978-3-540-75705-4

Vol. 88. Tina Yu, David Davis, Cem Baydar and Rajkumar Roy (Eds.)
Evolutionary Computation in Practice, 2008
ISBN 978-3-540-75770-2

Vol. 89. Ito Takayuki, Hattori Hiromitsu, Zhang Minjie and Matsuo Tokuro (Eds.)
Rational, Robust, Secure, 2008
ISBN 978-3-540-76281-2

Vol. 90. Simone Marinai and Hiromichi Fujisawa (Eds.)
Machine Learning in Document Analysis and Recognition, 2008
ISBN 978-3-540-76279-9

Vol. 91. Horst Bunke, Kandel Abraham and Last Mark (Eds.)
Applied Pattern Recognition, 2008
ISBN 978-3-540-76830-2

Vol. 92. Ang Yang, Yin Shan and Lam Thu Bui (Eds.)
Success in Evolutionary Computation, 2008
ISBN 978-3-540-76285-0

Vol. 93. Manolis Wallace, Marios Angelides and Phivos Mylonas (Eds.)
Advances in Semantic Media Adaptation and Personalization, 2008
ISBN 978-3-540-76359-8

Vol. 94. Arpad Kelemen, Ajith Abraham and Yuehui Chen (Eds.)
Computational Intelligence in Bioinformatics, 2008
ISBN 978-3-540-76802-9

Vol. 95. Radu Dogaru
Systematic Design for Emergence in Cellular Nonlinear Networks, 2008
ISBN 978-3-540-76800-5

Vol. 96. Aboul-Ella Hassanien, Ajith Abraham and Janusz Kacprzyk (Eds.)
Computational Intelligence in Multimedia Processing: Recent Advances, 2008
ISBN 978-3-540-76826-5

Vol. 97. Gloria Phillips-Wren, Nikhil Ichalkaranje and Lakhmi C. Jain (Eds.)
Intelligent Decision Making: An AI-Based Approach, 2008
ISBN 978-3-540-76829-9

Vol. 98. Ashish Ghosh, Satchidananda Dehuri and Susmita Ghosh (Eds.)
Multi-Objective Evolutionary Algorithms for Knowledge Discovery from Databases, 2008
ISBN 978-3-540-77466-2

Vol. 99. George Meghabghab and Abraham Kandel
Search Engines, Link Analysis, and User's Web Behavior, 2008
ISBN 978-3-540-77468-6

Vol. 100. Anthony Brabazon and Michael O'Neill (Eds.)
Natural Computing in Computational Finance, 2008
ISBN 978-3-540-77476-1

Anthony Brabazon
Michael O'Neill
(Eds.)

Natural Computing in Computational Finance

With 79 Figures and 61 Tables

Springer

Dr. Anthony Brabazon
Head of Research - School of Business
Quinn School
University College Dublin
Belfield, Dublin 4
Ireland
anthony.brabazon@ucd.ie

Dr. Michael O'Neill
Director - Natural Computing Research
and Applications
School of Computer Science and Informatics
University College Dublin
Belfield, Dublin 4
Ireland
m.oneill@ucd.ie

ISBN 978-3-540-77476-1 e-ISBN 978-3-540-77477-8

Studies in Computational Intelligence ISSN 1860-949X

Library of Congress Control Number: 2008922057

Cover design: Deblik, Berlin, Germany

Printed on acid-free paper

9 8 7 6 5 4 3 2 1

springer.com

To Maria
Tony

To Gráinne and Aoife
Michael

Preface

The inspiration for this book stemmed from the success of EvoFin 2007, the first European Workshop on Evolutionary Computation in Finance and Economics, which was held as part of the EvoWorkshops at Evo* in Valencia, Spain in April 2007. The range and quality of papers submitted for the workshop underscored the significant level of research activity which is taking place at the interface of natural computing and finance. After the workshop, a call for papers was issued for this volume and following a rigorous, peer-reviewed, selection process a total of fourteen chapters were finally selected. The chapters were selected on the basis of technical excellence and as examples of the application of a range of natural computing and agent-based methodologies to a broad array of financial domains. The book is intended to be accessible to a wide audience and should be of interest to academics, students and practitioners in the fields of both natural computing and finance.

We would like to thank all the authors for their high-quality contributions and we would also like to thank the reviewers who generously gave of their time to peer-review all submissions. We also extend our thanks to Dr. Thomas Ditzinger of Springer-Verlag and to Professor Janusz Kacprzyk, editor of this book series, for their encouragement of and their support during the preparation of this book.

Dublin
December 2007

Anthony Brabazon
Michael O'Neill

Contents

1

Natural Computing in Computational Finance: An Introduction

Anthony Brabazon and Michael O'Neill

Natural Computing Research and Applications Group, UCD CASL, University College Dublin, Dublin, Ireland. anthony.brabazon@ucd.ie, m.oneill@ucd.ie

1.1 Introduction

Natural computing can be broadly defined as the development of computer programs and computational algorithms using metaphorical inspiration from systems and phenomena that occur in the natural world. The inspiration for natural computing methodologies typically stem from real-world phenomena which exist in high-dimensional, noisy and uncertain, dynamic environments. These are characteristics which fit well with the nature of financial markets. Prima facie, this makes natural computing methods interesting for financial modelling applications. Another feature of natural environments is the phenomenon of emergence, or the activities of multiple individual agents combining to create their own environment.

This book contains fourteen chapters which illustrate the cutting-edge of natural computing and agent-based modelling in modern computational finance. A range of methods are employed including, Differential Evolution, Genetic Algorithms, Evolution Strategies, Quantum-Inspired Evolutionary Algorithms, Bacterial Foraging Algorithms, Genetic Programming, Agent-based Modelling and hybrid approaches including Fuzzy-Evolutionary Algorithms, Radial-Basis Function Networks with Kalman Filters, and a Multi-Layer Perceptron-Wavelet hybrid. A complementary range of applications are addressed including Fund Allocation, Asset Pricing, Market Prediction, Market Trading, Bankruptcy Prediction, and the agent based modelling of payment card and financial markets.

The book is divided into three sections each corresponding to a distinct grouping of chapters. The first section deals with optimisation applications of natural computing in finance, the second section explores the use of natural computing methodologies for model induction and the final section illustrates a range of agent-based applications in finance.

A. Brabazon and M. O'Neill: *Natural Computing in Computational Finance: An Introduction*, Studies in Computational Intelligence (SCI) **100**, 1–4 (2008)
www.springerlink.com

1.2 Optimisation

A wide variety of natural computing methodologies including genetic algorithms, evolutionary strategies, differential evolution and particle swarm optimisation have been applied for optimisation purposes in finance. A particular advantage of these methodologies is that, applied properly, they can cope with difficult, multi-modal, error surfaces. In the first six chapters, a series of these algorithms are introduced and applied to a variety of financial optimisation problems.

Passive portfolio management strategies have become very common in recent decades. In spite of the apparent simplicity of constructing an asset portfolio in order to track an index of interest, it is difficult to do this in practice due to the dynamic nature of the market and due to transactions constraints. As discussed in chpt. 2 (*Constrained Index Tracking under Loss Aversion Using Differential Evolution* by Dietmar Maringer), the solution space is non-convex suggesting a useful role for population-based, global optimisation, heuristics like differential evolution. This chapter applies differential evolution for asset selection in a passive portfolio.

The issue of optimal asset allocation for defined contribution pension funds is addressed in chpt. 3 (*An Evolutionary Approach to Asset Allocation in Defined Contribution Pension Schemes* by Kerem Senel, Bulent Pamukcu and Serhat Yanik). To date, there have been few examples of applications of natural computing in the pensions domain. Chpt. 3 shows the application of the genetic algorithm for asset allocation in a pension fund.

The classical portfolio optimisation problem is tackled using evolutionary strategies in chpt. 4 (*Evolutionary Strategies for Building Risk-Optimal Portfolios* by Piotr Lipinski). A particular advantage when using evolutionary algorithms for this task is that a modeller can easily employ differing risk measures and real-world investment constraints when determining optimal portfolios. This chapter provides a clear illustration of how this can be done.

A variant on classical portfolio optimisation is provided in chpt. 5 (*Evolutionary Stochastic Portfolio Optimization* by Ronald Hochreiter). This chapter focusses on stochastic portfolio optimisation and combines theory from the fields of stochastic programming, evolutionary computation, portfolio optimisation, as well as financial risk management in order to produce a generalised framework for computing optimal portfolios under uncertainty for various probabilistic risk measures.

A considerable amount of research has been undertaken in recent years in order to improve the scalability of evolutionary algorithms. This has led to the development of several new algorithms and methodological approaches including the compact genetic algorithm and estimation of distribution algorithms (see chpt. 13 for an introduction to EDAs). One interesting avenue of this work has seen the metaphorical combination of concepts from evolution and quantum mechanics to form the subfield of quantum-inspired evolutionary algorithms (QIEA). Chpt. 6 (*Non-linear Principal Component Analysis of the Implied Volatility Smile using a Quantum-inspired Evolutionary Algorithm* by Kai Fan et al) provides an introduction to this area and illustrates the application of a QIEA for the purposes of undertaking a non-linear principal component analysis of the implied volatility smile of stock options.

The last chapter in the first section, chpt. 7 (*Estimation of an EGARCH Volatility Option Pricing Model using a Bacterial Foraging Optimisation Algorithm* by Jing Dang et al) explores the bacteria-foraging optimisation algorithm (BFOA). The BFOA is a natural computing algorithm loosely inspired by the foraging behaviour of E. coli bacteria. The chapter illustrates the algorithm by applying it to estimate the parameters of an EGARCH model which can be used to price volatility options.

1.3 Model Induction

While optimisation applications of natural computing are important and plentiful, in many real-world applications the underlying model or data generating process is not known. Hence, the task is often to 'recover' or discover an underlying model from a dataset. This is usually a difficult task as both the model structure and associated parameters must be found. The five chapters in this section demonstrate a variety of approaches for this task.

The first chapter in this section, chpt. 8 (*Fuzzy-Evolutionary Modeling for Single-Position Day Trading* by Célia da Costa Pereira and Andrea G. B. Tettamanzi), combines an evolutionary algorithm with a fuzzy predictive model in order to construct an automated day-trading system. The developed trading model is expressed as a set of fuzzy IF-THEN rules. A particular aspect of fuzzy systems, is that they allow for the incorporation of expert domain knowledge even when that knowledge can only be expressed in general terms by the expert.

A major applications area of model induction techniques in finance is that of credit scoring / risk-assessment. Bankruptcy prediction is an example of this application. In chpt. 9 (*Strong Typing, Variable Reduction and Bloat Control for Solving the Bankruptcy Prediction Problem Using Genetic Programming* by Eva Alfaro-Cid, Alberto Cuesta-Cañada, Ken Sharman and Anna I. Esparcia-Alcázar), genetic programming (GP) is applied to uncover bankruptcy prediction models. The chapter also embeds a useful discussion on bloat control, an often overlooked issue which is important when seeking to promote the development of parsimonious models in GP.

Kalman filters are a recursive estimation methodology and provide a powerful tool for estimation of the internal state of a process given a series of noisy observations. Traditionally, the Kalman filter was based on the assumption that the system of interest was a linear dynamical system. Chpt. 10 (*Using Kalman-filtered Radial Basis Function Networks for Index Arbitrage in the Financial Markets* by David Edelman) combines the non-linear modelling capabilities of a radial basis function network with a Kalman filter to produce a index arbitrage trading model.

The study in chpt. 11 (*On Predictability and Profitability: Would GP Induced Trading Rules be Sensitive to the Observed Entropy of Time Series?* by Nicolas Navet and Shu-Heng Chen) examines whether there is a clear relationship between the entropy rates of stock price time series and the ability of a GP system to evolve good rules for trading that stock. The results reported in the chapter suggest that contrary to common wisdom, predictability is neither a necessary nor a sufficient condition for profitability. While tests of predictability may suggest the existence of temporal patterns in a financial time series, they do not necessarily provide information on how easy or difficult it is, for GP or any other machine learning technique, to discover this pattern.

The model induction part of the book concludes with chapter 12 (*Hybrid neural systems in exchange rate prediction* by Andrzej Bielecki, Pawel Hajto and Robert Schaefer), which adopts a hybrid approach to foreign exchange rate prediction through the combination of Wavelet Analysis with a Multi-Layer Perceptron Artificial Neural Network.

1.4 Agent-based Modelling

Agent-based modelling (ABM) has become a fruitful area of financial and economic research in recent years. ABM allows the simulation of markets which consist of heterogeneous agents, with differing risk attitudes and differing expectations to future outcomes, in contrast to traditional assumptions of investor homogeneity and rational expectations. ABMs attempt to explain market behaviour, replicate documented features of real-world markets, and allow us to gain insight into the likely outcomes of different policy choices.

Three chapters in this book adopt an ABM approach. Chpt. 13 (*Evolutionary Learning of the Optimal Pricing Strategy in an Artificial Payment Card Market* by Biliana Alexandrova-Kabadjova, Edward Tsang and Andreas Krause) illustrates a novel agent-based model of a payment card market. Payment card markets are very large internationally and are hard to examine analytically due to their rich set of agent interactions. This renders them very suitable for examination using ABM. In the simulations in the chapter, the authors derive the demand function for payment cards as well as the profit function of card issuers, observing that the fixed fees charged by the card issuers are a vital driver of demand and profits. The chapter provides insights into the optimal pricing strategy for payment card issuers and also has implications for regulators of these markets.

Chpt. 14 (*Can Trend Followers Survive in the Long-Run? Insights from Agent-Based Modeling* by Xue-Zhong He, Philip Hamill and Youwei Li) employs an ABM based on a market fraction asset pricing model in order to investigate the market dominance, the profitability, and the survival rates of both fundamental and trend-following investors across varying time scales. The results indicate that in contrast to the prediction of traditional financial theory, trend-followers can survive in the market in the long run and in the short run they can outperform fundamentalists. The chapter also investigates the effect of the composition of the initial population of investors on market dynamics.

A critical aspect of real-world financial markets is that they are co-evolutionary environments, with individual agents adapting to the actions of others. A well-known co-evolutionary model from the ecological domain which has not yet been widely applied in ABM in finance, the predator-prey model, is introduced in chpt. 15 (*Co-Evolutionary Multi-Agent System for Portfolio Optimization* by Rafał Dreżewski and Leszek Siwik) and is applied to the important portfolio optimisation problem.

Part I

Optimisation

2

Constrained Index Tracking under Loss Aversion Using Differential Evolution

Dietmar Maringer

Centre for Computational Finance and Economic Agents (CCFEA), University of Essex, UK.
dmaring@essex.ac.uk

Summary. Index tracking is concerned with forming a portfolio that mimics a benchmark index as closely as possible. Traditionally, this implies that the returns between the index and the portfolio should differ as little as possible. However, investors might happily accept positive deviations (ie, returns higher than the index's) while being particularly concerned with negative deviations. In this chapter, we model these preferences by introducing loss aversion to the index tracking problem and analyze the financial implications based on a computational study for the US stock market. In order to cope with this demanding optimization problem, we use Differential Evolution and investigate some calibration issues.

2.1 Introduction

Over recent years, passive portfolio management strategies have seen a remarkable renaissance. Assuming that the market cannot be beaten on the long run (in particular after transaction costs), these strategies aim to mimic a given market (or sector) index by investing either into a replication of the benchmark, or by selecting a portfolio which exhibits a behavior as similar to the benchmark's as possible. The market share of products such as exchange traded funds (ETFs) has increased significantly, and it is argued that passive portfolio management is becoming predominant. According to (5), almost half the capital in the Tokyo Stock exchange is subject to passive trading strategies, and (2) report that assets benchmarked against the S&P 500 exceed US$1 trillion.

In contrast, active portfolio management tries to generate excess returns by picking stocks which are expected to outperform the market and avoiding assets that are expected to underperform. Both approaches have their advantages and disadvantages: active strategies rely heavily on superior predictions while passive strategies require few assumptions about future price movements. Passive strategies will also copy the benchmark's poor behavior (in particular when segments of the market drag down the overall performance) while active strategies can react more flexibly in bear markets; etc. If investments are benchmarked against the index, a fund that aims to replicate this benchmark will, by definition, have a lower likelihood to severely

D. Maringer: *Constrained Index Tracking under Loss Aversion Using Differential Evolution*, Studies in Computational Intelligence (SCI) **100**, 7–24 (2008)
www.springerlink.com

fall below this benchmark. It is often argued that it is the actively managed funds' poor performance that makes passive funds all the more attractive (see, e.g., (2)). An alternative, more fundamental argument against active funds is the basis for works including (15) or (3), stating that all investments ought to be efficient *ex ante*, while deviations *ex post* should be transitory. The rational behind this argument is, simply speaking, that if all market participants have the same access to the market and to information, then the market index should be some sort of average of the individual investments. By definition and in the absence of transaction costs, over- and underperformers should then balance. In particular, no participant should be able to produce results that are persistently on just one side of the average. Outperforming the index could therefore be just a matter of good luck rather than skill. Constantly remaining in this position (in particular when transaction costs are incurred) becomes an increasingly rare experience the longer a period of time is considered, as empirical studies confirm.[1]

Arguably the most common approach in passive portfolio management is index tracking. In this strategy, investors select portfolios that mimic the behavior of an index representing the market or a market segment as closely as possible. To find the optimal combination, a distance measure between tracking portfolio and benchmark index is defined which is to be minimized. This so-called Tracking Error (TE) is typically defined as the mean squared deviation between the returns. Akin to volatility, this measure is more sensitive to bigger deviations, but it is oblivious to the sign of the deviation; hence, it does not distinguish between under- or over-performance of the tracking portfolio. Other things equal, investors appear to be mainly concerned with losses rather than any deviation, positive and negative, from their expected outcome. While some authors find that this might even lead to preference functions contradicting the traditional utility analysis for rational risk aversion (see (4) and (18)) the only assumption about investors' preferences made in this contribution is that they want to copy the market and that negative deviations are more "hurtful" than positive ones. Translated to index tracking, investors will try to avoid falling below the benchmark, hence they might want to particularly avoid negative deviations.

This chapter investigates how different levels of loss aversion affect the choice in asset selection for index tracking under realistic constraints. These constraints include an integer constraint on the number of assets, and that initial weights of included assets must fall within certain limits to avoid dominating assets as well as excessive fragmentation and data-fitting. As a consequence, the solution space becomes discrete, exhibits frictions and has multiple optima. Standard optimization techniques cannot deal with these problems satisfactorily and therefore tend to simplify the optimization problem. Heuristic methods, on the other hand, can deal with

[1] As (1, p. 133) put it in their opening statement: "There is overwhelming evidence that, post expenses, mutual fund managers on average underperform a combination of passive portfolios of similar risk. [...] Of the few studies that find that managers or a subset of managers with a common objective (such as growth) outperform passive portfolios, most, if not all, would reach opposite conclusions when survivorship bias and/or correct adjustment for risk are taken into account."

these situations (see, e.g., (8)). In particular, it has been found in (10) and (9) that Differential Evolution is well suited to solve this type of optimization problem.

The rest of this chapter is organized as follows. After formalizing the decision problem in section 2.2, section 2.3 will present how it can be solved numerically using Differential Evolution. Section 2.4 presents the results of a computational study, and section 2.5 concludes.

2.2 The Asset Selection Problem under Loss Aversion

2.2.1 Passive Asset Management and Preferences

The simplest way of tracking a given index would be to replicate it by investing into the assets included in it with their exact corresponding weights. This approach, however, is hampered by several practical limitations. For one, when the index is large and has a high number of different assets, monitoring a correspondingly large tracking portfolio becomes not only cumbersome but also rather costly. Furthermore, if the index is computed in a way that the required number of stocks is not constant over time (e.g., when the relative weights are constant), a perfect replication would require frequent portfolio revision which to some extent defeats the purpose of passive portfolio management. Finally, practical or institutional reasons can impede a perfect replication: If the investor's budget is limited and only whole-numbered quantities or lots of assets can be bought, or if there are upper and/or lower limits on the weight an individual asset can have within the portfolio, deviations of the benchmark's structure can become inevitable. Finally, this one-to-one replication strategy requires that the assets are available in the first place (which might be a problem not only for illiquid assets, but also if these assets are not easily accessible, e.g., in commodity indices). Hence, a trade-off between the revision frequency and the magnitude of the tracking error has to be accepted, and in many cases using a subset of the index's component will be preferred to a full replication.

Deviating from the index's structure might be reasonable if the investor does not want to copy all the market movements: in line with the usual non-satiation assumption of investor preferences, he might be willing to accept a (slightly) less well diversified portfolio with a higher risk if the average return is higher. On a similar note, a symmetric measure for the tracking error will not necessarily capture the investor's actual preferences: while the returns of the tracking portfolio should not fall below the index's, the investor will probably not object if the portfolio outperforms the benchmark. Putting more emphasis on losses lessens the impact of positive deviations when evaluating the TE. Explicitly aiming at outperforming the benchmark blends the index tracking approach with active strategies where, typically, predictions about individual assets' future returns are required. In this contribution, however, the original paradigm of passive strategies is maintained where no specific predictions are required and the only assumption about returns is that they follow some but (more or less) stable distribution without further specification.

2.2.2 Risk Aversion and Loss Aversion

Traditionally, the measure for the tracking error is the (root of the) mean squared difference between portfolio and index returns. This measure does not capture the investor's preferences for positive deviation, nor does it distinguish bullish (i.e., increasing) and bearish (i.e., decreasing) markets, and minimizing the tracking error will not maximize the investor's utility.

Findings in behavioral finance suggest that an individual's utility is not only affected by the future level of wealth, but also whether this level is above or below a certain threshold which is usually the current level of wealth. The same level of future wealth provides a higher utility if it represents an increase in wealth rather than a decrease in wealth. The marginal utility will be lower (higher) if this given level of wealth depending on it represents a profit (loss). In other words, decision makers exhibit loss aversion by reacting more sensitively to losses than suggested by the traditional assumptions on risk aversion.[2] A simple way to model this behavior is to introduce a measure of loss aversion, λ, and transform the actual terminal levels of wealth, w_T, into "perceived" wealth, $\widetilde{w}_T$, where losses are amplified while profits are kept unchanged:

$$\widetilde{w}_T = \begin{cases} w_0 \cdot \exp(r) & r \geq 0 \\ w_0 \cdot \exp(r \cdot \lambda) & r < 0 \end{cases} \tag{2.1}$$

$$= w_0 \cdot \exp\left(r \cdot (1 + (\lambda - 1)\Im_{r<0})\right) \tag{2.2}$$

where r is the log return, $r = \ln(w_T / w_0)$ and $\Im_{r<0}$ is a binary indicator for losses. If $\lambda = 1$, the decision maker has no extra attitude to losses beyond the (still intact) risk aversion, while $\lambda > 1$ models additional loss aversion. Cases where $\lambda < 1$ indicate loss seeking behavior which contradicts the other assumptions and can therefore be neglected.

One could find several ways to translate this into an index tracking framework. However, an investor following a passive strategy can be assumed to accept losses if this also reflects the development in the benchmark index. It therefore appears more plausible to regard losses in terms of opportunity costs, i.e., if the tracking portfolio's returns fall below the index's. This has the advantage that it also covers situations where the tracking portfolio loses not as much as the benchmark which, *ceteris paribus*, can be assumed favorable, while being outperformed in a bullish market is not favorable. Hence, in this contribution the decision maker is concerned with perceived deviations of the tracking portfolio's returns, r_P, from the index's, r_I,

[2] Kahneman and Tversky (4) found in their experiments that, as a consequence of loss aversion, individuals are reluctant to realize losses and therefore tend to stick to assets that have generated losses, making their behavior that of an irrational risk seeker (prospect theory). Since this chapter is not concerned with repeated investment decisions and, more importantly, assumes that decision makers comply to the usual assumptions of rationality and risk aversion, their model will not be applied in this contribution. For a critical assessment of prospect theory, see, e.g., (6) and (7).

$$\widetilde{\Delta r} = \begin{cases} r_P - r_I & r_P \geq r_I \\ (r_P - r_I) \cdot \lambda & r_P < r_I \end{cases} \tag{2.3}$$

$$= (r_P - r_I) \cdot (1 + (\lambda - 1)\Im_{r_P < r_I}) \tag{2.4}$$

where $\Im_{r_P < r_I}$ indicates outperformance (and hence opportunity costs). The parameter λ can be interpreted akin to loss aversion for levels of wealth: If $\lambda > 1$, then losses are amplified, and their effects are bigger than under risk aversion alone. The more λ exceeds one, the more harmful losses are perceived and the more keen the decision maker will be to avoid, in particular, these deviations. As a consequence, asymmetric preferences are introduced and reinforced. Furthermore, high loss aversion reduces the (relative) contribution of positive deviations to the TE.

2.2.3 The Constrained Index Tracking Problem under Loss Aversion

Under real life conditions, index tracking is hampered by several practical limitations. Following (10), it can be assumed that an investor has a certain initial budget B_0 and that only a non-negative and integer quantity of n_i of stock i can be bought. If this stock has an initial price of $S_{0,i}$, then the initial fraction invested in this stock is $x_{0,i} = (n_i \cdot S_{0,i})/B_0$. As prices will change over time while quantities are kept constant, the fractions of asset i will also change over time. In the presence of lower and upper limits on the initial weights for included assets, x^ℓ and x^u, respectively, and the absence of short selling, the quantities n_i must either fall within certain bandwidths or be equal to zero. By assumption, these limits must be kept only at time $t = 0$. If price changes over the holding period lead to violations of these limits, portfolio revisions are not required. Note that introducing lower and upper limits also introduces implicit cardinality constraints: if no asset must exceed an initial weight of x^u, then at least $k^{\min} = \lceil 1/x^u \rceil$ must be included, also the lower weight limit allows for at most $k^{\max} = \lfloor 1/x^\ell \rfloor$ positive weights.

The value of the portfolio at this time is $P_t = \sum_{i=1}^{N} n_i \cdot S_{t,i}$ where $S_{t,i}$ denotes the price of stock i at time t. In this contribution, tracking an index I with a portfolio P requires that over a given holding period their daily returns $r_{t,P}$ and $r_{t,I}$, respectively, are as similar as possible. In addition, negative deviations are perceived less favorable than positive ones. Hence, the investor wants to minimize the perceived differences where, depending on the level of loss aversion, there is different additional concern with losses. A popular measure for the tracking error, TE, is the root of the mean squared deviations of returns.[3] This measure is adopted and applied to perceived deviations, $\widetilde{\Delta r}$. The optimization problem for the constrained index tracking portfolio under loss aversion can therefore be summarised as follows:

$$\min TE = \sqrt{\frac{1}{T}\sum_{t=1}^{T} \widetilde{\Delta r_t}^2} \tag{2.5a}$$

[3] See, e.g., (11) or (14); this definition is also widely used in the industry. Alternatively, authors including (13) define the tracking error as the difference between returns.

subject to

$$\widetilde{\Delta r_t} = (r_{t,P} - r_{t,I})(1 + (\lambda - 1)\Im_{r_{t,P}<r_{t,I}}) \tag{2.5b}$$

$$r_{t,P} = \ln(P_t/P_{t-1}), \qquad r_{t,I} = \ln(I_t/I_{t-1}) \tag{2.5c}$$

$$\Im_{r_{t,P}<r_{t,I}} = \begin{cases} 1 & \text{if } r_{t,P} < r_{t,I} \\ 0 & \text{if } r_{t,P} \geq r_{t,I} \end{cases} \tag{2.5d}$$

$$P_t = \sum_{j=1}^{N} n_i \cdot S_{t,i} \tag{2.5e}$$

$$n_i \in \mathbb{N}_0^+ \tag{2.5f}$$

$$n_i : \begin{cases} x^\ell \leq \frac{n_i \cdot S_{0,i}}{B_0} \leq x^u & \text{if asset } i \text{ is included} \\ 0 & \text{otherwise} \end{cases} \tag{2.5g}$$

$$n_i \cdot S_{0,i} = B_0 \tag{2.5h}$$

The solution space for this problem is non-convex, and the integer constraint on the number of assets makes it a discrete problem, traditional numerical methods can therefore not be employed. Heuristic methods, on the other hand, are more flexible with respect to the shape of the solution space and the constraints that have to be met, as will be shown in the following section.

2.3 Differential Evolution for Portfolio Selection

2.3.1 The Principle and Application to Asset Selection Problems

Differential Evolution (DE) is a population based optimization heuristic for continuous search spaces, suggested by Storn and Price (16, 17); a comprehensive presentation can be found in (12). The basic idea is to generate new solutions by linearly combining three distinct current solutions plus crossing over with a fourth; a tournament principle is then employed to decide over replacement. In the course of iterations, the population should evolve and converge towards the (global) optimum.

More specifically, the typical DE implementation is structured as follows. The algorithm starts by generating P random initial solutions where P is the population size. DE is designed for continuous problems, and solutions are represented as vectors $\mathbf{v}_p, p = 1, \ldots, P$, containing the values of the decision variables, which, for the Index Tracking problem at hand, will be the asset weights. $v_p[i]$ therefore represents the weight of asset i in p's solution. Each of the subsequent iterations comprises of the following steps. First, for each of current solution p, one new solution $\widetilde{\mathbf{v}}_p$ is generated. This is done by randomly picking three further distinct current members of the population, $c_1 \neq c_2 \neq c_3$, and linearly combining their corresponding solution vectors. To do so, the first solutions is chosen as the base vector, $\mathbf{v}_{c_1}$ to which the weighted difference of the other two solutions, $F \cdot (\mathbf{v}_{c_2} - \mathbf{v}_{c_3})$, is added: $\mathbf{v}_c := \mathbf{v}_{c_1} + F \cdot (\mathbf{v}_{c_2} - \mathbf{v}_{c_3})$. Next, this combined solution is crossed over with the original candidate solution, p, $\widetilde{\mathbf{v}}_p = \text{crossover}(\mathbf{v}_p, \mathbf{v}_c, \pi)$, where π is the cross over probability that element i comes from parent p. The i-th element of the new solution is therefore computed as follows:

$$\widetilde{v}_p[i] := \begin{cases} v_p[i] & \text{with probability } \pi \\ v_{c_1} + F \cdot (v_{c_2}[i] - v_{c_3}[i]) & \text{otherwise} \end{cases} \quad (2.6)$$

Graphically speaking, the difference vector "moves" the base solution within the solution space. The larger the difference in the i-th element, the larger the move in this dimension, while no (or small) differences in other elements preserve the current position in those dimensions. The latter is the case in particular when the population has converged and "flocks" around a certain point in the solution space. In order to avoid premature convergence to local optima, it is common practise to add noise ("jitter"). In this case, new solutions are generated according to

$$\widetilde{v}_p[i] := \begin{cases} v_p[i] & \text{with probability } \pi \\ v_{c_1} + (F + z_1[i]) \cdot (v_{c_2}[i] - v_{c_3}[i] + z_2[i]) & \text{otherwise} \end{cases} \quad (2.6^*)$$

where the vectors $\mathbf{z}_j, j = 1,2$ are vectors with either zero (with probability π_j) or normally distributed values with expected value zero and some predefined standard deviation σ_j. Further extensions can contain a second difference vector where another two current solutions are picked randomly or where the distance from the best solution so far is introduced.

Once a new solution $\widetilde{\mathbf{v}}_p$ has been generated for each current solution $\mathbf{v}_p$, a tournament is run where $\widetilde{\mathbf{v}}_p$ replaces $\mathbf{v}_p$ if it has a better fitness value. The updated population of candidate solutions then enters the next iteration, and the process is repeated until some halting criterion is met, e.g., when the population has converged or a given number of iterations has been passed.

An important aspect in the implementation of any heuristic is constraint satisfaction. For the given asset selection problem, the asset weights have to be chosen such that (ideally) they add up to one and must fall within certain ranges; furthermore the integer constraint on number of assets makes it a discrete problem. In the suggested implementation, these constraints are met by introducing an interpretation or mapping function that converts a candidate solution $\mathbf{v}$ into a valid solution of asset weights $\mathbf{x}$. This function comprises of the following steps. Due to undesirable properties as well as minimum weight constraints, it might be favorable not to include all of the available assets in the tracking portfolio. Hence, it must first be decided which assets shall be assigned positive weights. This is done by finding the elements i where the corresponding element in the solution vector is positive, $v[i] > 0$. If this applies to fewer than $k^{\min} = \lceil 1/x^u \rceil$ assets, the budget could not be spent without violating the upper weight limit; if it applies to more than $k^{\max} = \lfloor 1/x^\ell \rfloor$ then the lower weight limit would be exceeded. In these cases, the elements of $\mathbf{v}$ with the $k^{\min}$ and $k^{\max}$ largest values, respectively, are picked. Next, the included assets are assigned the minimum weight while all the other weights are set equal to zero. Finally, the weights of the included assets are increased proportional to their values in $\mathbf{v}$ and the

Algorithm 1: Pseudocode for tracking error (TE) minimization with Differential Evolution

```
1  randomly initialize population of vectors v_p, p = 1...P;
2  repeat
3      %% generate new solutions ṽ_p;
4      for current solutions v_p, p = 1...P do
5          randomly pick c_1 ≠ c_2 ≠ c_3 ≠ p;
6          for all elements i in the solution vector do
7              with probability π_1 : z_1[i] ← N(0,σ_1) else z_1[i] ← 0 ;
8              with probability π_2 : z_2[i] ← N(0,σ_2) else z_2[i] ← 0 ;
9              randomly pick u[i] ~ U(0,1);
10             if u[i] < π then
11                 ṽ_p[i] ← v_p[i];
12             else
13                 ṽ_p[i] ← v_c1[i] + (F + z_1[i])·(v_c2[i] − v_c3[i] + z_2[i]);
14     %% select new population;
15     for current solutions v_p, p = 1...P do
16         if TE(ṽ_p) < TE(v_p) then
17             v_p ← ṽ_p;
18 until halting criterion met ;
```

Fig. 2.1. Differential evolution algorithm

weights add up to one. If for some asset(s) the upper weight limit is violated then the excess weight is redistributed proportionally to the remaining included assets until this constraint is also met. The actual number of stocks is determined according to $n_i \leftarrow \langle x_i V_0 / S_{0,i} \rangle$ where $\langle \cdot \rangle$ is the rounding operator.[4] This approach allows the algorithm to operate in the continuous space, but encourages convergence to solutions with suitable discrete counterparts.

2.3.2 Calibrating the Heuristic

Considerations and Experimental Setting

Unlike traditional deterministic methods, heuristics use stochastic ingredients in their search process. As a consequence, they are non-deterministic and independent restarts can lead to different reported solutions. It is therefore common practise to solve a problem repeatedly in independent runs and use the best of these results for

[4] A more strict version would round up (down) if this simple rounding operator violated the lower (upper) weight limit. Given the granularities $S_{0,i}/V_0$ for the chosen assets and initial budget, however, these violations are rather small. Preliminary experiments showed that this more sophisticated method is computationally more expensive, but has only negligible effects on the quality of results.

the subsequent analysis. To investigate the properties of the heuristic, however, the reported results have to be looked at in more detail. Due to their stochastic ingredients, the results reported by a heuristic must be considered random with a certain distribution, and the quality of a heuristic method can be described by the statistical properties of these reported results. These statistical properties can then be used to derive convergence proofs to show that a heuristics is capable of identifying the optimum with a given probability. These properties can also be used to find technical parameters that favor the heuristic's convergence to the global optimum by searching for parameters where the optimum itself or a sufficiently close solution is found with a given probability.

Ideally, the heuristic ought to find the global optimum in each run, resulting in a degenerated distribution with just one realization. In practical applications, however, the reported results can and will differ. Though the resulting distribution is (theoretically) truncated at the global optimum, it might well be that the empirical distribution of reported results from a series of restarts is not: when the global optimum is never found then this bound is never reached.

If a heuristic is to be evaluated for several test problems, it might be beneficial to "standardize" the reported results in a way that they become comparable. As will be presented in Section 2.4, the main computational study distinguishes six different time windows and three different levels of loss aversion. For each of these 18 cases, the optimal tracking error will differ. To evaluate the heuristic, a total of 28 075 independent restarts were performed with different combinations of parameters, resulting in approximately 1550 reported solutions per case c. From these reported solutions $TE_{r,c}$, the best was chosen, $TE_c^* = \min_r TE_{r,c}$, and the relative deviations of reported from optimal solutions, $D_{r,c} = TE_{r,c}/TE_c^* - 1$, computed. While the distributions of the $D_{r,c}$'s might still differ between cases, they are more similar than that of the $TE_{r,c}$'s; moreover, by definition they are all truncated at the same value, namely 0.

An indicator for the reliability of the implementation is how often this limit is actually reached. More recently, sometimes statistics such as the mean and standard deviation of the analyzed indicator are reported; this makes sense only when the assumption of a Gaussian distribution is reasonable (and the truncation at the global optimum is considered in the estimation process); otherwise they are not very illuminating. Usually statistically more stable and more meaningful is information on how likely a certain deviation is to be reached or exceeded. In the lack of a suitable parametric distribution, the latter can be measured by the quantiles of the empirical distribution of the $D_{r,c}$. These quantiles represent which values are exceeded with the respective frequency; they therefore indicate which values are also likely to be exceeded with a certain probability in future restarts. These quantiles can also be used to compare the results of different experimental settings: in superior settings, exceedances beyond a given deviation should happen less frequently, and for a given confidence level, the maximum exceedance should be smaller. Finally, these quantiles allow some indication on how many restarts are advisable.

The Trade-off Between Computational Time, Population Size and Number of Generations

One of DE's advantages is the low number of technical parameters it requires. Furthermore, it is often found that DE requires less tuning than comparable methods and that standard values often yield reasonable stable results already. Since the solution space for the given problem is multi-modal with frictions and interpretation function adds further complexity, experiments for different parameter settings appear reasonable.

The crucial ingredient to any heuristic search method is the number of function evaluations (FEs) during the optimization process. If the algorithm stops not when a convergence criterion is met but when an exogenously given number of FEs has been reached, then for population based methods, a decision has to be made whether these FEs should rather be spent on small populations with many generations or a larger, more diverse population with less iterations to converge. The number of conceded FEs for this implementation are randomly set to 10 000, 25 000, 50 000, 100 000 or 250 000. Next, alternatives for the population size P are 50, 75, 100 and 150. Depending on the combination of these parameter values, the number of generations ranges from 67 (FEs = 10 000, $P = 150$) to 5 000 (FEs = 250 000, $P = 50$).

Table 2.1 lists the respective 1%, 10% and 50% quantiles for the different combinations. Other things equal, the figures suggest that increasing the number of function evaluations is highly beneficial: if the FEs are doubled, the deviation often reduces by half or more, in some cases even by more, in magnitude. When there are at least 100 000 FEs, then one in ten restarts is likely to find a solution with a Tracking Error at most 1% above the optimal one's. With 250 000 FEs, every other run can be expected to find a solution which deviates by at most 5% from the global one, and one in 100 restarts is likely to end in the global optimum.

Table 2.1. 1% (10%, 50%) Quantiles for $D_{r,c}$ for different numbers of function evaluations (FEs) and population size (P)

	population size, P			
FEs	50	75	100	150
10 000	0.0334 (0.1189, 0.3083)	0.0613 (0.1483, 0.3446)	0.0941 (0.1738, 0.4153)	0.1751 (0.2728, 0.5264)
25 000	0.0126 (0.0335, 0.1912)	0.0075 (0.0484, 0.1764)	0.0118 (0.0638, 0.1855)	0.031 (0.0604, 0.2335)
50 000	0.0023 (0.0133, 0.1474)	0.0036 (0.0182, 0.128)	0.0015 (0.0223, 0.1254)	0.0033 (0.0278, 0.1357)
100 000	0.0003 (0.0087, 0.1139)	0.0005 (0.0072, 0.1011)	0.0005 (0.0098, 0.1067)	0.0009 (0.0123, 0.1011)
250 000	0 (0.0027, 0.0537)	0.0001 (0.0019, 0.0287)	0 (0.0011, 0.017)	0 (0.0024, 0.0205)

These figures also shed light on another relevant question: other things equal, should a large number of function evaluations be used on one (or a few) run(s) with many iterations or on more frequent restarts with fewer iterations per run? If a total of 1 million FEs can be used for 10 restarts with 100 000 FEs each or 100 restarts with 10 000 FEs each, then 10 restarts of the short runs are equally expensive as one long run. Hence, the 1% quantile of the short runs can be compared to the 10% quantile of the long ones. For all of the tested combinations, having longer runs yields better results than having more restarts with more reported results to choose from.

When looking at the different population sizes, the effect is less consistent. In general smaller populations can converge faster as they have less diversity within them. This might be an advantage when the number of generations is small and converging to a local optimum might be better than not converging at all. With more generations available, however, slower convergence reduces the chances of getting stuck in local optima, and it is the larger populations that benefit. Not surprisingly, it is the latter that also see the strongest improvements in reported results, and one can expect that all the quantiles will be superior to those of smaller population sizes if FEs were increased further. It is particularly noteworthy that increasing the population while keeping the number of generations constant is beneficial. Since in these experiments, the number of generations equals FEs divided by population size, doubling both FEs and population size, e.g., leaves the number of generations untouched. Comparing the quantiles for the different settings favors larger population sizes – at least for the tested ranges. This suggests that larger (and hence more diverse) populations have an advantage over smaller ones when conceded the same time to evolve in terms of generations. A general rule of thumb suggests the population size should be about three times the number of variables; with 64 (65) asset weights to optimize, this would correspond to $P \approx 200$. The empirical results from this study suggest that this would require a substantially higher number of FEs than the ones investigated. From a practical point of view, population sizes of 75 and 100 appear to work sufficiently reliably (provided a sufficiently high number of FEs).

Using Additional Noise in the Optimization Process

When generating new solutions, the scaling factor F and the cross over probability π play an important role (see equations (2.6) and (2.6*)). In most implementations, they are both often chosen in the range between 0.5 and 0.9. In this study, either parameter can take the values 0.25, 0.5 or 0.75. In addition, the cross over probability can be zero, implying that the new solution is equal to the linear combination of three current solutions, but inherits no element of the solution against which it is compared in the subsequent tournament. Table 2.2 summarizes the quantiles of deviations from the global optimum under different parameter constellations. Introducing crossover is beneficial as it increases the probability of finding good solutions. This is particularly true for when small populations and low numbers of FEs are allowed (top half of the table). It is noteworthy, however, that low values for F have a positive effect. In combination with high values for π, this implies that the base vector is not dramatically changed by adding the weighted difference vector, and that the new candidate solution is mainly inheriting properties of its future opponent in the tournament, crossed over with the base vector. The figures also confirm that the results are more sensitive to F than to π.

Table 2.2. 1% (10%, 50%) Quantiles for $D_{r,c}$ for different values of the scaling factor F and cross over probability π for all reported results (top half) and results where FEs $\geq$ 50000 and $P \geq 75$ (bottom half)

F	cross over probability π 0	0.25	0.5	0.75
0.25	0.1153 (0.3831, 0.8095)	0.0088 (0.0331, 0.1439)	0.0006 (0.0098, 0.0588)	0.0001 (0.0041, 0.0286)
0.5	0.0093 (0.0918, 0.5506)	0.0002 (0.0053, 0.04)	0.0002 (0.0043, 0.0464)	0.0003 (0.0054, 0.0543)
0.75	0.0224 (0.1149, 0.6762)	0.0172 (0.0655, 0.2003)	0.018 (0.0477, 0.1475)	0.0131 (0.0325, 0.1053)
0.25	0.0828 (0.2648, 0.6864)	0.0044 (0.0193, 0.08)	0.0002 (0.0049, 0.0203)	0 (0.0009, 0.0103)
0.5	0.0053 (0.0427, 0.3913)	0.0001 (0.0027, 0.0178)	0.0003 (0.0028, 0.0178)	0.0004 (0.0043, 0.0215)
0.75	0.0207 (0.0777, 0.3913)	0.0132 (0.0448, 0.1331)	0.0155 (0.0361, 0.0895)	0.0099 (0.0258, 0.0565)

Tables 2.3 and 2.4 summarize the quantiles for the different setting for the noise terms when equation (2.6*) is used to generate new solutions. Note that $\pi_1 = \pi_2 = 0$ reduces this model to the basic version without noise. Differences in the quantiles for different values of σ_i when $\pi_i = 0$ are due to sampling errors and give some indication about the Monte Carlo error for the estimated quantiles. Noting this, the results suggest that adding some noise to the scaling factor F ($\pi_1 > 0$) improves the distribution of reported results; the actual magnitude of the noise, however, seems less important. Adding noise to the difference vector (π_2), on the other hand, has hardly any noticeable effect unless small populations and low numbers of FEs might also be used.

Table 2.3. 1% (10%, 50%) Quantiles for $D_{r,c}$ for different probabilities that the noise term $z_1[i]$ is non-zero (π_1) and its standard deviation in that case (σ_1) for all reported results (top half) and results where FEs $\geq$ 50000 and $P \geq 75$ (bottom half)

π_1	standard deviation for the noise term, σ_1 0.05	0.25	0.5
0	0.0012 (0.0205, 0.2316)	0.0011 (0.0198, 0.2561)	0.001 (0.019, 0.2442)
0.025	0.0009 (0.0179, 0.2512)	0.0007 (0.0158, 0.2)	0.001 (0.014, 0.159)
0.1	0.0011 (0.0177, 0.2245)	0.0007 (0.0167, 0.1673)	0.0005 (0.0131, 0.1307)
0	0.0002 (0.0086, 0.0985)	0.0004 (0.0083, 0.1058)	0.0004 (0.0083, 0.0937)
0.025	0.0002 (0.0077, 0.1049)	0.0004 (0.0073, 0.0782)	0.0004 (0.0071, 0.0497)
0.1	0.0004 (0.0089, 0.094)	0.0003 (0.0088, 0.0711)	0.0003 (0.0067, 0.0427)

Comparing all of the results so far, one can confirm the often purported claim that Differential Evolution is usually stable with respect to the chosen parameters. Once

Table 2.4. 1% (10%, 50%) Quantiles for $D_{r,c}$ for different probabilities that the noise term $z_2[i]$ is non-zero (π_1) and its standard deviation in that case (σ_2) for all reported results (top half) and results where FEs ≥ 50000 and $P \geq 75$ (bottom half)

	standard deviation for the noise term, σ_2		
π_2	0.005	0.025	0.05
0	0.0007 (0.0179, 0.2163)	0.0012 (0.0182, 0.1992)	0.0009 (0.0165, 0.1938)
0.025	0.0006 (0.018, 0.2067)	0.0009 (0.0143, 0.1912)	0.0012 (0.0146, 0.1824)
0.1	0.0008 (0.0192, 0.1983)	0.001 (0.0176, 0.1991)	0.0007 (0.0156, 0.2003)
0	0.0004 (0.0089, 0.0846)	0.0004 (0.0089, 0.0769)	0.0004 (0.0078, 0.0722)
0.025	0.0002 (0.0085, 0.0831)	0.0003 (0.0081, 0.0661)	0.0005 (0.0072, 0.0685)
0.1	0.0003 (0.0089, 0.0766)	0.0003 (0.0075, 0.0776)	0.0003 (0.0061, 0.0706)

a sufficiently large number of function evaluations is chosen, together with a reasonable number of restarts and a populations not too small in size, the algorithm is hardly affected by the values of the remaining parameters as long as they fall within certain (but generally broad) bandwidths.

2.4 Computational Study

2.4.1 The Data

For the computational study, the Dow Jones Industrial Average (DJIA64) is to be tracked by using a subset of the stocks included in it. Adjusted daily prices for 65 stocks[5] were downloaded from `finance.yahoo.com` for the period March 2000 to November 2006, leading to a total of 1648 days with observations. Nine missing data are replaced by the averages of the prices of the adjacent days, and one stock has to be excluded for all windows preceding 2004 due to missing data.

For the financial analysis, the in sample periods consist of 500 observations each, representing about two years; the out of sample tests are performed on the subsequent 250 trading days (i.e., the subsequent year). The initial budget is set to 100 000, and the weight limits are $x^{\ell} = 0.01$ and $x^{h} = 0.5$.

2.4.2 Financial Results

The returns of an ideal tracking portfolio should show no deviations from the index's returns. Given the real world constraints, however, which have been considered in this contribution, this is not achievable, yet the decision maker will aim to come as close to this ideal as possible. Traditionally, this means to minimize the root of the

[5] Note that the composition of the DJIA64 has changed during the observed period.

mean squared deviations between portfolio and index returns. Under loss aversion, investors are more sensitive towards losses. This means that the reaction to losses appears exaggerated when compared to a traditional risk aversion setting and can manifest itself in two ways: the investor will try reduce losses either in magnitude or in frequency, and the investor will request even bigger profits on the positive side to restore an acceptable balance between (expected) return and risk. In either case, the investor will develop a stronger preference for positive skewness, for which he even might accept a (slight) increase in volatility.[6]

For the index tracking problem for the given DJIA data set, loss aversion ($\lambda > 1$) shows predominately in the changed magnitude in returns when compared optimal solutions for investors who are loss neutral ($\lambda = 1$). While the frequency of losses remains more or less unchanged, the mean of the deviations between the tracking portfolio's and the index's returns increases because the mean return increases on days where the tracking portfolio outperforms the index ($r_D > 0$) while it decreases otherwise. As a consequence, the skewness of these deviations increases (as predicted for loss averse investors) (see table 2.5). The standard deviation of r_D remains virtually unchanged, while their kurtosis increases in two years, decreases in another two years and remains constant in the remaining two years.

The financial results presented so far are, strictly speaking, in sample results: the optimized assets weights would have been achievable only under perfect foresight. A more realistic approach is to use a history of returns to optimize the weights and then form a portfolio with exactly these weights. Table 2.6 reports the statistics for the differences between the returns of tracking portfolio and index when decision makers invest for one year, do not readjust their portfolios and chose their asset weights such that it would have been the optimal choice for the preceding two year period. To some extent, these out of sample results reflect the in sample findings: the frequency of days with portfolio returns lower than the index's is hardly effected by the level of loss aversion, nor is the kurtosis. At the same time, with increasing loss aversion, the skewness tends to be higher, and the same is true for the Sharpe ratio. It is noteworthy, however, that the out of sample deviations do not show the high levels of kurtosis that could be observed for the in sample results in later years. The reasons for this are twofold: For one, out of sample results are – by definition – less prone to data fitting. Secondly and more important, the similarity of in and out of sample results for an optimization problem like the one considered here is also an indicator of the stability of the underlying assets' returns. Hence, it is not surprising that the actual tracking errors are bigger (in particular when the composition of the index changes and the decision maker does not adjust the tracking portfolio).

2.5 Conclusion

This chapter investigates the index tracking problem under realistic constraints where, in addition, decision makers can have different levels of loss aversion. Due to

[6] See also (9) for the effects in actively managed portfolios.

Table 2.5. Frequency of losses and statistics of differences in (conditioned) returns between tracking portfolio and index, $r_D = r_P - r_I$, under loss neutrality ($\lambda = 1$) and loss aversion ($\lambda > 1$)

λ	2000–02	2001–03	2002–2004	2003–05	2004–06	2005–2007
			frequency of losses, mean($\Im_{r_D<0}$)			
1.00	0.41483	0.45691	0.41884	0.44289	0.40281	0.45622
1.25	0.45892	0.45691	0.42084	0.44689	0.39479	0.46313
1.50	0.44890	0.45491	0.42285	0.43287	0.40481	0.46774
			mean(r_D)			
1.00	0.00020	0.00014	0.00013	0.00009	0.00009	0.00005
1.25	0.00021	0.00016	0.00013	0.00010	0.00010	0.00006
1.50	0.00022	0.00016	0.00014	0.00011	0.00012	0.00006
			mean($r_D \mid r_D < 0$)			
1.00	-0.00112	-0.00062	-0.00053	-0.00042	-0.00034	-0.00031
1.25	-0.00099	-0.00060	-0.00052	-0.00042	-0.00034	-0.00029
1.50	-0.00101	-0.00060	-0.00051	-0.00042	-0.00031	-0.00029
			mean($r_D \mid r_D > 0$)			
1.00	0.00114	0.00079	0.00060	0.00049	0.00038	0.00036
1.25	0.00124	0.00080	0.00060	0.00051	0.00038	0.00036
1.50	0.00123	0.00080	0.00061	0.00051	0.00040	0.00037
			standard deviation(r_D)			
1.00	0.001447	0.000896	0.000773	0.000661	0.000546	0.000527
1.25	0.001448	0.000896	0.000773	0.000662	0.000546	0.000527
1.50	0.001453	0.000899	0.000775	0.000666	0.000548	0.000529
			skewness(r_D)			
1.00	-0.05705	0.18258	-1.28895	-1.28101	-0.65755	-0.74783
1.25	0.04730	0.26553	-1.18525	-0.99069	-0.46092	-0.56830
1.50	0.10531	0.33182	-1.09341	-0.69125	-0.17956	-0.39504
			kurtosis(r_D)			
1.00	3.94	3.06	17.73	17.20	19.74	25.28
1.25	3.92	3.10	16.79	15.29	19.90	25.89
1.50	3.92	3.12	15.97	13.33	21.24	25.94
			Sharpe Ratio(r_D)			
1.00	0.139830	0.160291	0.162276	0.133908	0.162969	0.101599
1.25	0.147869	0.173980	0.169229	0.143540	0.178148	0.112160
1.50	0.154523	0.182161	0.174573	0.160756	0.211543	0.121664
			actual Tracking Error, mean(r_D^2)			
1.00	0.001460	0.000907	0.000782	0.000666	0.000552	0.000529
1.25	0.001463	0.000909	0.000783	0.000668	0.000554	0.000530
1.50	0.001469	0.000913	0.000786	0.000674	0.000560	0.000533

Table 2.6. Out of sample differences in returns between tracking portfolio and index, $r_D = r_P - r_I$, under loss neutrality ($\lambda = 1$) and loss aversion ($\lambda > 1$)

λ	2003	2004	2005	2006	2007
			frequency of losses, mean($\Im_{r_D<0}$)		
1.00	0.542169	0.461847	0.409639	0.502008	0.451087
1.25	0.534137	0.461847	0.413655	0.481928	0.451087
1.50	0.526104	0.461847	0.417671	0.477912	0.467391
			mean(r_D)		
1.00	-0.000023	0.000778	0.000112	0.000040	0.000129
1.25	-0.000027	0.000769	0.000110	0.000048	0.000127
1.50	-0.000019	0.000766	0.000107	0.000057	0.000128
			std(r_D)		
1.00	0.002271	0.005331	0.000851	0.001048	0.001103
1.25	0.002272	0.005269	0.000839	0.001040	0.001069
1.50	0.002233	0.005258	0.000830	0.001035	0.001057
			skewness(r_D)		
1.00	0.076169	0.482286	-0.410592	0.179581	0.273688
1.25	0.134023	0.481608	-0.404146	0.142729	0.265465
1.50	0.161411	0.490486	-0.364425	0.109294	0.110819
			kurtosis(r_D)		
1.00	4.18	3.74	6.89	7.35	3.03
1.25	4.28	3.74	6.84	7.66	3.02
1.50	4.25	3.77	6.69	8.07	3.02
			Sharpe Ratio(r_D)		
1.00	-0.010261	0.145876	0.131729	0.038443	0.116941
1.25	-0.012062	0.145945	0.130525	0.046456	0.118676
1.50	-0.008420	0.145654	0.128627	0.055289	0.120888
			actual Tracking Error, mean(r_D^2)		
1.00	0.002266	0.005377	0.000857	0.001047	0.001107
1.25	0.002267	0.005315	0.000845	0.001039	0.001074
1.50	0.002228	0.005303	0.000835	0.001034	0.001061

the nature of the constraints, the solution space exhibits frictions and is non-convex and discrete; traditional optimization methods based on first order conditions are therefore not appropriate. Heuristic optimization methods such as Differential Evolution (DE), on the other hand, can deal with such demanding solution spaces. It was discussed how constraint satisfaction can be dealt, and experiments were performed to test different variants of DE and find values for the required technical parameters. The main findings of these experiments shows that (in line with the literature) DE is considerably stable with respect to its parameters. It was also found that for this problem the population size needs not to be substantially higher than the number of decision variables and that large populations are favorable merely with large number of generations.

Meanwhile, there exist numerous variants of DE. A common extension is to add noise which has an effect similar to mutation in other evolutionary methods and which should help to reduce the likelihood of getting stuck in local optima. For the given problem, the experiments suggested that this is not necessary when the other parameters are chosen appropriately. Further variants such as using several difference vectors or enforcing the elitist were not considered here but shall be investigated in future studies.

From a financial point of view, the main result is that the presence of loss aversion has an effect on the investment choice, albeit a small one. As predicted, decision makers will accept slightly bigger deviations from the benchmark if this allows for a higher positive (or a less negative) skewness. Typically, these are matched with slightly higher mean returns because both negative deviations are lowered while positive ones are increased. At the same time, the frequency of falling below the benchmark, however, is hardly affected. Out of sample, these effects mostly persist; however, with neither asset returns nor the index's composition being as stable as assumed by the optimization model, imprecisions are inevitable, and a closer analysis had to be omitted. Extensions to the index tracking model could account for these aspects and include stability measures; furthermore opportunities to readjust the portfolio and the inclusion of transaction costs are of great practical interest, yet have to be left to future research.

Acknowledgements

Valuable comments by the editors, two anonymous referees, and seminar participants at the Universities of Graz and Essex are gratefully acknowledged.

References

[1] Elton EJ, Gruber MJ, Blake CR (1996) The persistence of risk-adjusted mutual fund performance. The Journal of Business 69(2):133–157
[2] Frino A, Gallagher DR, Oetomo TN (2005) The index tracking strategies of passive and enhanced index equity funds. Australian Journal of Management 30(1):23–56
[3] Jensen M (1968) The performance of mutual funds in the period 1945–1964. The Journal of Finance 23(2):389–416
[4] Kahneman D, Tversky A (1979) Prospect theory: An analysis of decision under risk. Econometrica 47:263–291
[5] Konno H, Hatagi T (2005) Index-plus-alpha tracking under concave transaction cost. Journal of industrial and management optimization 1(1):87–98
[6] Levy M, Levy H (2002) Prospect theory: Much ado about nothing? Management Science 48(10):1334–1349
[7] List JA (2004) Neoclassical theory: Evidence from the market place. Econometrica 72(2):615–625
[8] Maringer D (2005) Portfolio Management with Heuristic Optimization. Springer-Verlag

[9] Maringer D (forthcoming) Computational Methods in Financial Engineering, Springer, chap Risk Preferences and Loss Aversion in Portfolio Optimization
[10] Maringer D, Oyewumi O (2007) Index tracking with constrained portfolios. Intelligent Systems in Accounting, Finance and Management 15:57–71
[11] Pope P, Yadav PK (1994) Discovering error in tracking error. Journal of Portfolio Management 20:27–32
[12] Price K, Storn R, Lampinen J (2005) Differential Evolution: A Practical Approach to Global Optimization. Springer
[13] Roll R (1992) A mean/variance analysis of tracking error. Journal of Portfolio Management 18(4):12–22
[14] Rudolf M, Wolter H, Zimmerman H (1999) A linear model for tracking error minimization. Journal of Banking and Finance 23:85–103
[15] Sharpe WF (1966) Mutual fund performance. Journal of Business 39:119–138
[16] Storn R, Price K (1995) Differential evolution – a simple and efficient adaptive scheme for global optimization over continuous spaces. Technical report, International Computer Science Institute, Berkeley
[17] Storn R, Price K (1997) Differential evolution – a simple and efficient heuristic for global optimization over continuous spaces. Journal of Global Optimization 11(4):341–359
[18] Thaler R (1980) Toward a positive theory of consumer choice. Journal of Economic Behaviour and Organization 1:39–60

3

An Evolutionary Approach to Asset Allocation in Defined Contribution Pension Schemes

Kerem Senel[1], A. Bulent Pamukcu[2], and Serhat Yanik[3]

[1] Bilgi University, Istanbul Commerce University, Turkey. ksenel@iticu.edu.tr
[2] Istanbul Commerce University, Turkey. abpamukcu@iticu.edu.tr
[3] Istanbul University, Turkey. syanik@istanbul.edu.tr

Summary. With the increasing popularity of defined contribution pension schemes, the related asset allocation problem has become more prominent. The usual portfolio asset allocation approach is far from being appropriate since the asset allocation problem faced by defined contribution pension schemes is fundamentally different. There have been many attempts to solve the problem analytically. However, most of these analytical solutions fail to incorporate real world constraints such as short selling restrictions for the sake of mathematical tractability. In this chapter, we present an evolutionary approach to the asset allocation problem in defined contribution pension schemes. In particular, we compare the simulation results from a genetic algorithm with the results from an analytical model, a simulated annealing algorithm, and two asset allocation strategies that are widely used in practice, namely the life cycle and threshold (funded status) strategies.

3.1 Introduction

The pension fund allocation problem deserves special treatment in its own right since it is quite different than the usual asset allocation paradigm. The goal of pension fund management is to provide a reasonable retirement income with minimum risk. Minimum risk does not and should not mean investing in low risk assets only. The pension portfolio should include high risk assets to utilize their high expected return, at least for some period during the investment horizon. Since the investment horizon is very long, occasional low performance of high risk assets during some of the years are expected to be more than compensated by high performance in other years.

The Markowitz Portfolio Theory, or the so-called Modern Portfolio theory, attempts to minimize risk (the variance of portfolio return) given a target return or, equivalently, maximize return given a target risk level in a single period (17). The multi-period nature of the pension fund management problem is where the complication arises. It is obvious that a pension portfolio will need rebalancing as time passes and actual asset returns are revealed (24). If actual returns turn out to be higher than expected and the current size of the fund portends an ultimate surplus, then, there is no point in taking extra risk by investing heavily in high risk assets for the next period. On the other hand, if actual returns turn out to be low and the current size of

K. Senel et al.: *An Evolutionary Approach to Asset Allocation in Defined Contribution Pension Schemes*, Studies in Computational Intelligence (SCI) **100**, 25–51 (2008)
www.springerlink.com

the fund suggests an ultimate deficit, then, it may be a good idea to increase exposure to high risk assets in order to be able to catch up with the long term target. Therefore, in stark contrast to a single period setting where the Markowitz portfolio theory is utilized for return maximization given a target risk level, pension fund allocation is about maximizing the probability of attaining a long term target return in a multi period setting where the size of the fund should be taken into account at the end of each period and the portfolio should be rebalanced accordingly.

In other words, the within-period standard deviation is not really relevant to long-term institutional investors such as pension funds, who are more concerned with the variability of the terminal size of their portfolios, from which pensioners will derive their benefits (16). Hence, pension funds cannot simply strive to maximize risk-adjusted returns by utilizing a standard portfolio optimization approach, since this may decrease the chances of attaining the long term target, i.e. a reasonable retirement income.

In this chapter, we demonstrate that an evolutionary approach may provide an alternative solution for optimal investment allocation decision in defined contribution pension schemes. Most of the previous research papers attempt to solve the problem analytically (3, 11, 20, 27). The problem with analytical solutions is that they make a lot of restricting assumptions such as lognormal distributions, time-invariant covariance matrices, or avoidance of short selling restrictions that are not (or, rather, that cannot be) incorporated into the model for the sake of mathematical tractability. Although some of these restricting assumptions can be relaxed, as previously demonstrated by relaxing the assumption of time-invariant covariance matrix (26), such improvements come at the expense of increased mathematical complexity.

Genetic algorithms provide numerical solutions that are not bound by such restricting assumptions. For instance, asset returns can be simulated via a bootstrap method so that the genetic algorithm can work with any distribution and not just with a lognormal distribution. Similarly, short selling restrictions can easily be incorporated into the genetic algorithm. In a general review of genetic algorithms and soft computing techniques in the field of insurance (22), a number of studies with genetic algorithms are mentioned (9, 14, 28). In one of these studies, which investigates the asset allocation problem, a comparison of genetic algorithm with the Newton's method demonstrates that the genetic algorithm is more robust to discontinuities in the search space and is not as sensitive to starting values as is Newton's method (14).

Thus, genetic algorithms emerge as a potentially appropriate solution for the highly nonlinear search space of a portfolio optimization problem with discontinuities. Nonlinearity usually arises due to quadratic terms in portfolio variance for classical portfolio optimization or nonlinear cost (disutility) functions, which are commonly used for pension portfolio optimization. Discontinuities are generally due to short selling restrictions or other potential constraints such as upper limits for individual assets. Therefore, this chapter examines the relative performance of genetic algorithms in solving the asset allocation problem for defined contribution pension schemes. In particular, we compare the simulation results from a genetic algorithm with results from an analytical model, a simulated annealing algorithm, and two asset allocation strategies that are widely used in practice, namely the life cycle and threshold (funded status) strategies.

The layout of the rest of this chapter is as follows. First, we give a brief description of defined contribution pension schemes and, then, focus on the asset allocation practice in defined contribution plans. This section provides the rationale behind the so-called life cycle funds, which automatically take care of rebalancing pension portfolios. This is followed by the model for the defined contribution pension scheme we use in our simulations. The next section focuses on the genetic algorithm we utilize and the alternative solutions we employ as a benchmark. Then, we present the experiment setup, which also includes a subsection on what the optimal number of assets in a well diversified portfolio should be and how we decide on the assets to be included in the pension portfolio. The subsequent section discusses the simulation results. Finally, we present our conclusions together with suggestions for future studies.

3.2 Defined Contribution Pension Plans

A defined contribution pension plan provides each participant with an individual account into which plan contributions are paid. Most defined contribution plans render tax advantages in exchange for certain restrictions such as penalties for withdrawal before a certain age. Contributions may be in the form of employee contributions (salary deferral), employer contributions, and employer matching. The funds in the individual's account are then invested in a number of assets, which are mostly mutual funds designed to match different risk attitudes and investment horizons. Generally, the employee has a certain degree of discretion and responsibility in the asset allocation process, whereas the employer has fiduciary responsibility over the selection of investment choices and product providers. Retirement benefits are provided through the individual's account, mostly through purchase of an annuity at retirement.

Defined contribution schemes are radically different to defined benefit schemes. As the name suggests, the participant's contribution is fixed in a defined contribution pension plan, whereas the exact level of future retirement income is uncertain due to a number of risk factors such as the future return on investment and the annuity rate at retirement. The amount of contribution and charges deducted by the product provider also affect the terminal value of funds and hence, the retirement income (3).

In other words, in contrast to defined benefit schemes, the investment risk is assumed by the plan participant, not the plan sponsor, in a defined contribution scheme. Hence, a poor return on investment may directly result in a lower retirement income unless contribution rates are increased accordingly. In this chapter, we focus on the investment risk during the accumulation phase, which is the most important factor determining the level of retirement income.

3.3 Asset Allocation Practice and Life Cycle Funds

In the last decade, the responsibility of asset allocation has greatly shifted to individuals with the increasing popularity of defined contribution plans. According to a study by Employee Benefit Research Institute, the percentage of family heads with a defined contribution plan (typically a 401(k)-type plan) has increased strikingly in the US. In 1992, the percentage of family heads having a defined benefit plan only was 42%, whereas the percentage of those having a defined contribution plan only was 41%. In 2004, these figures have drastically changed into 26% and 56% for defined benefit and defined contribution, respectively (6). However, it is hard to say that there is a great deal of sophistication in the investment process. It has been demonstrated that most individuals follow quite a naive investment strategy by dividing their contributions evenly across the funds offered in the plan (1). Hence, it follows that the proportion invested in stocks turns out to be high if there are a large number of stock funds available to the investor, or vice versa.

Pension companies offer advisory services that try to match investment portfolios with different risk attitudes and investment horizons. Risk attitudes and investment horizons are investigated via risk tolerance questionnaires which often use simple scoring systems that try to fit the individual in one of the predetermined asset allocation structures such as conservative, moderate, or aggressive. Investors are advised and expected to review the allocation regularly and make the necessary adjustments based on their risk tolerance and financial goals.

Individuals are naturally expected to shift their funds toward less risky assets as they approach retirement. For instance, GFOA (Government Finance Officers Association) recommends that public employers assure that adequate investment education and asset allocation information be provided to participating employees in defined contribution plans that permit participants to self-direct their investments (10). In order to accomplish this objective, they suggest some recommendations and guidelines. In one of these guidelines, they say "Participants should be systematically reminded of their potential need to change their asset allocations as they age or experience various life events. For example, a reminder might be issued to participants as they cross certain age levels, as well as in conjunction with a change in employment status."

Nevertheless, experience shows that many participants n ever reallocate their investments or even review their allocation. A smart solution for this problem has been the introduction of life cycle funds. These funds set a target maturity date and allocate the investments over time, from aggressive to conservative. Typically, a fund will start with a heavy allocation to equities, and, over time, the allocation will move from primarily equities to a balanced portfolio of equity and bonds to primarily fixed income. In this type of fund offering, the participant would select the fund that has a maturity date similar to the participant's own investment horizon, often the participant's retirement age.

In a prospectus describing the funds that are available to the employees of the University of California for defined contribution plans, 5 out of the available 19 funds are life cycle funds that are designed for accommodating the needs of individuals that are expected to retire in different years (13). For instance, it is quoted in the description for the "UC Pathway 2030 Fund" that "As the Fund moves toward its target date, its asset allocation becomes more conservative...The UC Pathway 2030 Fund is appropriate for those investors planning to retire between 2025 and 2035."

Life cycle allocation and threshold (funded status) allocation are two common strategies that are employed for the asset allocation of life cycle funds. In a typical life cycle allocation, funds are 100% invested in equities for, say, the first 10 years. Then, this is followed by, for instance, a 10% per annum switch into bonds during the last 10 years. Although this is a dynamic strategy, it includes no feedback. The threshold allocation strategy, on the other hand, incorporates a form of feedback control. In this strategy, 100% of the funds are invested in equities if the size of the fund is below a lower threshold so that the fund can benefit from higher expected return of equity investing. If the size of the fund is above a higher threshold, then 100% of the funds are invested in bonds with the aim of preserving the status quo with lower risk of fixed income investing. In between, i.e. when the size of the fund is between the lower and higher thresholds, the proportion of funds invested in bonds increases linearly as the size of the fund increases (3). Simulations of analytical models also verify the appropriateness of life cycle strategies; i.e., the analytical models suggest a gradual switch into low risk assets as the individual approaches retirement (11, 27).

In this chapter, we focus on the asset allocation problem of a life cycle fund. We compare the simulation results from a genetic algorithm with results from an analytical model, the life cycle, and threshold strategies. On the other hand, we also test the efficacy of the genetic algorithm by making comparisons with results from another stochastic search technique, namely the simulated annealing algorithm.

3.4 The Model

In our defined contribution pension scheme, funds can be invested in n assets with different levels of risk. Contributions, which are assumed to be a fixed percentage of salary, are paid at the beginning of each period. The only decrement from accumulated funds is assumed to be retirement. Taxation is not taken into account; i.e., contributions and investment income are assumed to be exempt from tax. We assume that the final fund (the actual level of accumulated funds at retirement), f_{final}, is converted into a whole life annuity due. Using the retiree's expected mortality and the return of the low-risk asset as the discount rate, the annual retirement income is given by

$$Annual\,Retirement\,Income = \frac{f_{final}}{\ddot{a}_x} \tag{3.1}$$

where $\ddot{a}_x$ is the actuarial present value of a whole life annuity of 1 payable at the beginning of each year (starting immediately after retirement) as long as the retiree who is at the age of x at retirement survives. The actuarial present value of a whole life annuity of 1 payable at the beginning of each year, $\ddot{a}_x$, is given by

$$\ddot{a}_x = \sum_{k=0}^{\omega} {}_k p_x v^k \tag{3.2}$$

where ω is the maximum age, ${}_k p_x$ is the probability that a retiree who is at the age of x (at retirement) survives for k years, and v is the discount factor that uses the return of the low-risk asset. v is given by

$$v = E(\mathrm{e}^{-X}), \qquad X \sim N(\mu_{low-risk\,asset}, \sigma^2_{low-risk\,asset}) \tag{3.3}$$

or

$$v = exp(-\mu_{low-risk\,asset} + 0.5\sigma^2_{low-risk\,asset}) \tag{3.4}$$

where $\mu_{low-risk\,asset}$ and $\sigma^2_{low-risk\,asset}$ are the mean and variance of the low-risk asset. Since the final fund is assumed to be converted into an annuity, the retiree will be more concerned with the net replacement ratio than the final fund. The net replacement ratio is defined as the ratio of retirement income to final salary. Assuming that the real salary growth rate is assumed to be zero and the real salary level is set equal to 1, the net replacement ratio is given by

$$Net\,Replacement\,Ratio = \frac{Annual\,Retirement\,Income}{Final\,Salary} = \frac{\frac{f_{final}}{\ddot{a}_x}}{1} = \frac{f_{final}}{\ddot{a}_x} \tag{3.5}$$

3.5 The Genetic Algorithm and Alternative Solutions

3.5.1 The Genetic Algorithm

A genetic algorithm can briefly be described as a search technique inspired by biological evolution. A typical genetic algorithm imitates evolutionary mechanisms such as "selection", "crossover", "mutation", and "survival of the fittest". Potential solutions to the problem at hand, which comprise the "population", evolve into the next generation of solutions by an evaluation of how fit they are via a "fitness function".

The genetic algorithm and the alternatives have been implemented in R 2.5.1 (2007-06-27) (19). R includes an "R Based Genetic Algorithm" for binary and floating point chromosomes (29). We have used the floating point chromosome version to minimize the expected cost (disutility) function for the next period. Following (11, 26, 27), the cost function is defined as

$$C_t = (F_t - f_t)^2 + \alpha(F_t - f_t) \tag{3.6}$$

C_t : Cost incurred at the end of period t
F_t : Target level for accumulated funds at the end of period t
f_t : Actual level of accumulated funds at the end of period t
α : Risk aversion parameter

where $\alpha \geq 0$.

When the actual level of funds is below the target level, the cost function is positive. Hence, negative deviations from the target level are always penalized. On the other hand, when the actual level of funds exceeds the target level, the cost function first takes negative values. Then, after a certain point, which depends on the risk aversion parameter α, the cost function becomes positive as the positive deviation increases. In other words, positive deviations from the target level may be rewarded or penalized depending on the size of the deviation. Large positive deviations are penalized in order to increase the probability of attaining the target level by preventing excessive risk exposure.

For the above cost function, risk aversion is decreasing with increasing risk aversion parameter α. In the limit when α approaches infinity, positive deviations are never penalized. Hence, greater values of α are associated with less risk-averse individuals. The target level for each period is given a priori. Hence, the optimization problem can be stated as

$$\min E[C_{t+1}|I_t(information\, available\, at\, time\, t)] \tag{3.7}$$

We have provided the following arguments to the genetic algorithm:

stringMin: (vector with minimum values for each gene) Since each gene corresponds to the weight of an asset in the portfolio, we have set stringMin to a vector of 0s. In other words, short selling is not allowed. The length of the vector is equal to the number of assets in the portfolio.
stringMax: (vector with maximum values for each gene) Similar to stringMin, we have set stringMax to a vector of 1s.
popSize: (population size) The population size is set equal to 50.
iters: (number of iterations) The number of iterations is set equal to 200.

The parameters stringMin and stringMax take care of the short selling restriction. However, there is one more restriction to account for; i.e., the sum of portfolio weights should add up to 1. In order to handle this constraint, the cost function includes a transformation from the original weight vector that is used by the genetic algorithm to a new weight vector that satisfies the constraint for the sum of portfolio weights. The transformation simply divides the original weight vector by the sum of its constituents. The weight vectors used in simulations are the transformed vectors. We have used the default values of the following arguments for the "R Based Genetic Algorithm":

mutationChance: (chance that a gene in the chromosome mutates) The default value is 1/(length of the chromosome+1).
elitism: (number of chromosomes that are kept into the next generation) The default value is 20% of the population size.

The floating point chromosome version of the "R Based Genetic Algorithm" randomly generates an initial population using uniform distribution for obtaining values as dictated by stringMin and stringMax. An iteration starts by producing a fitness score for each member of the population. As elitism is employed, the best members of the population are saved for the next generation. The other members of the next generation are generated via single-point crossover and a nonuniform mutation operator. The nonuniform mutation operator with a decreasing dampening factor over time provides an opportunity for fine tuning as the potential solutions approach a global optimum.

3.5.2 Alternative Solutions

The Analytical Model

In our simulations, we use the analytical model in (26) (see Appendix 3.8). It minimizes the same expected cost (disutility) function as the genetic algorithm. The problem with analytical solutions is that they make a lot of restricting assumptions such as lognormal distributions, time-invariant covariance matrices, or avoidance of short selling restrictions that are not (or, rather, that cannot be) incorporated into the model for the sake of mathematical tractability (11, 27). Although some of these restricting assumptions can be relaxed, as previously demonstrated by relaxing the assumption of time-invariant covariance matrix (26), such improvements come at the expense of increased mathematical complexity.

Since the analytical model does not take into account the short selling restriction, the portfolio weights should be truncated to [0, 1]. This is a straightforward procedure for two assets. If one of the portfolio weights is less than 0, the other portfolio weight must be greater than 1, since the sum of portfolio weights is equal to 1. Hence, if one of the portfolio weights is less than 0, it is set equal to 0 and the other portfolio weight, which is greater than 1, is set equal to 1 if we have only two assets.

The problem becomes complicated for more than two assets. If the above algorithm is used, the sum of portfolio weights may not be equal to 1. For instance, assume that we have three assets for which the respective portfolio weights are -0.5, 0.3, and 1.2. If we set the portfolio weights for the first and third assets equal to 0 and 1, respectively, the sum of the portfolio weights is going to be equal to 1.3. Therefore, we use the following algorithm for the case of n assets:

- Find the maximum distance between the portfolio weights and the violated boundaries for those portfolio weights outside [0, 1].
- – If the portfolio weight corresponding to the maximum distance is greater than 1:
    - Set that portfolio weight equal to 1.
    - Set the other portfolio weights equal to 0.
  – If the portfolio weight corresponding to the maximum distance is less than 0:
    - Set that portfolio weight and the other negative portfolio weights equal to 0.
    - Reduce the positive portfolio weights so that they are proportional to their original sizes and the total reduction is equal to the total increase for the previously negative portfolio weights.

If this algorithm is used for the above mentioned example, the curtailed portfolio weights become 0, 0.2, and 0.8.

The Simulated Annealing Algorithm

The simulated annealing algorithm has originally been developed as a modified Monte Carlo scheme for high dimensional integrals in statistical physics problems (18). Later, it was generalized to solve nonlinear problems (15) and evolved into a popular stochastic search technique that is especially useful for dealing with cost functions possessing quite arbitrary degrees of nonlinearities, discontinuities, and stochasticity (12). The algorithm is also able to handle arbitrary boundary conditions and constraints imposed on these cost functions.

There is a direct analogy between annealing, i.e. the cooling of a metal into a minimum energy structure, and simulated annealing. The version we use in our simulations is also known as Boltzmann annealing (12) as it utilizes a probability function similar to the probability of increase in energy due to the law of thermodynamics. Our general algorithmic flow is given below:

- specify initial temperature and cooling schedule
- randomly initialize solution vector
- evaluate cost
- initialize best cost = cost
- repeat for prespecified number of times
  - obtain new solution vector (by randomly changing solution vector)
  - evaluate new cost
  - · if new cost < best cost
      (accept new solution vector unconditionally)
      solution vector = new solution vector; best cost = new cost
    · else
      (accept new solution vector conditionally)
      · compute probability of increase in energy (cost)
      · generate random number between 0 and 1
      · if random number < probability of increase in energy
        solution vector = new solution vector; best cost = new cost
  - compute new temperature

In our simulations, we minimize the same expected cost (disutility) function used by the genetic algorithm and the analytical model. We set the initial temperature equal to 1 and use an exponential cooling scheme to compute the new temperature; i.e.,

$$T[i+1] = cT[i], \qquad 0 < c < 1 \tag{3.8}$$

where $T[i]$ is the temperature of the i-th loop and c is a constant. We choose c to be equal to 0.99. When initializing the solution vector or generating a new solution vector, only the first n-1 elements are considered to be orthogonal. The n-th element, or the weight of the n-th asset, is simply found by subtracting the sum of the weights of the first n-1 elements from 1. This ensures that the sum of portfolio weights add up to 1. Nevertheless, generating a new solution vector randomly or calculating the n-th element by subtraction will result in the violation of short selling restriction. This is taken care of by using the same algorithm used by the analytical model for restricting the portfolio weights within [0, 1].

To obtain a new solution vector, we change only one randomly selected element of the current solution vector. If the temperature is high, it may be favorable to obtain more distant new solutions to search untapped regions of the solution space. However, as the temperature cools down and the solution approaches the global minimum, the new solution should better be in the proximate vicinity of the current solution so as to speed up convergence. To ensure this gradual narrowing of the search diameter, we use current temperature as the standard deviation of the Gaussian distribution for the random change in the solution vector (12):

$$g(x) = \frac{1}{\sqrt{2\pi T}} \exp[\frac{(x-x_0)^2}{2T}] \tag{3.9}$$

where x and x_0 are the new and current values for the randomly selected element of the solution vector. If the new solution is not accepted unconditionally, then we check whether it is accepted conditionally (probabilistically). First, we compute the probability of an increase in energy (cost) (12):

$$h(new\,cost - cost) = \exp[\frac{-(new\,cost - cost)}{T}] \tag{3.10}$$

By generating a random number between 0 and 1 and comparing it with the probability of an increase in energy, we decide if we should accept the new solution. The conditional acceptance of a worse solution helps to prevent getting stuck at local minima. It is important to note that, as the deviation from the current cost gets larger and/or the temperature cools down, the probability of an increase in energy or, equivalently, the probability of acceptance of a worse solution diminishes.

In order to compare the performances of the genetic and simulated annealing algorithms on a fair ground, we repeat the previous loop as much as the number of iterations for the genetic algorithm, i.e. 200 times. Running the simulated annealing algorithm for 50 times, i.e. as much as the population size of the genetic algorithm, and choosing the best solution vector among the 50, we provide equal chances to both algorithms.

The Life Cycle Strategy

Life cycle allocation is a common strategy that is employed for the asset allocation of life cycle funds. In a typical life cycle allocation, funds are 100% invested in equities for, say, the first 10 years. Then, this is followed by, for instance, a 10% per annum switch into bonds during the last 10 years (3). We follow exactly the same strategy in our simulations. We divide the period before retirement into two equal subperiods. In the first subperiod, 100% of funds are invested in an equity portfolio whereas, in the second period, the exposure to equities is reduced linearly so that 100% of funds are switched to bonds at retirement.

In constructing the equity portfolio, we use the regular (Markowitz) portfolio optimization by using the available stocks. We set the target return at a rate that is equal to the mean of the arithmetic stock returns. Although this is a dynamic strategy, it includes no feedback. The weight of each asset in the portfolio is predetermined and does not depend on the current size of the fund.

The Threshold (Funded Status) Strategy

The threshold allocation strategy, on the other hand, incorporates a form of feedback control (3). In this strategy, 100% of the funds are invested in equities if the size of the fund is below a lower threshold so that the fund can benefit from higher expected return of equity investing. If the size of the fund is above a higher threshold, then 100% of the funds are invested in bonds with the aim of preserving the status quo with lower risk of fixed income investing. In between, i.e. when the size of the fund is between the lower and higher thresholds, the proportion of funds invested in bonds increases linearly as the size of the fund increases. The equity portfolio used in this strategy is e xactly the same as the Markowitz portfolio mentioned in the previous section. We set the lower and upper threshold levels at 0.40 and 0.80, respectively.

3.6 Experiment Setup

3.6.1 Optimal Number of Assets for a Well-Diversified Portfolio

Before we present our selection of assets a nd the relevant data, we need to explain how we decide on the number of assets to be included in the portfolio. The optimal number of assets for a well diversified portfolio remains to be a controversial subject. The general belief that 8 to 10 assets are sufficient to construct a well diversified portfolio is based on an earlier study (8). This belief has been critically challenged by subsequent studies (2, 5, 7, 25). In these studies, the suggested number of assets for a well diversified portfolio ranges from 30 (25) to as high as 500 (5).

Nevertheless, it is very important not to overlook the potential costs due to over-diversification (23). These potential costs include increased monitoring costs, dilution of a portfolio manager's best investment ideas when a portfolio's value is spread across a large number of stocks, and increased transaction costs due to a higher fixed component of commission fees when multiple trades are made. Besides, there is another theoretical justification behind imperfect diversification (4). An investor preference for positive skewness may be the rationale behind holding a limited number of assets. The simple fact that a stockholder can never lose more than 100% whereas gains are potentially unlimited, i.e. positive skewness, is naturally appealing for an investor. Furthermore, for investors with positive marginal utility of wealth, consistent risk aversion, and strict consistency of moment preference, it can be shown that investors will have preference for positive skewness (21). Diversification is desired since it reduces the variance of portfolio return. However, it also reduces the positive skewness, which is something undesirable (4).

3.6.2 Asset Selection and Data

Regarding the number of assets in our portfolio, we took the middle road by setting the number of assets to 10, but also choosing one of the assets to be an index tracking portfolio so that a sufficient degree of diversification can be achieved. Setting aside another asset as the low risk asset, there remains 8 more assets for return enhancement. The data of our first asset belongs to the Dow Jones Industrial Average, which we use as a substitute for an index tracking portfolio. The Dow Jones Industrial Average is comprised of 30 of the largest and most widely held public companies in the US. Hence, holding such an index should provide at least a reasonable degree of diversification. The next eight assets are randomly selected stocks from the constituents of the S&P 500 index, which are not included in the Dow Jones Industrial Average. They represent the assets chosen for return enhancement. These stocks include Apple Inc., Bank of America Corp., Chevron Corp., Ford Motor Co., Goodyear Tire & Rubber Co., Kellogg Co., Eli Lilly & Co., and 3M Co. The last asset is the 10-Year US Treasury Note, which represents the low risk asset. We use daily price data from the period Jan. 2, 2001 to Jun. 29, 2007. This corresponds to a total of 1630 daily returns.

3.6.3 Simulations

For our simulations, we set the initial fund level to zero and the contribution rate to 8% of the salary. Contributions are paid at the beginning of each year. We make our simulations for a total of 20 years and assume that the individual retires at the age of 65. The 1983 US GATT (unisex) mortality table is used for calculating the individual's expected mortality. The annual target rate of return for calculating interim fund level targets is set equal to the Chisini average of asset returns.[1] We do not allow short selling; i.e., portfolio weights are limited within [0, 1].

For our base case scenario, the risk aversion parameter α is set equal to 2.[2] By using the genetic algorithm and alternative methods, 1,000 simulations are carried out to simulate the portfolio weights, the interim and final fund levels, and the net replacement ratio, which may be considered as the single most important indicator for the success of a retirement fund. Then, we compute a number of risk measures, i.e. probability of failing the target, mean shortfall, and 5th percentile (Value at Risk):

- Probability of failing the target is computed by dividing the number of failures (those simulations for which the actual net replacement ratio is below the target level) by the total number of simulations.

[1] Chisini Average of Gross Asset Returns $exp(X_{it}) = exp[(\frac{1}{n})X_{it}]$; $X_{it} \sim N(\mu_i, \sigma_i^2)$; i = 1, 2, ..., n; n is the number of available assets.

[2] A range of α values between 0 and 30 are used in (11) (for the same cost function we use and for 20 years to retirement).

- Mean shortfall is the average absolute difference between the actual and target net replacement ratios for failures. For example, if the target net replacement ratio is 0.80 (the target retirement income is 80% of the final salary) and the actual net replacement ratio is 0.75 (the actual retirement income is 75% of the final salary), the shortfall is 0.05. The mean shortfall is the conditional expected value of shortfall given that the actual net replacement fails to meet the target.
- 5th percentile is the lowest 5th percentile of the simulated distribution of net replacement ratio. In other words, the probability of attaining a lower net replacement ratio than the 5th percentile is only 5%. If 1,000 simulations are carried out, the 5th percentile is the 50th lowest net replacement ratio.

Fig. 3.1 depicts a basic flowchart of the simulation process. We repeat the first 100 simulations for different values of α (10, 20 and 50) to analyze the impact of risk aversion on relative performances of different methods. We use the same simulated path of asset returns for sensitivity analysis so as to ensure that the differences in simulation results do not stem from different asset returns.

3.7 Simulation Results

3.7.1 Base Case Scenario ($\alpha = 2$)

Optimal Asset Allocation - Portfolio Weights

Figs. 3.2–3.6 give the mean of simulated portfolio weights for equities generated by different asset allocation methods. Since there is an index tracking portfolio and a total of 8 stocks, which all qualify as equity investments, the mean portfolio weight for equities is found by deducting the mean portfolio weight for the low risk asset, i.e. the 10 year US Treasury Note, from 1. It has been previously demonstrated that simulations of analytical models verify the appropriateness of life cycle strategies (11, 27). Our simulations of the analytical model also confirm this result. The mean portfolio weight for equities gradually declines from a level close to 1 to around 0.4 and is stabilized at that level. The stabilization occurs in year 14.

For both the genetic and simulated annealing algorithms, the mean portfolio weight for equities is similarly stabilized at around 0.4. On the other hand, much earlier than the analytical model, it reaches that level in year 7 for the simulated and genetic algorithms. This clearly indicates that the rebalancing process is much faster for both algorithms, compared to the analytical model.

For the threshold strategy, the shape of the mean portfolio weight for equities resembles that of the life cycle strategy with one distinction. The mean portfolio weights for equities are significantly higher for the threshold strategy; or, equivalently, they decline more slowly compared to the life cycle strategy. Compared to the analytical model, genetic, and simulated annealing algorithms, both the threshold and life cycle strategies seem to be taking more risk with higher portfolio weights for equities.

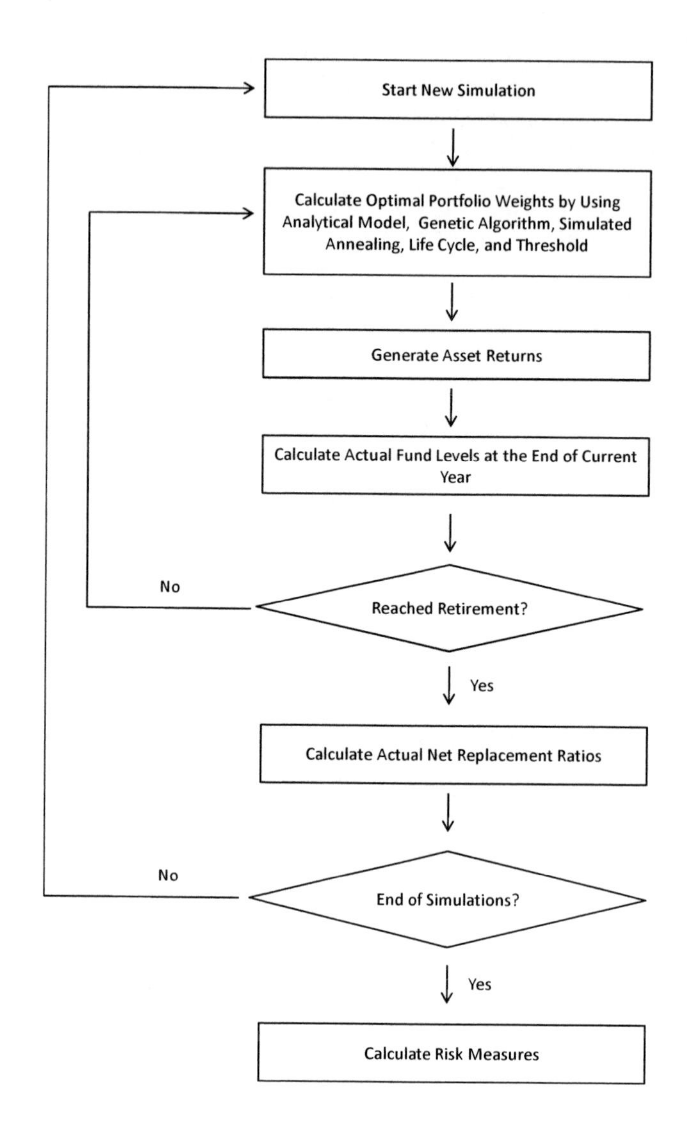

Fig. 3.1. Basic Flowchart of the Simulation Process

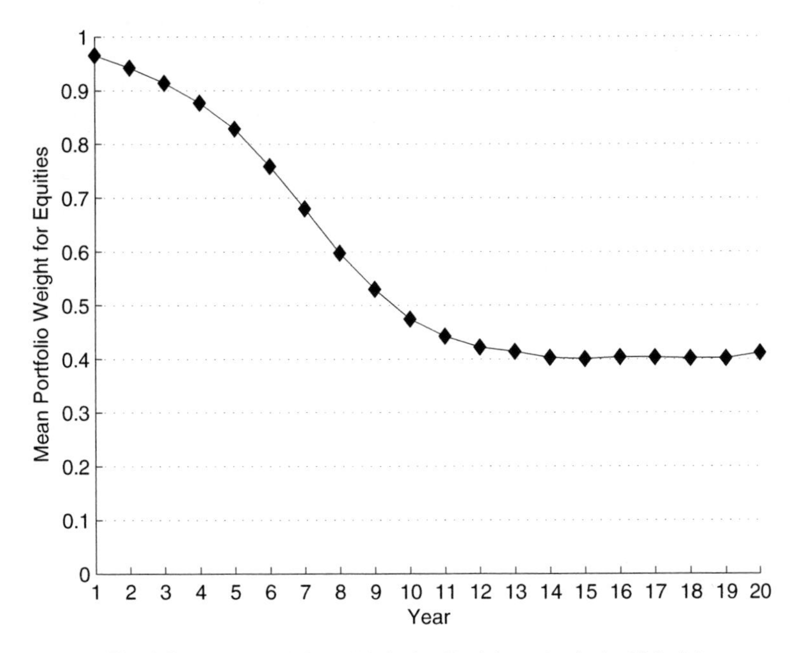

Fig. 3.2. Mean Portfolio Weight for Equities - Analytical Model

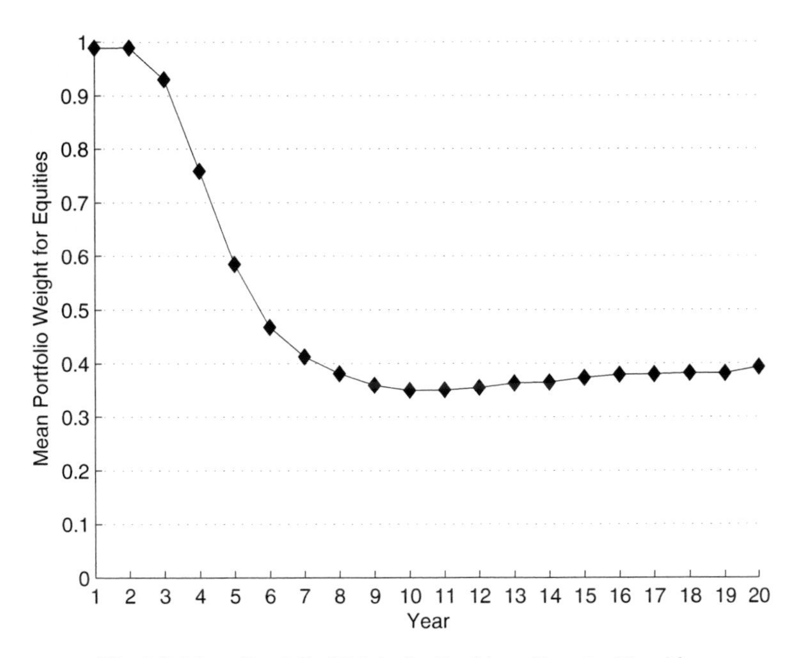

Fig. 3.3. Mean Portfolio Weight for Equities - Genetic Algorithm

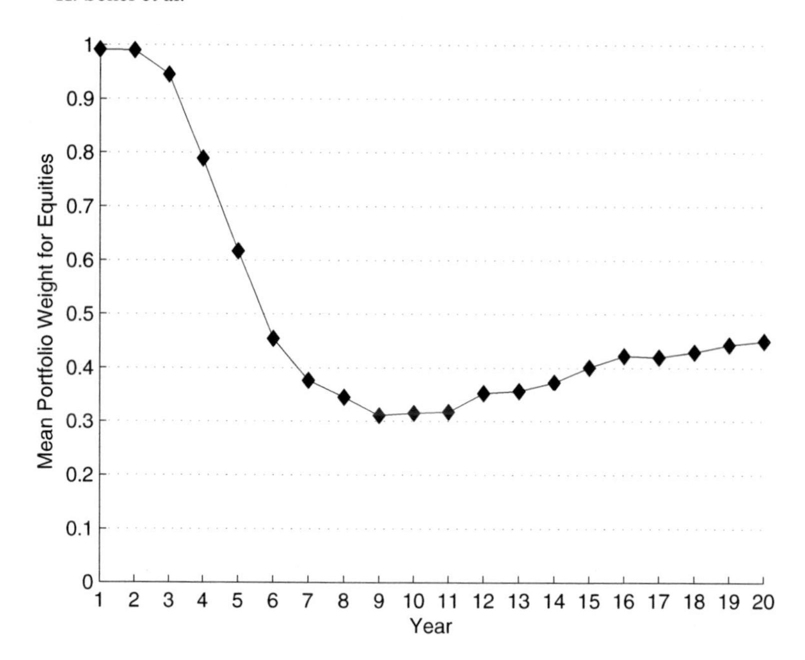

Fig. 3.4. Mean Portfolio Weight for Equities - Simulated Annealing

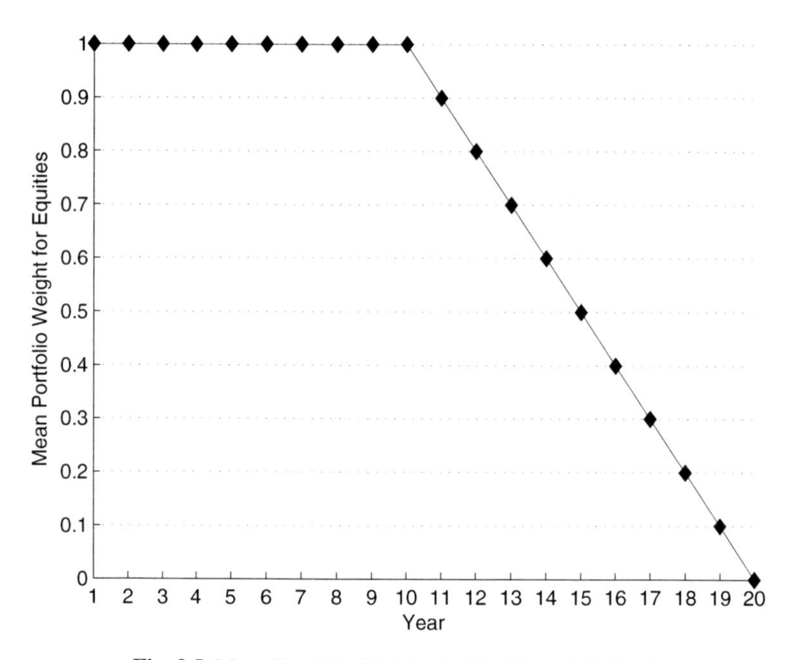

Fig. 3.5. Mean Portfolio Weight for Equities - Life Cycle

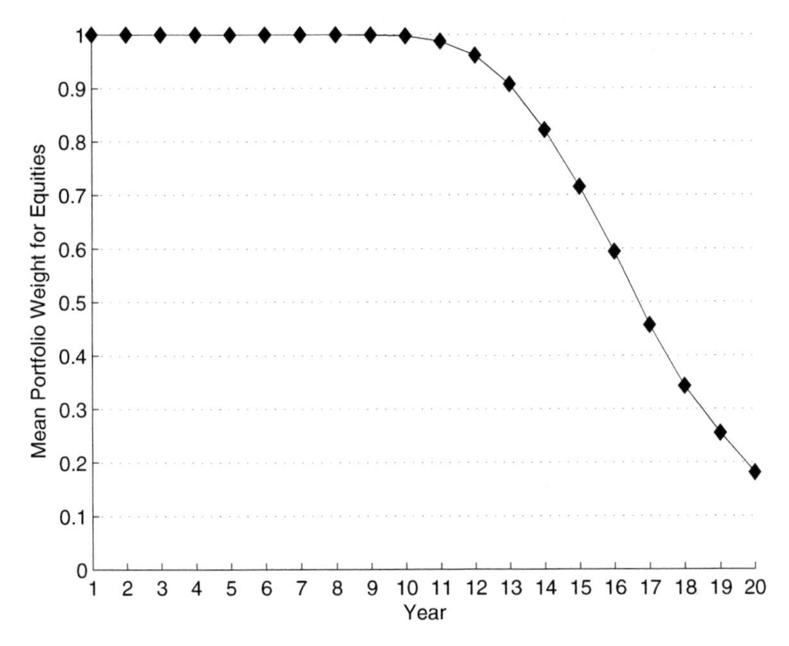

Fig. 3.6. Mean Portfolio Weight for Equities - Threshold

Net Replacement Ratios

Since we assume that the final fund will be converted into an annuity, the retiree will be more concerned with the net replacement ratio than the final fund. The net replacement ratio is defined as the ratio of retirement income to final salary. Figs. 3.7–3.11 give the histograms of simulated net replacement ratios using different asset allocation methods whereas table 3.1 summarizes the distributional properties.

Table 3.1. Distributional Properties of Net Replacement Ratios

	Analytical	Gen.Alg.	Sim.Ann.	Life Cycle	Threshold
Mean	0.6967	0.7101	0.7069	0.7621	0.8669
Standard Deviation	0.0914	0.0815	0.1061	0.3807	0.2455
Skewness	0.3010	0.4453	0.8456	1.6641	1.2166
Kurtosis	2.3447	1.9430	3.3381	4.2436	4.9628

The histograms depict that the analytical model, the genetic, and the simulated annealing algorithms do a better job in terms of reducing the uncertainty regarding net replacement ratio. Their common denominator is the existence of a cost function. Although the threshold strategy includes some form of feedback control, it does not seem to be as effective as the cost function.

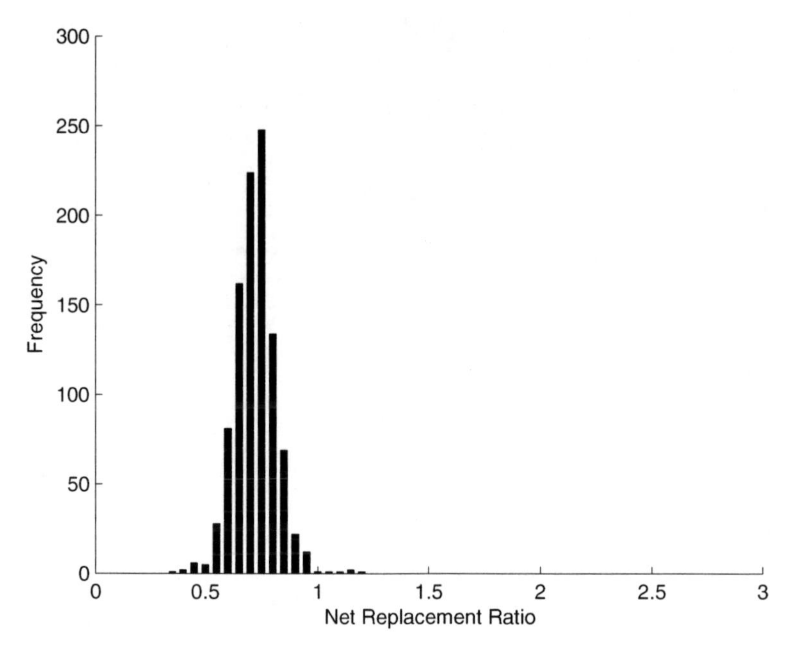

Fig. 3.7. Net Replacement Ratio - Analytical Model

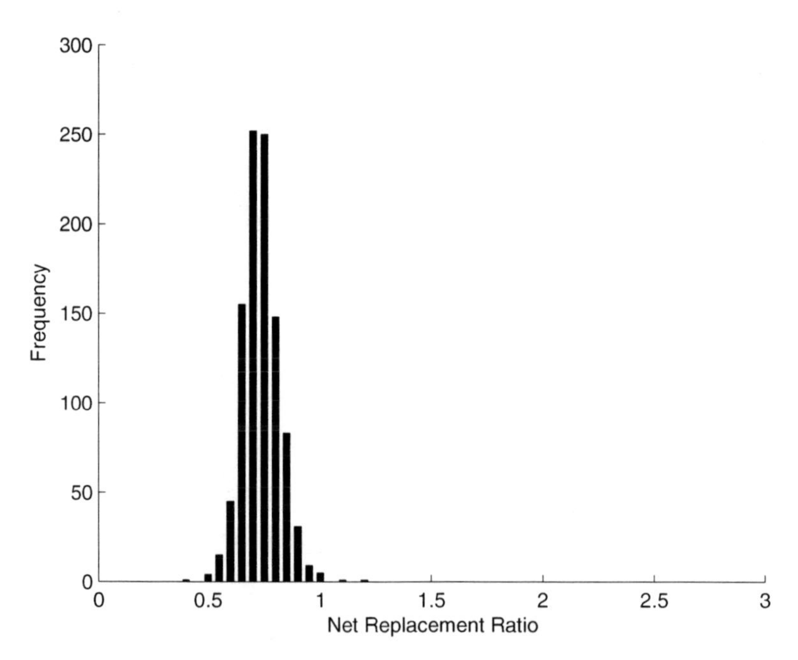

Fig. 3.8. Net Replacement Ratio - Genetic Algorithm

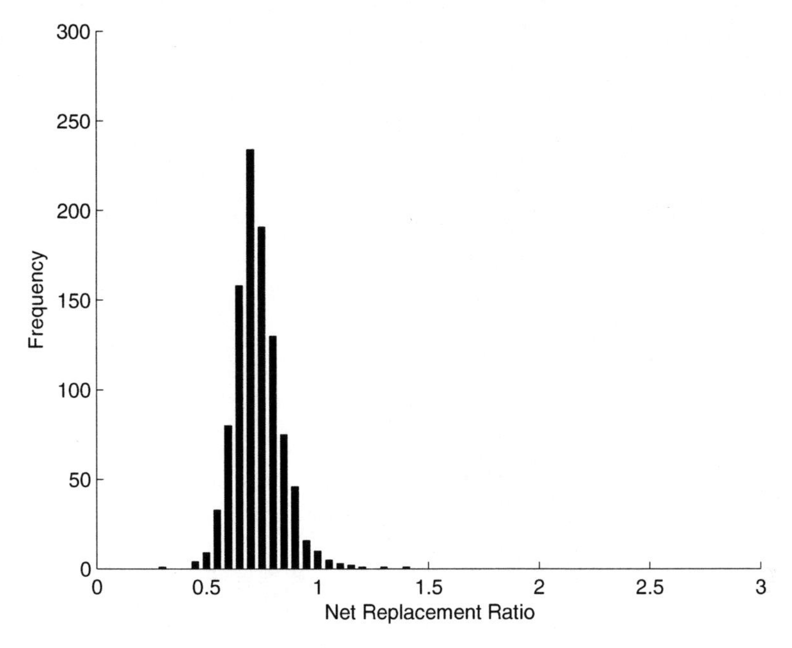

Fig. 3.9. Net Replacement Ratio - Simulated Annealing

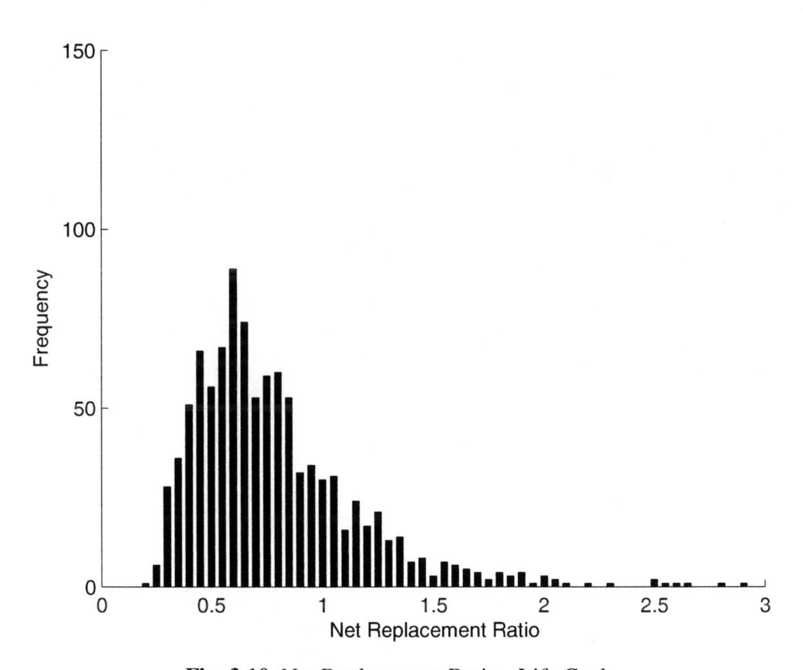

Fig. 3.10. Net Replacement Ratio - Life Cycle

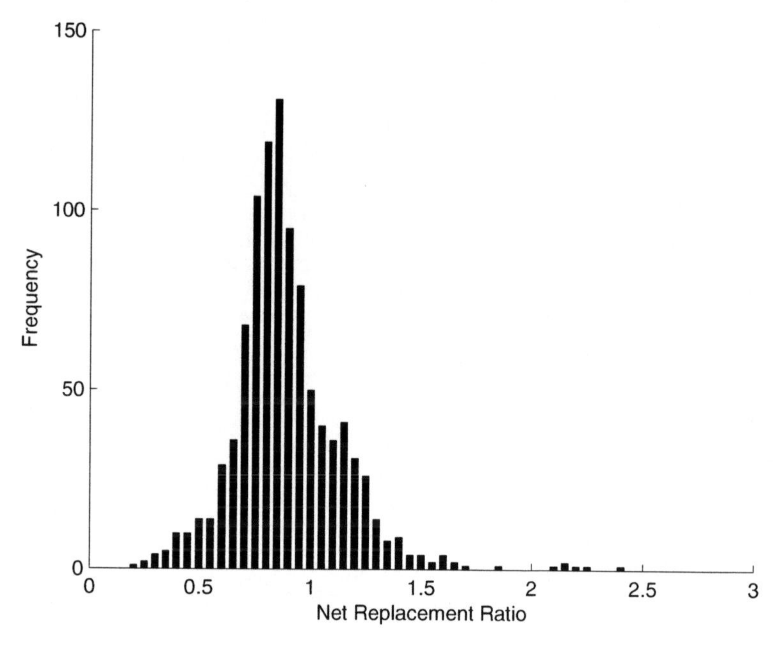

Fig. 3.11. Net Replacement Ratio - Threshold

The threshold strategy has the highest mean net replacement ratio among the five methods, but its standard deviation is also high. If the domination criterion is selected as "higher mean with lower standard deviation", it only dominates the life cycle strategy. In terms of the same criterion, the genetic algorithm slightly dominates the analytical model and the simulated annealing algorithm.

However, it is impossible to tell if the threshold strategy should be preferred over the genetic algorithm, or vice versa. Despite having a significantly higher mean, the standard deviation of simulated net replacement ratios for the threshold strategy is three times that of the genetic algorithm. Furthermore, the kurtosis levels imply a much higher probability of extreme values for the threshold strategy compared to the other methods.

It should be emphasized once again that the goal in this portfolio optimization problem is not return or net replacement ratio maximization. Naturally, it is good to have a higher net replacement ratio on average. Nevertheless, a higher net replacement ratio should not be compromised with higher risk. The goal of portfolio optimization for pension funds is to increase the probability of attaining, or at least approaching, a reasonable retirement income, i.e. a target net replacement ratio. Therefore, although the distributional properties help us in evaluating different methods, the acid test for relative performances will be risk measures, i.e. the probability of failing the target, mean shortfall, and 5th percentile.

Risk Measures

Risk measures computed for the simulated net replacement ratios for different asset allocation methods are presented in table 3.2. The threshold strategy has the lowest probability of failure whereas the genetic algorithm has the best scores in terms of mean shortfall and fifth percentile.

The simulation results reveal that there is an evident tradeoff between the size and probability of failure. The threshold strategy failed only 118 times in 1,000 simulations and the shortfall averaged at 13.68% (of the final salary). Meanwhile, the genetic algorithm failed 197 times whereas the mean shortfall realized at only 4.30%. In other words, the threshold strategy fails rarely; but, when it does, it misses the target substantially.

In terms of the third risk measure, i.e. the 5th percentile, the genetic algorithm outperforms the threshold strategy with a 7.6% difference. In other words, compared to the threshold strategy, the ratio of retirement income to final salary will be 7.6% higher for the genetic algorithm in the case of a really undesirable outcome.

Table 3.2. Risk Measures

	Analytical	Gen.Alg.	Sim.Ann.	Life Cycle	Threshold
Probability of Failure	0.2680	0.1970	0.2700	0.4700	0.1180
Mean Shortfall	0.0565	0.0430	0.0553	0.1673	0.1368
5th Percentile	0.5559	0.5890	0.5537	0.3224	0.5131

Thus, in terms of downside risk, especially in terms of the size of shortfall from the target level in the case of a failure, the risk measures for our base case scenario indicate that the genetic algorithm outperforms the threshold strategy. On the other hand, in terms of the mean-variance measure, there was a relatively small difference in favor of the genetic algorithm compared to the analytical model and simulated annealing algorithm. When risk measures are considered, the genetic algorithm outperforms these two methods significantly.

3.7.2 Sensitivity Analysis (α = 10, 25, and 50)

Net Replacement Ratios

Tables 3.3–3.5 summarize the distributional properties of simulated net replacement ratios for different levels of the risk aversion parameter α.
The first three methods, i.e. the analytical model, the genetic, and the simulated annealing algorithms, attempt to minimize the same expected cost function that incorporates the risk aversion parameter α. Therefore, as risk aversion decreases, the mean and standard deviation of net replacement ratios increase, as expected.[3] On the

[3] Risk aversion decreases as α increases (see sect. 3.5.1).

Table 3.3. Distributional Properties of Net Replacement Ratios (α=10)

	Analytical	Gen.Alg.	Sim.Ann.	Life Cycle	Threshold
Mean	1.0694	1.1028	1.0857	0.7944	0.8752
Standard Deviation	0.1803	0.1476	0.1596	0.3994	0.2635
Skewness	-0.1371	1.1603	1.8022	1.1749	1.1325
Kurtosis	3.0827	4.0671	7.7428	1.2573	5.5661

Table 3.4. Distributional Properties of Net Replacement Ratios (α=25)

	Analytical	Gen.Alg.	Sim.Ann.	Life Cycle	Threshold
Mean	1.6675	1.8874	1.8760	0.7944	0.8752
Standard Deviation	0.3732	0.3795	0.3837	0.3994	0.2635
Skewness	-0.6392	2.8702	3.0741	1.1749	1.1325
Kurtosis	2.4946	13.8618	15.7853	1.2573	5.5661

Table 3.5. Distributional Properties of Net Replacement Ratios (α=50)

	Analytical	Gen.Alg.	Sim.Ann.	Life Cycle	Threshold
Mean	2.4177	3.2044	3.1502	0.7944	0.8752
Standard Deviation	0.8009	0.6510	0.6487	0.3994	0.2635
Skewness	-0.0831	1.8322	3.3758	1.1749	1.1325
Kurtosis	1.0573	4.4008	18.8976	1.2573	5.5661

other hand, the life cycle and threshold strategies are independent of the risk aversion parameter α.

For α equals 10, all of the first three methods outperform the life cycle and threshold strategies with higher mean and lower standard deviation. Among these three methods, the genetic algorithm yields the best mean variance combination. Although this outcome fortifies the results of our base case scenario, risk measures for different values of α will shed further light on the relative performances of alternative methods, particularly from the viewpoint of downside protection.

Risk Measures

Risk measures computed for the simulated net replacement ratios for different values of the risk aversion parameter α are presented in Tables 3.6–3.8.
The results manifest three important points. First, the analytical model, the genetic, and the simulated annealing algorithms dominate the life cycle and threshold strategies at all α levels in terms of all risk measures. This demonstrates the value of using a cost (disutility) function in portfolio optimization. Second, despite using the same expected cost function, the genetic and simulated annealing algorithms outperform the analytical model drastically. This provides further evidence of the weakness of

Table 3.6. Risk Measures (α=10)

	Analytical	Gen.Alg.	Sim.Ann.	Life Cycle	Threshold
Probability of Failure	0.0300	-	-	0.4200	0.1000
Mean Shortfall	0.1137	NA	NA	0.1808	0.1900
5th Percentile	0.6994	0.9028	0.8735	0.2983	0.4689

Table 3.7. Risk Measures (α=25)

	Analytical	Gen.Alg.	Sim.Ann.	Life Cycle	Threshold
Probability of Failure	0.0300	-	-	0.4200	0.1000
Mean Shortfall	0.1039	NA	NA	0.1808	0.1900
5th Percentile	0.8521	1.4880	1.4558	0.2983	0.4689

Table 3.8. Risk Measures (α=50)

	Analytical	Gen.Alg.	Sim.Ann.	Life Cycle	Threshold
Probability of Failure	0.0300	-	-	0.4200	0.1000
Mean Shortfall	0.0994	NA	NA	0.1808	0.1900
5th Percentile	0.8896	2.4962	2.4454	0.2983	0.4689

analytical models in the realm of nonlinear search spaces with discontinuities. Third, the genetic algorithm yields (slightly) better risk measures than the simulated annealing algorithm. Although both algorithms utilize some form of mutation, the genetic algorithm also uses a crossover phase that may potentially enhance the probability of generating better solutions. Considering that both algorithms have been given equal chances, the results can be considered as an indication for the utility of crossover phase.[4]

3.8 Conclusions and Future Studies

In this chapter, we present an evolutionary approach to the asset allocation problem in defined contribution pension schemes. In proposing an evolutionary approach, we consider the highly nonlinear search space with discontinuities. Nonlinearities arise due to quadratic terms in portfolio variance for classical portfolio optimization and nonlinear cost (disutility) functions, which are commonly used for pension portfolio optimization. Discontinuities are due to short selling restrictions and other potential constraints such as upper limits for individual assets. In particular, we compare the simulation results from a genetic algorithm with results from an analytical model, a simulated annealing algorithm, and two asset allocation strategies that are widely used in practice, namely the life cycle and threshold (funded status) strategies.

[4] See Sect. 3.5.2.

The results of our simulations, which are shown to be robust to changing risk aversion, presage a significant potential for the use of evolutionary algorithms in the asset allocation problem of pension funds. The genetic algorithm generally outperforms the other methods both in terms of the classical mean-variance measure and three other risk measures that are thought to be more relevant in the case of portfolio optimization for pension funds.

For future studies, we shall consider the distribution phase, i.e. the post-retirement phase. In particular, we will focus on the timing of conversion of funds into income via purchase of annuities and the investment of funds during the drawdown period. These problems appear to be even more challenging and provide an important test ground for using evolutionary approaches in the field of insurance.

Appendix

The actual level of funds at time t+1 can be expressed as

$$f_{t+1} = (f_t + c)(\mathbf{y}'\mathbf{W}) \tag{3.11}$$

or

$$f_{t+1} = (f_t + c)(\mathbf{W}'\mathbf{y}) \tag{3.12}$$

where c is the contribution rate, $\mathbf{y}$ is the vector of portfolio weights and $\mathbf{W}$ is the vector of real (gross) asset returns. Here, the real salary growth rate is assumed to be zero and the real salary level is set equal to 1. Therefore, the annual contribution by the individual is assumed to be equal to a constant contribution rate, c, throughout the whole period. Although $\mathbf{y}$ and $\mathbf{W}$ are dependent on time, the subscript t is dropped for convenience. $\mathbf{y}$ and $\mathbf{W}$ can explicitly be written as

$$\mathbf{y}' = \begin{bmatrix} y_{1t} \; y_{2t} \; \dots \; y_{nt} \end{bmatrix} \tag{3.13}$$

and

$$\mathbf{W}' = \begin{bmatrix} W_{1t} \; W_{2t} \; \dots \; W_{nt} \end{bmatrix}, \qquad W_{it} = \exp(X_{it}), \qquad X_{it} \sim N(\mu_i, \sigma_i^2) \tag{3.14}$$

where X_{it} is the real rate of interest for the i-th asset in period t, which is assumed to be constant throughout the period, and n is the number of available assets. The asset returns are assumed to be lognormally distributed. If (3.11) is rewritten explicitly,

$$f_{t+1} = (f_t + c) \begin{bmatrix} y_{1t} \; y_{2t} \; \dots \; y_{nt} \end{bmatrix} \begin{bmatrix} \exp(\mu_1 + V_{1t}) \\ \exp(\mu_2 + V_{2t}) \\ \dots \\ \exp(\mu_n + V_{nt}) \end{bmatrix}, \qquad V_{it} \sim N(0, \sigma_i^2) \tag{3.15}$$

or

$$f_{t+1} = (f_t + c)[y_{1t}\exp(\mu_1)\exp(V_{1t}) + \ldots + y_{nt}\exp(\mu_n)\exp(V_{nt})] \tag{3.16}$$

Define

$$\mathbf{y}'_T = \begin{bmatrix} y_{1t}\exp(\mu_1) & y_{2t}\exp(\mu_2) & \ldots & y_{nt}\exp(\mu_n) \end{bmatrix} \tag{3.17}$$

and

$$\mathbf{W}'_T = \begin{bmatrix} \exp(V_{1t}) & \exp(V_{2t}) & \ldots & \exp(V_{nt}) \end{bmatrix} \tag{3.18}$$

where $\mathbf{y}_T$ is the transformed vector of portfolio weights and $\mathbf{W}_T$ is the transformed vector of (gross) asset returns. Similar to $\mathbf{y}$ and $\mathbf{W}$, $\mathbf{y}_T$ and $\mathbf{W}_T$ are also dependent on time, but the subscript t is dropped for convenience. Then, (3.16) can be rewritten as

$$f_{t+1} = (f_t + c)(\mathbf{y}'_T\mathbf{W}_T) \tag{3.19}$$

or

$$f_{t+1} = (f_t + c)(\mathbf{W}'_T\mathbf{y}_T) \tag{3.20}$$

If the constraint that the portfolio weights should add up to 1 is imposed upon, the optimization problem can be stated as

$$\min E[C_{t+1}|I_t(information\,available\,at\,time\,t)], \qquad s.t.\ \mu'\mathbf{y}_T = 1 \tag{3.21}$$

where μ is a vector comprised of all $\exp(-\mu_i)$'s. μ can explicitly be written as

$$\mu' = \begin{bmatrix} \exp(-\mu_1) & \exp(-\mu_2) & \ldots & \exp(-\mu_n) \end{bmatrix} \tag{3.22}$$

In order to solve the optimization problem in (3.21), a Lagrange multiplier κ is associated with the above constraint and the Lagrangian

$$E(C_{t+1}|I_t) - \kappa(\mu'\mathbf{y}_T - 1) \tag{3.23}$$

is formed. Now, the optimization problem in (3.21) becomes

$$\min[E(C_{t+1}|I_t) - \kappa(\mu'\mathbf{y}_T - 1)] \tag{3.24}$$

In (26), $\mathbf{y}_T$, the transformed optimal portfolio weights that minimize the expected cost function at t+1, are derived as

$$\mathbf{y}_T = \frac{(2F_{t+1} + \alpha)(f_t + c)\mathbf{H}^{-1}E(\mathbf{W}_T|I_t) + \kappa\mathbf{H}^{-1}\mu}{2(f_t + c)^2} \tag{3.25}$$

where,

$$\mathbf{H} = E(\mathbf{W}_T\mathbf{W}'_T|I_t) \tag{3.26}$$

and

$$\kappa = \frac{2(f_t + c)^2 - (2F_{t+1} + \alpha)(f_t + c)\mu' \mathbf{H}^{-1} E(\mathbf{W}_T | I_t)}{\mu' \mathbf{H}^{-1} \mu} \tag{3.27}$$

References

[1] Benartzi S, Thaler R H (2001) Naive Diversification Strategies in Defined Contribution Saving Plans. American Economic Review 91(1):79–98
[2] Bird R, Tippett M (1986) Notes, Naive Diversification and Portfolio Risk - A Note. Management Science 32(2):244–251
[3] Blake D, Cairns A, Dowd K (1999) PensionMetrics: Stochastic Pension Plan Design and Value at Risk During the Accumulation Phase. BSI-Gamma Foundation, Working Paper Series, 11:77 pages
[4] Conine T E Jr, Tamarkin M J (1981) On Diversification Given Asymmetry in Returns. Journal of Finance 36(5):1143–1155
[5] De Wit D P M (1998) Naive Diversification. Financial Analysts Journal 54(4):95–100
[6] EBRI Employee Benefit Research Institute Notes, 28(2):12 pages, Feb. 2007
[7] Elton E J, Gruber M J (1977) Risk Reduction and Portfolio Size: An Analytical Solution. Journal of Business 50(4):415–437
[8] Evans J L, Archer S H (1968) Diversification and the Reduction of Dispersion: An Empirical Analysis. Journal of Finance 23(5):761–767
[9] Frick A, Herrmann R, Kreidler M, Narr A, Seese D (1996) Genetic-Based Trading Rules - A New Tool to Beat the Market with? - First Empirical Results. Proceedings of AFIR Colloquium, 1-3 Oct. 1996, 997–1017
[10] GFOA (Government Finance Officers Association) Asset Allocation - Guidance for Defined Contribution Plans (1999), 5 pages, http://www.gfoa.org/services/rp/corba/corba-asset-allocation-benefit.pdf
[11] Haberman S, Vigna E (2002) Optimal Investment Strategies and Risk Measures in Defined Contribution Pension Schemes. Insurance: Mathematics and Economics 31:35–69
[12] Ingber L (1993) Simulated Annealing: Practice versus Theory. Mathematical and Computer Modelling 18(11):29–57
[13] Investments Managed by the University of California Office of the Treasurer, 4 pages, http://www.ucop.edu/treasurer/invinfo/Descriptions.pdf
[14] Jackson A (1997) Genetic Algorithms for Use in Financial Problems. Proceedings of AFIR Colloquium, 13-15 Aug. 1997, 481–503
[15] Kirkpatrick S, Gelatt C D Jr, Vecchi M P (1983) Optimization by Simulated Annealing. Science 220(4598):671–680
[16] Lee S, Byrne P (2001) The Impact of Portfolio Size on the Variability of the Terminal Wealth of Real Estate Funds. RICS Foundation, The Cutting Edge 2001, 16 pages
[17] Markowitz H (1952) Portfolio Selection. Journal of Finance 7(1):77–91

[18] Metropolis N, Rosenbluth A W, Rosenbluth M N, Teller A H, Teller E (1953) Equation of State Calculations by Fast Computing Machines. Journal of Chemical Physics 21(6):1087–1092
[19] R Development Core Team (2007) R: A language and environment for statistical computing. R Foundation for Statistical Computing, Vienna, Austria, ISBN 3-900051-07-0, http://www.R-project.org
[20] Romaniuk K (2005) The Optimal Asset Allocation of the Main Types of Pension Funds: a Unified Framework. Proceedings of 2005 FMA Annual Meeting, 13-15 Oct. 2005, 57 pages
[21] Scott R C, Horvath P A (1980) On the Direction of Preference for Moments of Higher Order than the Variance. Journal of Finance 35(4):915–919
[22] Shapiro A F (2002) The Merging of Neural Networks, Fuzzy Logic, and Genetic Algorithms. Insurance: Mathematics and Economics 31:115–131
[23] Shawky H A, Smith D M (2005) Optimal Number of Stock Holdings in Mutual Fund Portfolios Based on Market Performance. Financial Review 40:481–495
[24] Sherris M (1993) Portfolio Selection Models for Life Insurance and Pension Funds. Proceedings of AFIR Colloquium, 2 Apr. 1993, 915–930
[25] Statman M (1987) How Many Stocks Make a Diversified Portfolio? Journal of Financial and Quantitative Analysis 22(3):353–363
[26] Tuncer R, Senel K (2004) Optimal Investment Allocation Decision in Defined Contribution Pension Schemes with Time-Varying Risk. Proceedings of the 8th Congress on Insurance: Mathematics and Economics, 43 pages
[27] Vigna E, Haberman S (2001) Optimal Investment Strategy for Defined Contribution Pension Schemes. Insurance: Mathematics and Economics 28:233–262
[28] Wendt R Q (1995) Build Your Own GA Efficient Frontier. Risks and Rewards 24:1,4–5
[29] Willighagen E (2005) genalg: R Based Genetic Algorithm, R package version 0.1.1

4

Evolutionary Strategies for Building Risk-Optimal Portfolios

Piotr Lipinski

Institute of Computer Science, University of Wroclaw, Poland.
lipinski@ii.uni.wroc.pl
http://www.ii.uni.wroc.pl/~lipinski

Summary. This chapter describes an evolutionary approach to portfolio optimization. It rejects some assumptions from classic models, introduces alternative risk measures and proposes three evolutionary algorithms to solve the optimization problem. In order to validate the approach proposed, results of a number of experiments using data from the Paris Stock Exchange are presented.

4.1 Introduction

Evolutionary Algorithms have been successfully incorporated into many fields of science and technology, notably including the domains of economics and finance (5), (7), (10), (14). This chapter presents another application of evolutionary algorithms in this domain, namely an evolutionary approach to the problem of portfolio optimization. Although some analytical methods are well-known for classic versions of the problem (1), (4), an extension of the problem by introducing more complex risk measures and loosening several artificial assumptions requires a new efficient approach, which cannot be developed on the basis of classic methods due to the irregularity of the objective function and the search space. However, the opportunities provided by evolutionary algorithms (2), (12) may lead to an efficient optimization of portfolio structures.

Moreover, apart from theoretical constraints, the approach presented in this chapter focuses also on a few practical constraints such as budget constraints, which means that the user of the system has only finite amount of money, as well as investor capabilities and preferences, which means that the user has to obey various market regulations and pay transaction fees. Moreover, an important constraint is constituted by time restrictions and hardware limits.

4.2 Classic Approach to Portfolio Optimization

In this chapter, we focus on the main goal of investors, which is to optimally allocate their capital among various financial assets. Searching for an optimal portfolio of

P. Lipinski: *Evolutionary Strategies for Building Risk-Optimal Portfolios*, Studies in Computational Intelligence (SCI) **100**, 53–65 (2008)
www.springerlink.com

stocks, characterized by random future returns, seems to be a difficult task and is usually formalized as a risk-minimization problem under a constraint of expected portfolio return (4). The risk of portfolio is often measured as the variance of returns, but many other risk criteria have been proposed in the financial literature (1) (3). Portfolio theory may be traced back to the seminal Markowitz paper (11) and is presented in an elegant way in (4). Consider a financial market on which n risky assets are traded. Let

$$\mathbf{R} = (R_1, R_2, \ldots, R_n)' \tag{4.1}$$

be the square-integrable random vector of random variables representing their return rates. Denote as $\mathbf{r} = (r_1, r_2, \ldots, r_n)' \in \mathbb{R}^n$ the vector of their expected return rates

$$\mathbf{r} = (\mathbf{E}[R_1], \mathbf{E}[R_2], \ldots, \mathbf{E}[R_n])' = \mathbf{E}[\mathbf{R}] \tag{4.2}$$

and as $\mathbf{V}$ the corresponding covariance matrix which is assumed positive definite. A portfolio is a vector $\mathbf{x} = (x_1, x_2, \ldots, x_n)' \in \mathbb{R}^n$ verifying

$$x_1 + x_2 + \ldots + x_n = 1. \tag{4.3}$$

Hence x_i is the proportion of capital invested in the i-th asset. Denote as X the set of all portfolios. For each portfolio $\mathbf{x} \in X$, we define

$$R_{\mathbf{x}} = x_1 R_1 + x_2 R_2 + \ldots + x_n R_n = \mathbf{x}'\mathbf{R} \tag{4.4}$$

as the random variable representing the portfolio return rate and then

$$\mathbf{E}[R_{\mathbf{x}}] = x_1 r_1 + x_2 r_2 + \ldots + x_n r_n = \mathbf{x}'\mathbf{r} \tag{4.5}$$

is the portfolio expected return rate. For a fixed level $e \in \mathbb{R}$ of expected return rate, let

$$X_e = \{\mathbf{x} \in X : \mathbf{E}[R_{\mathbf{x}}] = e\} \tag{4.6}$$

be the set of all portfolios leading to the desired expected return rate e. Therefore, the classic Markowitz's problem of portfolio optimization may be formulated as finding $\tilde{\mathbf{x}} \in X_e$ such that:

$$\mathbf{Var}[R_{\tilde{\mathbf{x}}}] = \min\{\mathbf{Var}[R_{\mathbf{x}}] : \mathbf{x} \in X_e\}, \tag{4.7}$$

where the variance is considered as the risk measure. Such a problem, defined in the classic portfolio theory, may be solved using analytical methods (4). Although the approach has very strong mathematical foundations and completed theoretical models, it requires some artificial and unreal assumptions, so there appear some competitive practical approaches, which extend these theoretical models to the real investment market (6) (9) (10).

4.3 Competitive Approaches to Portfolio Optimization

In spite of its wide diffusion in the academic and professional worlds, the classic approach is often criticized for its artificial assumptions (3). Therefore, there are an increasing number of competitive approaches to portfolio optimization, which are based rather on heuristic descended from empirical observations than on theoretical models. Unfortunately, loosing some theoretical assumptions, the problem becomes more and more complex, hence it cannot be solved in the classic way (3).

Many competitive approaches are based on artificial intelligence (6), (9), (10). They extend the classic model by introducing several alternative risk measures instead of variance. Such extended problems cannot be solved by analytical methods, because of their complexity, irregularity and lack of proper optimization tools. However, evolutionary algorithms used as an optimization method can do with such problems returning satisfying results. In order to extend the classic problem of portfolio optimization, let us replace the criteria (4.7) with the criteria

$$\rho(\tilde{\mathbf{x}}) = \min\{\rho(\mathbf{x}) : \mathbf{x} \in X_e\}, \tag{4.8}$$

where $\rho : X \to \mathbb{R}$ is a risk measure, i.e. a function which assign to each portfolio $\mathbf{x} \in X$ its risk $\rho(\mathbf{x}) \in \mathbb{R}$. In this chapter, we consider a few risk measures, which are the most popular alternatives to variance, such as semivariance of the return rate (3)

$$\rho(\mathbf{x}) = \mathbf{SVar}[R_{\mathbf{x}}] = \mathbf{E}[(R_{\mathbf{x}} - \mathbf{E}[R_{\mathbf{x}}])_-^2] \tag{4.9}$$

where

$$(R_{\mathbf{x}} - \mathbf{E}[R_{\mathbf{x}}])_- = \begin{cases} 0, & \text{if } \mathbf{E}[R_{\mathbf{x}}] \le R_{\mathbf{x}} \\ R_{\mathbf{x}} - \mathbf{E}[R_{\mathbf{x}}], & \text{if } R_{\mathbf{x}} < \mathbf{E}[R_{\mathbf{x}}] \end{cases} \tag{4.10}$$

or the downside risk of the return rate (3)

$$\rho(\mathbf{x}) = \mathbf{DSR}_q[R_{\mathbf{x}}] = \mathbf{E}[(R_{\mathbf{x}} - \mathbf{E}[R_{\mathbf{x}}])^q] \tag{4.11}$$

for a given $q \ge 0$ (clearly, semivariance is the downside risk with $q = 2$). These two measures take into consideration the fact that investors usually consider the risk of an investment rather as the probability that the value of the investment will decrease below the expected level than as the probability that it will increase above the expected level (so-called *asymmetric* risk).

In order to construct a more objective risk measure, we combine a few different measures introducing a new risk measure, a composition of variance, semivariance and downside risk, in the following form

$$\rho(\mathbf{x}) = \alpha_V \mathbf{Var}[R_{\mathbf{x}}] + \alpha_S \mathbf{SVar}[R_{\mathbf{x}}] + \sum_{i=1}^{n} \alpha_i \mathbf{DSR}_{q_i}[R_{\mathbf{x}}] \tag{4.12}$$

for a given n, given $q_1, q_2, \ldots, q_n$ and $\alpha_V, \alpha_S, \alpha_1, \alpha_2, \ldots, \alpha_n$, chosen in a separate optimization process described further.

4.4 Classic and Practitioner Performance Measures

In portfolio optimization, there are two factors taken into consideration: the expected return rate and the risk, defined either by the variance of the return rate or by another measure. According to the model, investors always deal with the trade off between a high expected return rate and a low risk. However, in practice, investors often look for a universal measure to compare investments with different expected return rates and different risks, as they are willing to accept a higher risk if the expected return rate would be extremely high, or to accept a lower expected return rate if the risk would be low. Such a universal measure is also necessary to finally assess an investment, compare it with other investments built on different criteria, and to calculate how well the return of an investment compensates the risk taken.

There are a number of such measures, called performance measures (1), used by financial analysts and stock market investors. One of the most classic performance measures is the Sharpe ratio (1), denoted here as η_{Sh}, defined as the ratio of the return rate minus the return rate of the risk free investment to the standard deviation of the return rate. Another popular performance measure is the Treynor ratio (1), denoted here as η_{Tr}, which is the ratio of the return rate minus the return rate of the risk free investment to the beta coefficient (from the CAPM model described in (4)). There are also some performance measures with asymmetric preferences, such as the Sortino ratio (1), denoted here as η_{So}, defined as the ratio of the return rate minus the return rate of the risk free investment to the standard semideviation of the return rate, as well as some practitioner performance measures, such as the Sterling ratio (1), denoted here as η_{St}, which is the ratio of the return rate minus the return rate of the risk free investment to the maximum drawdown.

$$\eta_{Sh}(\mathbf{x}) = \frac{\mathbf{E}[\mathbf{R_x}] - \mathbf{r_0}}{\sqrt{\mathbf{Var}[\mathbf{R_x}]}}, \qquad \eta_{Tr}(\mathbf{x}) = \frac{\mathbf{E}[\mathbf{R_x}] - \mathbf{r_0}}{\beta_{\mathbf{x}}},$$

$$\eta_{So}(\mathbf{x}) = \frac{\mathbf{E}[\mathbf{R_x}] - \mathbf{r_0}}{\sqrt{\mathbf{SVar}[\mathbf{R_x}]}}, \qquad \eta_{St}(\mathbf{x}) = \frac{\mathbf{E}[\mathbf{R_x}] - \mathbf{r_0}}{\mathbf{MDD}[\mathbf{R_x}]},$$

where $\mathbf{MDD}[\mathbf{R_x}]$ denotes the maximum drawdown in the value of the portfolio $\mathbf{x}$, usually calculated in the context of a specific period. In this chapter, performance measures are used in the final evaluation of the discovered portfolios and in the comparison of portfolios built using different risk measures.

4.5 Data Description and Problem Definition

All the experiments are performed using data from the Paris Stock Exchange consisting of financial time series of price quotations of 40 different stocks, from the CAC40 index, over a specific time period. On the basis of these data, the return rates are estimated. Let

$$(\xi_k^{(1)}), (\xi_k^{(2)}), \ldots, (\xi_k^{(n)})$$

denote time series representing prices of stocks $A_1, A_2, \ldots, A_n$ respectively, i.e. for each $i = 1, 2, \ldots, n$ the sequence

$$\xi_0^{(i)}, \xi_1^{(i)}, \ldots, \xi_m^{(i)}$$

contains prices of the stock A_i in consecutive time instants of the specific period and $m+1$ denotes the length of the time period (the same for all the stocks A_i). Further, let

$$r_1^{(i)}, r_2^{(i)}, \ldots, r_m^{(i)}$$

denote a time series of return rates of the stock A_i in consecutive time instants of the specific period, i.e.

$$r_j^{(i)} = \frac{\xi_j^{(i)} - \xi_{j-1}^{(i)}}{\xi_{j-1}^{(i)}}, \quad \text{for } j = 1, 2, \ldots, m.$$

Therefore, for each $i = 1, 2, \ldots, n$, the expected return rate $r_i = \mathbf{E}[R_i]$ and the variance $\mathbf{Var}[R_i]$ may be computed respectively as

$$r_i = \frac{1}{m}\sum_{j=1}^{m} r_j^{(i)}, \quad \mathbf{Var}[R_i] = \frac{1}{m-1}\sum_{j=1}^{m}(r_j^{(i)} - r_i)^2. \tag{4.13}$$

Similarly, one may compute the correlation matrix Σ. For any portfolio $\mathbf{x} \in \mathbb{R}^n$, the expected return rate $r_{\mathbf{x}} = \mathbf{E}[R_{\mathbf{x}}]$ and the semivariance $\mathbf{SVar}[R_{\mathbf{x}}]$ may be computed respectively as

$$r_{\mathbf{x}} = \sum_{i=1}^{n} x_i r_i, \quad \mathbf{SVar}[R_{\mathbf{x}}] = \frac{1}{m-1}\sum_{j=1}^{m}(\sum_{i=1}^{n} x_i r_j^{(i)} - r_{\mathbf{x}})^2 \tag{4.14}$$

In our approach, the problem of portfolio optimization is expressed as finding a vector $\mathbf{x} \in \mathbb{R}^n$ minimizing a given risk measure $\rho : \mathbf{X} \to \mathbb{R}$ under the constraint that the expected return rate $r_{\mathbf{x}}$ is no lower than a given value $e \in \mathbb{R}$ (the value e is often defined by the expected return rate of a specific initial portfolio $\mathbf{x}_0$, given by the user instead of the value e explicite). It is considered in the context of a given set of stocks $A_1, A_2, \ldots, A_n$ with time series of its prices over a given time period. Such a problem with irregular risk measures constitutes a hard optimization problem.

4.6 Evolutionary Algorithms for Portfolio Optimization

In this chapter, we focus on solving the optimization problem defined in the previous section using evolution strategies. Three different algorithms are applied: a simple evolution strategy with the famous Rechenberg's *1/5 success rule* (called here **ES1**) (2), (12), a classic $ES(\mu, \lambda)$ evolution strategy with mutation parameters encoded in individuals (called here **ES2**) (2), (12) and a more advanced $ES(\mu, \lambda, \rho, \kappa)$ evolution strategy with mutation by multidimensional rotations (called here **ES3**) (13).

4.6.1 Search Space and Objective Function

In all the algorithms, the portfolio is encoded as an n-dimensional real number vector $\mathbf{x} \in \mathbb{R}^n$, where n is the number of stocks in the portfolio under consideration.

The search space is the entire n-dimensional real number space $\mathbb{R}^n$ – although some elements of the space may not represent portfolios, they may be normalized to fulfill the condition (4.3). The objective function is basically given by the risk measure $\rho : X \rightarrow \mathbb{R}$, but is slightly modified by some heuristic additional factors, which are also considered by certain financial experts and stock market analysts, such as the β coefficients of the portfolio evaluated and the specific initial portfolio. Therefore, the following objective function is studied

$$F(\mathbf{x}) = \frac{1}{1 + \varepsilon_1 \cdot \rho(\mathbf{x}) + \varepsilon_2 \cdot |\beta_{\mathbf{x}} - \beta_{\mathbf{x}_0}| + \varepsilon_3 \cdot \mathbf{Cov}[R_{\mathbf{x}}, R_{\mathbf{i}}]} \tag{4.15}$$

where $\mathbf{x}_0$ denotes the specific initial portfolio, usually given by the user, $R_{\mathbf{i}}$ denotes the return rate of the stock market index and $\beta_{\mathbf{x}}, \beta_{\mathbf{x}_0}$ denote the β coefficients of the portfolio evaluated $\mathbf{x}$ and the specific initial portfolio $\mathbf{x}_0$ respectively. Factors $\varepsilon_1, \varepsilon_2, \varepsilon_3$ are used to tune the algorithm and to adjust the importance of each component of the objective function.

These objective functions refer to some heuristics using parameters such as the β coefficient. By introducing the difference between the β_x of the generated portfolio and the β_{x_0} of the portfolio of reference, we penalize the portfolio having β_x far away from β_{x_0} of the reference. Nevertheless, the performance of a solution is defined in terms of expected return and risk of the portfolio over a test period as was mentioned in previous sections.

4.6.2 The ES1 Algorithm

Fig. 4.1 describes the **ES1** algorithm, based on a simple evolution strategy with mutation controlled by the Rechenberg's *1/5 success rule* and without recombination. First, the algorithm initializes the mutation parameter σ, being a real number, which defines the strength of the mutation, and creates an initial population $\mathscr{P}$, composed of N individuals, at random, either with uniform probability (each individual is drawn independently from the entire search space, and each point of the search space has equal probability of being drawn) or using a specific initial portfolio $\mathbf{x}_0$ (each individual is drawn independently from the neighborhood of the initial portfolio, usually by random noising its coefficients). Naturally, each individual in the initial population has to fulfill the financial constraints, thus, after being drawn, it undergoes a validation and if it does not satisfy all the constraints, it is drawn again or repaired. After creation, the population is evaluated according to the objective function F.

Each iteration of the algorithm consists of reproducing the old population, evaluating the new population and updating the mutation parameter. Reproduction begins with creating a parent population $\mathscr{P}^{(P)}$ of size N by random drawing N individuals from the original population $\mathscr{P}$, one by one, with the probability of being drawn for each individual in the original population $\mathscr{P}$ proportional to its value of the objective function (so-called *roulette wheel* method). After parent selection, each individual $\mathbf{x}$ from the parent population $\mathscr{P}^{(P)}$ undergoes the process of mutation, which consists of adding a random noise ε_i to each coefficient x_i. In the **ES1** algorithm, the random

```
ES1(F, N, σ0, θ1, θ2)
 1  σ ← σ0;
 2  𝒫 ← RANDOM-POPULATION(N);
 3  POPULATION-EVALUATION(𝒫, F);
 4  while not TERMINATION-CONDITION
 5  do
 6      𝒫^(P) ← PARENT-SELECTION(𝒫);
 7      𝒫^(C) ← MUTATION(𝒫^(P), σ);
 8      𝒫 ← REPLACEMENT(𝒫 ∪ 𝒫^(C));
 9      POPULATION-EVALUATION(𝒫, F);
10      SIGMA-UPDATING(σ, θ1, θ2);
```

Fig. 4.1. The **ES1** algorithm proposed to optimize an objective function F with a population $\mathscr{P}$ composed of N individuals, where $\sigma_0, \theta_1, \theta_2$ are algorithm parameters.

noise ε_i is drawn, separately for each individual and each coefficient, with gaussian distribution $\mathscr{N}(0,\sigma)$. Consequently, a new population $\mathscr{P}^{(C)}$ appears. Naturally, each individual in the children population $\mathscr{P}^{(C)}$ has to fulfill the financial constraints, thus, it undergoes a validation, as in creating the initial population. Finally, in replacement, the original population $\mathscr{P}$ is replaced with the best N individuals from the union of the original population $\mathscr{P}$ and the children population $\mathscr{P}^{(C)}$, in a deterministic manner. Afterwards, the population is evaluated according to the objective function F.

The parameter σ is updated according to the Rechenberg's *1/5 success rule*: it is increased by θ_1 when, in the last 5 iterations, the number of mutations leading to improvement of individuals exceeded 20% of total mutations, and is decreased by θ_2 when the number of mutations leading to deterioration of individuals exceeded 20% of total mutations. Normally, the algorithm terminates when it completes a specific number of iterations or when there is no increases in objective function values over a specific number of recent iterations.

4.6.3 The ES2 Algorithm

Fig. 4.2 describes the **ES2** algorithm, based on a classic evolution strategy with dynamic mutation using parameters encoded in individuals and with recombination. First, the algorithm creates an initial population $\mathscr{P}$, composed of N individuals, at random, as in the **ES1** algorithm. However, in this algorithm, each individual is equipped with one additional chromosome σ, being a real number vector of the same length as the main chromosome $\mathbf{x}$, used in the mutation, drawn with uniform probability over the $\mathbb{R}^n$ space. After creation, the population is evaluated according to the objective function F.
Afterwards, the population evolves under the influence of evolutionary operators, namely *parent selection*, *recombination*, *mutation*, and *replacement*, until a termination condition is satisfied. In parent selection, a new population $\mathscr{P}^{(P)}$ of size $4N$ appears by random drawing $4N$ individuals from the original population $\mathscr{P}$ (four

```
ES2(F, N, τ)
1  𝒫 ← RANDOM-POPULATION(N);
2  POPULATION-EVALUATION(𝒫, F);
3  while not TERMINATION-CONDITION
4  do
5      𝒫^(P) ← PARENT-SELECTION(𝒫);
6      𝒫^(C) ← RECOMBINATION(𝒫^(P));
7      𝒫^(C) ← MUTATION(𝒫^(C), τ);
8      𝒫 ← REPLACEMENT(𝒫 ∪ 𝒫^(C));
9      POPULATION-EVALUATION(𝒫, F);
```

Fig. 4.2. The **ES2** algorithm proposed to optimize an objective function F with a population $\mathscr{P}$ composed of N individuals, where τ is an algorithm parameter.

parents for each new descendant). Individuals are selected one by one with the probability of being drawn for each individual in the original population $\mathscr{P}$ proportional to its value of the objective function (so-called *roulette wheel* method). Next, in recombination, individuals from the parent population $\mathscr{P}^{(P)}$ are randomly matched in groups of four, and each four parents produce one descendant using one of two operators chosen randomly with equal probability: either the global intermediary recombination or the local intermediary recombination. In the first operator, coefficients of the descendant are arithmetic averages of coefficients of all the four parents. In the second operator, two parents are randomly chosen from these four parents chosen earlier, for each coefficient separately, and next, the coefficient of the descendant is the arithmetic average of coefficients of the two parents. Consequently, a new population $\mathscr{P}^{(C)}$ of size N appears. After recombination, each individual $\mathbf{x}$ from the children population $\mathscr{P}^{(C)}$ undergoes the process of mutation, which consists of adding a random noise ε_i to each coefficient x_i. In the **ES2** algorithm, the random noise ε_i is drawn, separately for each individual and each coefficient, with gaussian distribution $\mathscr{N}(0, \sigma_i)$, where σ_i denotes the i-th coefficient of the parameter σ encoded in the additional chromosome of the individual. The parameter σ undergoes the evolution as well: it is recombined together with the main chromosome $\mathbf{x}$ and mutated by adding to each coefficient σ_i a random noise ε_i drawn with gaussian distribution $\mathscr{N}(0, \tau)$. Finally, in replacement, the original population $\mathscr{P}$ is replaced with the best N individuals from the union of the original population $\mathscr{P}$ and the children population $\mathscr{P}^{(C)}$, in a deterministic manner. Afterwards, the population is evaluated according to the objective function F and the process of evolution is repeated until the algorithm completes a specific number of iterations or when there is no increases in objective function values over a specific number of recent iterations.

4.6.4 The ES3 Algorithm

Fig. 4.3 describes the **ES3** algorithm, based on an advanced evolution strategy with dynamic mutation by multidimensional rotations and with recombination. First, the algorithm creates an initial population, as in the **ES1** and **ES2** algorithm. However, in this algorithm, each individual is equipped with two additional chromosomes σ and α, used in the mutation, drawn with uniform probability over the $\mathbb{R}^n$ and $\mathbb{R}^{n(n+1)/2}$ space, respectively. Moreover, each individual has an associated parameter κ denoting the age of the individual (the number of iterations that the individual survive). After creation, the population is evaluated according to the objective function F.

ES3$(F, N, \tau_1, \tau_2, \kappa_0)$
1 $\mathscr{P} \leftarrow$ RANDOM-POPULATION(N);
2 POPULATION-EVALUATION$(\mathscr{P}, F)$;
3 **while not** TERMINATION-CONDITION
4 **do**
5 $\quad \mathscr{P}^{(P)} \leftarrow$ PARENT-SELECTION$(\mathscr{P})$;
6 $\quad \mathscr{P}^{(C)} \leftarrow$ RECOMBINATION$(\mathscr{P}^{(P)})$;
7 $\quad \mathscr{P}^{(C)} \leftarrow$ MUTATION$(\mathscr{P}^{(C)}, \tau_1, \tau_2)$;
8 $\quad \mathscr{P} \leftarrow$ REPLACEMENT$(\mathscr{P} \cup \mathscr{P}^{(C)}, \kappa_0)$;
9 $\quad$ POPULATION-EVALUATION$(\mathscr{P}, F)$;

Fig. 4.3. The **ES3** algorithm proposed to optimize an objective function F with a population $\mathscr{P}$ composed of N individuals, where τ_1, τ_2, κ_0 are algorithm parameters.

In general, the algorithm is similar to the **ES2** algorithm, but the difference lies in the mutation operator, described further, and in the replacement operator, where only the individuals with the age κ not exceeding the parameter κ_0 can survive.

In the **ES3** algorithm, mutation is controlled by two parameters σ and α, encoded in two additional chromosomes of each individual. The parameter α is used for drawing a random direction in the n-dimensional real number space and the parameter σ is used for drawing a random movement in this direction. As in the **ES2** algorithm, the additional chromosomes undergo the evolution as well. Details of the process of mutation may be found in (13).

4.7 Validation of the Approach

In order to validate our approach to portfolio optimization, we present the results of a large number of experiments performed on data from the Paris Stock Exchange, which include financial time series of daily price quotations of stocks constituting the CAC40 index, recorded over the period January 4, 1999 to June 29, 2007.

Before the actual experiments started, we performed some introductory experiments to calibrate the coefficients of the composite risk measure described in previous sections, i.e. n, $q_1, q_2, \ldots, q_n$ and $\alpha_V, \alpha_S, \alpha_1, \alpha_2, \ldots, \alpha_n$. A number of different settings was tried out and the settings chosen were those that produced the strongest correlation between the risk estimated a priori and the deviation of the future return rate evaluated a posteriori from the expected return rate estimated a priori. In spite of the composite risk measure, two other measures, namely the semivariance and the downside risk with $q = 0.5$ was applied in further experiments.

Each experiment began with choosing stocks $A_1, A_2, \ldots, A_n$ constituting financial instruments available for an investor. In each experiment, $n = 10$ or $n = 20$ stocks were randomly chosen from all the stocks in our financial database (including about 40 stocks). Next, an initial portfolio $\mathbf{x}_0$ was drawn corresponding to partitioning the investor's capital among the available stocks. Afterwards, a time instant t was chosen and the evolutionary algorithms presented in previous sections was applied to optimize the initial portfolio at the time t with respect to the risk measure ρ, i.e. to find an optimal portfolio $\mathbf{x}$ of equal or higher expected return rate $\mathbf{E}[\mathbf{x}] \geq \mathbf{E}[\mathbf{x}_0]$ and minimum risk measure $\rho(\mathbf{x})$. Each algorithm was run with each performance measure under consideration. All the computations concerning estimation of return rates were done over the period preceding the time instant t, i.e. over the time period $(t - \Delta t, t)$, where $\Delta t = 25$. Finally, all the optimal portfolios discovered were compared with respect to risk and performance measures.

In the validation, a few issues were studied. First, in order to assess the optimization quality, the risk $\rho(\mathbf{x}_0)$ of the initial portfolio $\mathbf{x}_0$ were compared with the risk $\rho(\mathbf{x})$ of the built optimal portfolio $\mathbf{x}$ and the risk $\rho(\mathbf{x}_*)$ of the reference portfolio $\mathbf{x}_*$ optimal according to the Markowitz model (it is worth noticing that the reference portfolio $\mathbf{x}_*$ need not be optimal according to the performance measure ρ different than variance). Next, in order to compare the portfolios optimal according to different risk measures, four performance measures were calculated for each of them.

Table 4.1 shows a summary of results concerning the risk and performance comparison. Each row corresponds to an experiment, repeated 10 times to avoid random influence, with different initial portfolio and different risk measure (in fact, the table shows the best of the 30 results obtained by running 10 times each of the three optimization algorithms). One may see that the built portfolio $\mathbf{x}$ had always lower risk than the initial portfolio $\mathbf{x}_0$. It is worth noticing that the built portfolio $\mathbf{x}$ also had always lower risk than the reference portfolio $\mathbf{x}_*$, which proved that a portfolio optimal according to variance is not optimal according other risk measures.

Next, the results of the experiments were studied with the aim to assess and compare the three evolutionary algorithms with respect to the quality of solutions and the computing time. In the validation, the quality of solutions was measured by a number ω of cases when the optimum found by the algorithm under study was better than the optima found by the other two algorithms.

Table 4.2 shows performances of the three evolutionary algorithms. The second column contains the number ω. The third column contains the average computing time. Not surprisingly, the last algorithm had the best performance, but it also required the longest computing time.

Finally, the optimal portfolios discovered were examined in order to assess and compare their profitability. For each experiment, a future time period $(t, t + \Delta t)$, where $\Delta t = 25$, unknown during the optimization process, were considered. Over that period, actual return rates of all the three portfolios, namely $\mathbf{x}_0, \mathbf{x}, \mathbf{x}_*$, were computed and compared. Table 4.3 shows the comparison of the actual return rates of the three portfolios. Each row corresponds to an experiment, repeated 30 times. In the first, second and third column, there is the number of cases where the portfolio $\mathbf{x}_0, \mathbf{x}, \mathbf{x}_*$, respectively, turned out to outperform the others. In the next two columns, there is

Table 4.1. Risk and performance measures for the initial portfolio $\mathbf{x}_0$, the built optimal portfolio $\mathbf{x}$, and the reference portfolio $\mathbf{x}_*$ optimal according to the Markowitz model for $n = 10$ and $n = 20$ stocks

n	ρ	$\rho(\mathbf{x}_0)$	$\rho(\mathbf{x})$	$\rho(\mathbf{x}_*)$	$\eta_{Sh}(\mathbf{x})$	$\eta_{Tr}(\mathbf{x})$	$\eta_{So}(\mathbf{x})$	$\eta_{St}(\mathbf{x})$
10	SVar	0.8374	0.4036	0.4463	0.9724	0.6148	0.7724	0.4714
10	DSR	0.8374	0.3482	0.4463	0.9874	0.5242	0.8141	0.4716
10	Comp.	0.8374	0.2567	0.4463	0.8643	0.5163	0.6814	0.3741
10	SVar	0.7464	0.4156	0.5183	0.9163	0.5916	0.8314	0.4147
10	DSR	0.7464	0.5012	0.5183	0.8914	0.6149	0.7913	0.3742
10	Comp.	0.7464	0.3043	0.5183	0.9012	0.5914	0.7831	0.4825
20	SVar	0.7913	0.4284	0.4384	0.9824	0.5817	0.6971	0.4279
20	DSR	0.7913	0.3817	0.4384	0.9305	0.6184	0.6814	0.4014
20	Comp.	0.7913	0.3784	0.4384	0.9745	0.6384	0.7194	0.4765
20	SVar	0.8012	0.4892	0.6103	0.9242	0.5942	0.8104	0.3975
20	DSR	0.8012	0.4356	0.6103	0.9847	0.6278	0.8042	0.4146
20	Comp.	0.8012	0.3874	0.6103	0.8942	0.6247	0.6792	0.4714

Table 4.2. Quality of solutions (a number of cases when the optimum found by the algorithm under study was better than the optima found by the other two algorithms) and computing time for the three evolutionary algorithms

Algorithm	ω	Computing Time
ES1 - SVar	28	17 s
ES1 - DSR	25	24 s
ES1 - Comp.	19	39 s
ES2 - SVar	34	18 s
ES2 - DSR	32	27 s
ES2 - Comp.	23	42 s
ES3 - SVar	38	33 s
ES3 - DSR	43	41 s
ES3 - Comp.	58	97 s

the number of cases where the portfolio $\mathbf{x}$ outperformed the buy & hold strategy and the stock market index, respectively. It can be seen that the built portfolio $\mathbf{x}$ and the reference portfolio $\mathbf{x}_*$ usually led to higher profits than the initial portfolio $\mathbf{x}_0$. Moreover, in some cases, the built portfolio $\mathbf{x}$ outperformed the stock market index.

4.8 Conclusions and Perspectives

This chapter concerns an evolutionary approach to portfolio optimization. It rejects some assumptions from classic models, introduces alternative risk measures such as semivariance, downside risk and a new composite risk measure, and proposes three evolutionary algorithms to solve the optimization problem. In order to validate the approach proposed, results of a number of experiments on real-life data from the Paris Stock Exchange are presented. Built investment strategies are compared with

Table 4.3. Comparison of actual future return rates of the initial portfolio $\mathbf{x}_0$, the built optimal portfolio $\mathbf{x}$, and the reference portfolio $\mathbf{x}_*$ optimal according to the Markowitz model (the number of cases where the portfolio outperformed the others) as well as the number of cases where the built optimal portfolio $\mathbf{x}$ outperformed the buy & hold strategy and the stock market index

$\mathbf{x}_0$	$\mathbf{x}$	$\mathbf{x}_*$	B&H	Index
0	27	3	30	4
1	27	2	30	6
0	28	2	30	7
4	23	3	29	5
5	24	1	30	4
0	28	2	29	3
2	25	3	30	8
0	27	3	30	9
3	25	2	28	4
0	29	1	30	6
2	26	2	30	6
4	25	1	30	5

competitive ones built with other risk measures as well as with the buy & hold strategy and the stock market index. Finally, the results show that the approach proposed is capable of investing more efficiently than the simple buy & hold strategy and the stock market index in some cases.

However, the approach proposed can be still improved by additional studies on the objective and fitness function as well as on risk measures. Moreover, modifying evolutionary operators, especially recombination, can increase the efficiency of the algorithms. Additional effort should be put on methods of portfolio validation in order to eliminate unacceptable solutions at the moment of its creation. It is also worth noting that the evolutionary approach in stock trading is still in an experimentation phase. Further research is needed, not only to build a solid theoretical foundation in knowledge discovery applied to financial time series, but also to implement an efficient validation model for real-life data. The presented approach seems to constitute a practical alternative to classical theoretical models.

References

[1] Aftalion F, Poncet P (2003) Les Techniques de Mesure de Performance. Economica

[2] Back T (1995) Evolutionary Algorithms in Theory and Practice. Oxford University Press, New York

[3] Harlow H V (1991) Asset allocation in a downside-risk framework. In: Financial Analysts Journal:30-40.

[4] Huang C F, Litzenberger R (1988) Foundations for Financial Economics. North-Holland

[5] Korczak J, Lipinski P (2004) Evolutionary Building of Stock Trading Experts in a Real-Time System. In: Proceedings of the 2004 Congress on Evolutionary Computation :940-947
[6] Korczak J, Lipinski P, Roger P (2002) Evolution Strategy in Portfolio Optimization. In: Collet P (eds) Artificial Evolution. Lecture Notes in Computer Science 2310:156-167
[7] Korczak J, Roger P (2002) Stock Timing using Genetic Algorithms. In: Applied Stochastic Models in Business and Industry :121-134
[8] Lipinski P, Korczak J (2004) Performance Measures in an Evolutionary Stock Trading Expert System. In: Bubak M, van Albada G, Sloot P, Dongarra J (eds) Proceedings of the International Conference on Computational Science. Lecture Notes in Computer Science 3039:835-842
[9] Lipinski P, Winczura K, Wojcik J (2007) Building Risk-Optimal Portfolio Using Evolutionary Strategies. In: Giacobini M (eds) Proceedings of EvoWorkshops 2007. Lecture Notes in Computer Science 4448:208-217
[10] Loraschi A, Tettamanzi A (1996) An Evolutionary Algorithm for Portfolio Selection Within a Downside Risk Framework. In: Forecasting Financial Markets. Wiley :275285.
[11] Markowitz H (1952) Portfolio Selection. In: Journal of Finance 7:77-91
[12] Schwefel H-P (1995) Evolution and Optimum Seeking. John Wiley and Sons
[13] Schwefel H-P, Rudolph G (1995) Contemporary Evolution Strategies. In: Advances in Artificial Life. Springer :893-907.
[14] Tsang E P K, Yung P, Li J (2004) EDDIE-automation, a decision support tool for financial forecasting. In: Journal of Decision Support Systems 37:559-565

5

Evolutionary Stochastic Portfolio Optimization

Ronald Hochreiter

Department of Statistics and Decision Support Systems, University of Vienna, Vienna, Austria ronald.hochreiter@compmath.net

Summary. In this chapter, the concept of evolutionary stochastic portfolio optimization is discussed. Selected theory from the fields of Stochastic Programming, evolutionary computation, portfolio optimization, as well as financial risk management is used to derive a generalized framework for computing optimal financial portfolios given an uncertain future using a probabilistic risk measure approach. A set of structurally different risk measures - Standard Deviation, Mean-absolute Downside Semi Deviation, Value-at-Risk, and Expected Shortfall - which are commonly used for practical portfolio management purposes have been selected to substantiate the approach with numerical results.

5.1 Introduction

During recent years, the increasing need for financial decision optimization algorithms for complex, and often non-convex optimization problems in the area of financial engineering led to a significant increase in the use of biologically inspired algorithms for practical financial management purposes, see e.g. (4). In this chapter, the well-known technique of Stochastic Programming is applied to solve financial portfolio optimization problems under uncertainty based on probabilistic risk measures. Evolutionary computation methods are exploited to allow for a generalization of the underlying problem structure and to solve the resulting optimization problems numerically in a systematic way. This chapter is organized as follows. The remainder of section 5.1 contains a short summary of Stochastic Programming, as well as an overview of previous evolutionary approaches to portfolio optimization. Section 5.2 surveys the field of (stochastic) portfolio optimization, and discusses the generalization of the well-known Markowitz portfolio approach to scenario-based Stochastic Programming. Furthermore, the loss-distribution based approach and its relation to probabilistic risk measures within the evolutionary optimization process is explained. Section 5.3 presents ideas and strategies for implementing a successful evolutionary portfolio optimization framework based on the discussion in section 5.2, and contains a section on different constraint handling techniques. Section 5.4 outlines details of the implementation, and presents a set of numerical results, while section 5.5 concludes this chapter.

R. Hochreiter: *Evolutionary Stochastic Portfolio Optimization*, Studies in Computational Intelligence (SCI) **100**, 67–87 (2008)
www.springerlink.com

5.1.1 Stochastic Programming

To summarize the concept of Stochastic Programming, consider the classical deterministic optimization problem, where a decision maker aims at finding an optimal (numerical) decision $x \in \mathbb{R}^n$ by minimizing a deterministic cost function $f(\cdot)$ (or by maximizing a profit function respectively) given a set $\mathscr{X}$ of constraints, which generally consists of various physical, organizational, and regulatory restrictions. The mathematical formulation of this problem can be simplified to the formulation shown in Equ. (5.1).

$$\begin{array}{ll} \text{optimize } x: & f(x) \\ \text{subject to} & x \in \mathscr{X}. \end{array} \tag{5.1}$$

During the 1950s Stochastic Programming was initiated by the seminal papers of Dantzig (6) and Beale (2). It is one of the main techniques for modeling and solving decision optimization problems under uncertainty, which is a class of optimization problems inherent to the application area of financial engineering. Due to the recent developments both from the computational and the algorithmic viewpoint, solutions of large stochastic programs are generally computable using standard computer hardware. The idea is to replace deterministic parameters by probability distributions on some probability space $(\Omega, \mathscr{F}, \mathbb{P})$, which will be denoted by Ξ in the following, and to optimize a stochastic cost (or profit) function $f(\cdot,\cdot)$ over some probability functional $\mathbb{F}$. A common choice regarding this functional is the expectation $\mathbb{E}$. As Rockafellar (23) points out, expectations are only suitable for situations where the interest lies in long-range operation, and stochastic ups and downs can safely average out, which is not the case for controlling financial market risk. The recent progress of unifying probabilistic risk measures, as presented in the seminal paper by Artzner et al. (1) on coherent risk measures, motivated for using probability functionals based on such financial risk measures. See the recent book (22) for more details on modeling, measuring, and managing risk for this class of optimization applications. A different view on the integration of risk measures using the concept of deviation measures is shown in (25). In summary, the resulting mathematical meta-formulation of a stochastic program for arbitrary probability functionals is shown in eq. (5.2).

$$\begin{array}{ll} \text{optimize } x: & \mathbb{F}(f(x, \Xi)) \\ \text{subject to} & (x, \Xi) \in \mathscr{X}. \end{array} \tag{5.2}$$

However, a concrete reformulation of this meta-model into some model, which can be solved with a numerical optimizer depends to a high degree on the chosen probability functional, as well as on the structure of the underlying probability space. The interested reader is referred to (26) for a recent theoretical overview of the area of Stochastic Programming, and to (37) for Stochastic Programming languages, environments, and applications. Interestingly, evolutionary approaches have not been applied to a wide range of real stochastic programming problems so far, with only a few examples available, e.g. recent works in the field of chemical batch processing, see (34) and (32).

5.1.2 Stochastic Portfolio Optimization

The analysis in this chapter is based on the stochastic single-stage (single-period) scenario-based risk-return portfolio optimization problem. This approach belongs to the class of static stochastic programming, as there are no dynamic asset rebalancing events taking place at intermediate time-stages between the day of the decision ($t = 0$) and the end of the investment horizon ($t = T$).

The scenario-based approach can be categorized as a generalization of the class of bi-criteria mean-variance portfolio optimization problems, which were initiated by the seminal work of Markowitz (see (17)) on portfolio theory. The main difference is that the whole underlying probability space is used for optimization purposes, such that a general class of risk measures can be applied. It is still straightforward to calculate parameters for the pure Markowitz approach by computing the mean vector and the correlation matrix of the given probabilistic model. This issue is discussed in section 5.2 in more detail.

However, the generality necessitates a reformulation of the stochastic meta-model, as shown in eq. (5.2), for each risk functional, e.g. if a deterministic-equivalent formulation is desired. To avoid this problem, an evolutionary approach may be used, such that intermediate reformulations are not necessary. This is a clear advantage considering that reformulations have to be built on a sound theoretical framework, and might need to be reconsidered, whenever the underlying structure changes slightly.

5.1.3 Evolutionary Portfolio Optimization

In the evolutionary computation approach, the underlying stochastic portfolio optimization problems will be solved by adapting a standard genetic algorithm, e.g. as surveyed by (3) and summarized in fig. 5.1, to handle the stochastic meta-model directly without any reformulation. Using such an evolutionary approach allows for application of the same optimization technique and decision support framework for every loss distribution-based risk measure, regardless of its underlying structure. This general idea was already outlined in (12).

The portfolio optimization problem naturally fits into the general problem structure usually handled with evolutionary algorithms. Evolutionary approaches have been successfully applied to different classes of portfolio optimization problems, see e.g. (28), (27), (31), (11), (15), as well as the references therein, or refer to (16). Some of the approaches mention the field of Stochastic Programming explicitly, e.g. (33), and (38). Most of the proposed methods put the main focus on multi-criteria optimization, see especially (7) for portfolio optimization from the viewpoint of Multiple Criteria Decision Making (MCDM) applying different heuristic solution techniques including genetic algorithms. This is clearly due to the multi-criteria nature of the portfolio optimization problem. However, most approaches are restricted to the pure Markowitz approach, i.e. using solely the expectation vector and correlation matrix of the financial assets under consideration, or use hybrid techniques, which are still

GA(F,N)
1 $\mathscr{P} \leftarrow$ GENERATE RANDOM-POPULATION(N);
2 POPULATION-EVALUATION($\mathscr{P},F$);
3 **while not** TERMINATION-CONDITION
4 **do**
5 $\mathscr{P}^{(P)} \leftarrow$ PARENT-SELECTION($\mathscr{P}$);
6 $\mathscr{P}^{(C)} \leftarrow$ RECOMBINATION($\mathscr{P}^{(P)}$);
7 $\mathscr{P}^{(C*)} \leftarrow$ MUTATION($\mathscr{P}^{(C)}$);
8 POPULATION-EVALUATION($\mathscr{P}^{(C*)},F$);
9 $\mathscr{P} \leftarrow$ REPLACEMENT($\mathscr{P} \cup \mathscr{P}^{(C*)}$);

Fig. 5.1. Meta-heuristic: Genetic Algorithm with a population N and a fitness function F.

conceptually based on the Markowitz case. One of the main problems from the viewpoint of financial engineers is the estimation of the correct correlation matrix, which is crucial for obtaining relevant results.

5.2 Portfolio Optimization

5.2.1 Mean-Variance Portfolio Optimization

The classical bi-criteria portfolio optimization problem based on Markowitz can be summarized as follows: An investor has to choose a portfolio from a set of (financial) assets $\mathscr{A}$ with finite cardinality $a = |\mathscr{A}|$ to invest her available budget. The bi-criteria problem stems from the fact that the investor aims at maximizing her return while aiming at minimizing the risk of the chosen portfolio at the same time. Markowitz initially proposed to minimize the variance of the portfolio subject to reaching at least an expected return of μ, i.e. given a correlation matrix C and an expectation vector $\bar{e}$, the resulting optimization problem can be written as a quadratic program, and is shown in eq. (5.3). The parameter μ is the minimal required amount of return specified by the investor.

$$\begin{array}{ll} \text{minimize } x: & x^T C x \\ \text{subject to} & \bar{e}^T x \geq \mu, \\ & \mathbf{1}^T x = 1, \\ & x \geq 0. \end{array} \tag{5.3}$$

This optimization problem can be seen as a stochastic program. It should also be noted, that Markowitz initially also suggested to maximize the expectation, and limit the variance by some upper limit. However, this optimization problem was not directly (re-)formulated for some time, because of the challenge of solving quadratically constrained optimization programs numerically.

5.2.2 Scenario-based Portfolio Optimization

The main advantage of scenario-based portfolio optimization is that the investor is able to choose from a large set of different methods to design and generate the representation of the future uncertainty. Thus, subjectivity can be integrated by applying different generation techniques. Historical data sets, time series estimation and simulation procedures, as well as various random sampling techniques are commonly used, and combined. The latter technique allows for integrating complex dependency structures between assets, e.g. copulas, which are an important tool for modern quantitative financial risk management, see (8). This is a clear advantage over the Markowitz case, where the investor is restricted to the first and second moment of the underlying probability space, and faces, in financial terms, the unrealistic restriction of linear dependence.

In the static stochastic programming case, the underlying uncertainty is represented as a multi-variate probability distribution on the respective probability space. The distribution may either be continuous or discrete. In most practical optimization models, the distribution has to be discrete in order to numerically compute a solution, i.e. continuous distributions have to be discretized. Analytical solutions based on continuous distributions can only be derived in unrealistically simple settings. In the application-centric scenario-based stochastic programming approach, the discretized probability distribution, which is used to compute the optimal decision is called *scenario set*. This discrete set of scenarios $\mathscr{S}$ has finite cardinality $s = |\mathscr{S}|$, where each s_i is equipped with a non-negative probability $p_i \geq 0$, and $\sum_{j=1}^{s} p_j = 1$. From the financial market viewpoint, each scenario contains one possible set of joint future returns of all a assets under consideration for the portfolio. Using the terminology of Markowitz, each scenario contains the discounted anticipated return of each asset.

Table 5.1. Example of return scenario matrix

	Asset 1	Asset 2	Asset 3	Asset 4	Asset 5	Asset 6	Asset 7	...
Scenario 1	1.0349	1.0332	1.0105	1.0097	1.0084	1.0302	1.0041	
Scenario 2	1.0117	1.0309	1.0141	1.0070	1.0260	1.0195	1.0181	
Scenario 3	0.9966	0.9853	0.9892	0.9947	0.9928	0.9827	0.9914	
Scenario 4	0.9617	0.9595	0.9782	0.9709	0.9790	0.9720	0.9857	
Scenario 5	1.0110	1.0059	1.0051	0.9996	1.0011	1.0038	1.0061	
Scenario 6	1.0071	1.0101	1.0045	1.0125	1.0080	1.0133	1.0038	
Scenario 7	1.0203	1.0207	1.0186	1.0212	1.0156	1.0409	1.0241	
...								

As mentioned above, there exists a plethora of techniques and tools to construct scenarios. The choice of the method is based the subjective taste of the respective investor, such that it is possible to adapt the stochastic decision framework to the beliefs of the investor by choosing an appropriate method to create scenarios. If a continuous (multi-variate) probability distribution is chosen, an discretization of the underlying continuous probability space, as shown e.g. by (21), has to be applied. The application of historical data is straight forward, while the use of other technique to generate the final set of scenarios, e.g. time-series estimation and simulation techniques, might need some further considerations. Finally, random sampling procedures can be used to integrate complex dependency structures. Additionally, there exists a set of methods, which aim at optimizing scenario sets, see e.g. (13) for techniques on how to generate and subsequently reduce the complexity of scenarios by minimizing probability metrics between the original set and the final set.

Table 5.1 shows a small part of one possible instance of a scenario matrix. In this particular example, historical data from selected Dow Jones STOXX Supersector indices are presented. When investors use plain historical data, it is common not to use recent historical data, but data from a time horizon, which is anticipated to represent the future.

Depending on the selected scenario generation method, the probabilities of each scenario may exhibit a different structure. Huge scenario sets, which are the result of large-scale simulations or sets based on historical data commonly consist of equiprobable scenarios. Non-equiprobable scenarios are often due to scenario reduction techniques, where similar scenarios are combined into one scenario to reduce the size of the scenario set, which may affect the computation time.

Using a representative scenario set eliminates the need for estimating the correct correlation matrix, and makes the optimization problem less sensitive to estimation errors. As already mentioned above, to obtain the input for the classical Markowitz approach, the expectation vector of size a, as well as the correlation matrix of size $a \times a$ can be calculated from the scenario set, which clearly leads to a loss of information.

5.2.3 Loss Distributions and Risk Mappings

Let $x \in \mathbb{R}^a$ be some portfolio. Without loss of generality, we use budget normalization, i.e. $\sum_{a \in \mathscr{A}} x_a = 1$. Each component x_i of the portfolio vector denotes the fraction of the available budget B invested into the respective asset i. We may now rewrite the scenario set $\mathscr{S}$ as a matrix S to calculate the discrete Profit & Loss (P&L) distribution ℓ for some portfolio x, which is simply the cross product $\ell_x = \langle x, S \rangle$. We will denote the loss distribution as ℓ in the following. Finally, let $x^*_\rho \in \mathbb{R}^a$ denote the optimal portfolio given some risk measure ρ and ℓ^*_ρ denote the respective ρ-optimal discrete loss distribution.

When we reconsider the bi-criteria aspect of this portfolio optimization problem, we can map the loss distribution ℓ to both dimensions, which are important for a successful financial portfolio management. The first dimension - the return (reward, value) dimension - is obtained by calculating the expectation $\mathbb{E}(\ell_x)$. The second dimension - the risk dimension - is calculated by applying a risk mapping $\rho(\ell_x) : \mathbb{R}^s \to \mathbb{R}$ of the chosen risk measure ρ to the loss distribution. The risk dimension received special importance due to enormous financial losses especially during the 1990s and regulatory frameworks like Basel II, as well as the academic discussion about coherence of risk measures. See also (18) for a in-depth discussion on quantitative risk management and how risk measures can be used for practical risk management purposes.

5.2.4 Risk Measures

In this loss distribution-based notation, risk measures are used to define specific risk mappings of loss distributions ℓ_x for some portfolio x as shown above. Each probabilistic risk measure can be classified by the structure of its underlying statistical measure. Thus, there are different classes of risk measures, e.g. deviation-based, and quantile-based. An example for a deviation-based risk measure is the Standard Deviation, which conforms to the Markowitz case. Examples for quantile-based risk measures include the quantile itself, which equals the Value-at-Risk (VaR) case, and the expectation conditional to the quantile, which results in the Expected Shortfall (ES). These measures are described below.

It should be noted, that the ES is also called Conditional Value-at-Risk (CVaR), because of the conditional expectation given the quantile, but should ideally be called Average Value-at-Risk (AVaR), as elaborated by (9), especially when multi-period, dynamic risk measures are considered.

For the purpose of this analysis, four of the most commonly applied risk measures ρ for portfolio management have been chosen to show comparative analysis for practical portfolio risk management - Standard Deviation, Mean Absolute Downside Semi Deviation (MADSD), Value at Risk (VaR), as well as Expected Shortfall (ES).

- **Standard Deviation.** Using the Standard Deviation for scenario-based portfolio optimization resembles the classical Markowitz approach by calculating the risk as shown in eq. (5.4).

$$\rho \equiv \sigma = \sqrt{\sum_{i \in \mathscr{S}} p_i(\ell_i - \mathbb{E}(\ell))^2}. \tag{5.4}$$

- **Mean Absolute Downside Semi Deviation.** The mean absolute deviation (MAD) has been used as a replacement for variance, because of its linearity. In our case, we consider the mean absolute downside semi-deviation, which only considers the lower part of the loss function, and thus eliminates one of the most significant criticism over two-sided deviation-based risk measures, and is shown in eq. (5.5).

$$\rho \equiv \text{MADSD} = \sum_{i \in \mathscr{S}} p_i |\max(\ell_i - \mathbb{E}(\ell), 0)|. \tag{5.5}$$

- **Value-at-Risk.** The Value-at-Risk at level (1-α) is the α-Quantile of the loss distribution. This risk measure gained significant importance, especially for regulatory purposes. For discrete distributions it is the $\alpha \cdot s$ smallest value of ℓ_x. While the Mean-VaR optimization problem is non-convex, an evaluation of the VaR (as shown in eq. (5.6)) of a distribution is straightforward.

$$\rho \equiv \text{VaR}_\alpha = \inf\{l \in \mathbb{R} : \mathbb{P}(l > \ell) \leq 1 - \alpha\} = \inf\{l \in \mathbb{R} : F_l(\ell) \leq \alpha\} \tag{5.6}$$

- **Expected Shortfall** The Expected Shortfall (ES) is used as a substitute for VaR mainly because a linear programming reformulation is available, and it additionally exhibits the property of being a coherent risk measure. It can be notated as the expectation over the quantile (VaR) of the loss distribution as shown in Equ. (5.7).

$$\rho \equiv \mathrm{ES}_\alpha = \mathbb{E}(\ell | \ell \leq \mathrm{VaR}_\alpha) \tag{5.7}$$

Even without considering additional portfolio constraints, the risk measures listed above result in completely different optimization program formulations, e.g. a quadratic optimization reformulation in the case of Markowitz, as shown in eq. (5.3). For ES, a linear programming reformulation for a finite set of scenarios exists, which has been presented in (24) and (35). Finally, there exists a variety of optimization heuristics to solve the non-convex Mean-VaR portfolio optimization problem, see (10) and the references therein. It should be noted that these reformulations are only valid if standard, convex, non-integer constraints are included. However, the scenario-based evolutionary optimization framework enables a common treatment of risk measures in one systematic way.

5.2.5 Portfolio Constraints

For practical purposes, it is crucial to include real-world constraints. Let $\mathscr{X}$ denote the set of organizational, regulatory and physical constraints. The list of constraints, which are commonly included into $\mathscr{X}$ are summarized below.

- **Upper and lower limits on asset weights.** Especially due to regulatory issues, the fractional amount of budget into each asset is limited, i.e. there may be individual lower and upper bounds on the portfolio weights.

  $$l_a \leq x_a \leq u_a \; \forall a \in \mathscr{A}$$

  The constraint of disallowing short selling is commonly implicitly added. This can be explicitly modeled by setting $l = 0$ in this formulation.
- **Minimum profit.** An important constraint is the requirement of a desired and constant minimal return M, i.e.

  $$\mathbb{E}(\ell_x) \geq M$$

  If it is not possible to reach this level of expected return with any portfolio given one specific scenario set, then the optimization problem has no feasible solution.
- **Maximum risk.** Likewise, the risk given a respective risk measure ρ can be limited by a constant level R, i.e.

  $$\rho(\ell_x) \leq R$$

  Different risk measures, i.e. even more than one, can be integrated and combined, to allow for a fulfillment of both regulatory needs and subjective perceptions of the decision maker.
- **Cardinality constraints.** Cardinality constraints limit the number of assets in the portfolio. They can either be strict

$$\#(x_a > 0) = K,\ 0 < K \leq a,\ K \in \mathbb{N},$$

or lower and upper bounds on the cardinality can be defined, i.e.

$$K_l \leq \#(x_a > 0) \leq K_u,\ 0 < K_l < K_u \leq a,\ K_l \in \mathbb{N}, K_u \in \mathbb{N}.$$

Prior to the optimization, a feasibility check has to be conducted to analyze, whether the set of constraints does lead to a feasible solution at all. Constraints are another problematic dimension for reformulating the stochastic meta-model. While some of the constraints and their extensions can be reformulated as convex optimization problems, more involved constraints like cardinality constraints, as well as non-linear, non-differentiable (i.e. combination of fixed and non-linear flexible cost) transaction cost structures, as well as buy-in thresholds, or round lots lead to non-convex, non-differential models, and have motivated the application of various heuristics such as evolutionary computation techniques. The handling of constraints is discussed in section 5.3.3 below.

An important visualization tool for financial managers is the so called efficient frontier depicting the trade-off between risk and return, to allow for picking a portfolio given the respective risk attitude. The minimum expected profit constraint is necessary to calculate these frontiers, i.e. by iterating over a set of minimal expected profits, which are necessarily calculated from the scenario set $\mathscr{S}$. These are shown and described in detail in the numerical results section 5.4.4 below.

5.2.6 Multi-criteria objective reformulations

As mentioned in the introduction, the portfolio optimization problem is clearly an MCDM problem, and can be notated as shown in eq. (5.8).

$$\begin{array}{ll} \text{maximize}\, x: & \mathbb{E}(\ell_x), \\ \text{minimize}\, x: & \rho(\ell_x), \\ \text{subject to} & x \in \mathscr{X}. \end{array} \tag{5.8}$$

In a non-multi-criteria setting, three different main portfolio optimization formulations are used. Either minimize the risk subject to reaching a minimum (expected) return M, i.e.

$$\begin{array}{ll} \text{minimize}\, x: & \rho(\ell_x) \\ \text{subject to} & \mathbb{E}(\ell_x) \geq M \\ x \in \mathscr{X}, & \end{array} \tag{5.9}$$

or maximize the value (expectation), subject to a maximum risk level R, i.e.

$$\begin{array}{ll} \text{maximize}\, x: & \mathbb{E}(\ell_x) \\ \text{subject to} & \rho(\ell_x) \leq R \\ & x \in \mathscr{X}, \end{array} \tag{5.10}$$

or apply the classical criteria-weighted model, where an additional risk-aversion parameter κ is defined, i.e.

$$\begin{array}{ll} \text{maximize } x: & \mathbb{E}(\ell_x) - \kappa\rho(\ell_x), \\ \text{subject to} & x \in \mathscr{X}. \end{array} \tag{5.11}$$

Equivalence of these three formulation has been proven for convex risk measures in (14). In the following, we will use the third formulation (5.11), which is a direct reformulation our bi-criteria problem into a single-criteria problem, and will be used in the evolutionary approach presented in the next Section. This mapping also allows for calculating the critical line conveniently, when the optimization is iterated over the parameter κ.

5.3 Evolutionary Portfolio Optimization

5.3.1 Genotype structure

In contrast to a portfolio picking problem, where an investor tries to select a sub-set of assets out of a large set of assets, and has to solve a combinatorial problem with a binary decision for each asset, we face a real-valued problem, because each asset decision denotes a fraction of the total budget to invest. There are basically two ways to define the genotype structure of the underlying portfolio problem. One may either use real-valued genes, or bit-encoding techniques. Most propositions apply the straight-forward real-value formulation. An advantage of a bit-encoding approach is that the problem can be solved using readily available software based on bit-encoded chromosomes. A comparison between discrete and continuous genotypes for the portfolio selection problem has been studied in (29). The scenario-based stochastic framework presented below makes use of a real-valued representation, for a number of reasons. First, the software has been developed from the scratch, so the restriction of available software does not need to be addressed. Secondly, this approach offers a natural scalability for large scenario sets with many assets, and finally it is possible to handle special operations like the δ-normalization directly using real-valued vector operations.

Table 5.2 displays the structure of the portfolio weight chromosome and the fitness calculation. For each chromosome (portfolio) x_c its respective loss distribution ℓ_{x_c} is calculated, and the expectation $\mathbb{E}(\ell_{x_c})$ as well as the associated risk $\rho(\ell_{x_c})$ is computed. These two dimensions (return and risk) are mapped to a single value via a risk-aversion parameter κ, which can be used to control the risk-aversion of the investor. The aim is to maximize the fitness value, i.e. to maximize the expected return and simultaneously minimize the expected risk. The example shows Value-at-Risk at 90% with $\kappa = -1$. In this example, it is noteworthy to observe the inversion of the sign of κ. This is due to the fact that we are considering loss functions, and a higher value of a quantile-based risk measure (such as VaR is this case) means lower risk in this case. The bottom row of table 5.2 shows one concrete portfolio from Section 5.4.4 ($\text{VaR}_{0.1}$, Data T_1), see also fig. 5.7 (left). It should be noted, that if the risk measure has a different unit system, a normalization factor has to be applied to ensure an equal numerical treatment of both dimensions, either by modifying κ, or integrating the normalization into the evaluation functions.

Table 5.2. Example chromosome - fitness: κ-weighted return and risk

	Asset 1	Asset 2	...	Asset a	Return $\mathbb{E}(\ell_x)$	Risk $\rho(\ell_x)$	Fitness $\mathbb{E}(\ell_x) - \kappa\rho(\ell_x)$
$x_c =$	$(x_1$	x_2	...	$x_a)$	$\mathbb{E}(\ell_{x_c})$	$\text{VaR}_{0.1}(\ell_{x_c})$	$\mathbb{E}(\ell_{x_c})+\text{VaR}_{0.1}(\ell_{x_c})$
	$(0$	0.1688	...	$0)$	1.0024	0.9852	1.9876

5.3.2 Evolutionary operators

The effect of using different crossovers for the Markowitz-based portfolio selection problem has been investigated in (30), where three types of crossovers have been compared: discrete N-point crossovers with $N = 3$, intermediate crossovers, as well as BLX-α crossovers. In this analysis, we use the following evolutionary operators:

- **Mutate by factor.** Randomly select n genes and multiply by factor f, which is either randomly selected, or set to some fixed level. Choosing a fixed factor of $f = 0$ represents a special case, which is useful for portfolio optimization - see below for more details.
- **N-point crossover.** Take two parent chromosomes and create two child chromosomes by crossing over at N positions, where N is selected depending on the number of assets under consideration.
- **Intermediate/blend crossover.** Take two parent chromosomes x_1, x_2 and create two child chromosomes y_1, y_2 by defining a weight-vector w with a weight $0 \leq w_i \leq 1$ for each gene, which may be either drawn randomly or pre-determined. The children are calculated using vector operations as shown in eq. (5.12).

$$\begin{aligned} y_1 &= w \cdot x_1 + (1-w) \cdot x_2 \\ y_2 &= (1-w) \cdot x_1 + w \cdot x_2 \end{aligned} \tag{5.12}$$

Most operators necessitate the need for a normalization after each operation to ensure that the basic budget constraint, i.e. $\sum x_{a \in \mathscr{A}} = 1$, is fulfilled. This can be done on the fly after each budget-constraint violating operation. Furthermore, to avoid numerical problems, a special δ-normalization is applied, which (re)sets every gene (asset weight) x_i to 0, if $0 < x_i \leq \delta$. This can be seen as a special operator for portfolio optimization, as the number of selected assets is usually small with respect to the amount of assets available. This is also the reason for using the factor $f = 0$ mutation, as well as the *realistic portfolio modifier* for adding random chromosomes to a population, which is described below. In terms of chromosomes it means that only a small number of (non-negative) genes should be positive.

5.3.3 Constraint handling

A survey of constraint handling techniques for evolutionary approaches can be found in (19), and (5), or see e.g. (36) for a generic framework for constrained optimization using genetic algorithms. Many approaches make use of a penalty-based approach, i.e. modifying the fitness, if a constraint is violated. The main advantage of such an approach is that the underlying algorithm does not need to be changed, see e.g. (20). Constraint handling methods, which are not based on penalty functions are often problem-dependent or are restricted to certain types of functions.

For general constraints, e.g. minimum return and maximum risk, we will use a penalty function approach. However, the portfolio optimization problem does also provide opportunities to implement special constraint handling techniques, especially for cardinality constraints, as well as upper asset investment limits.

Cardinality constraints

In the case of cardinality constraints, we have to consider two cases. First, when there are too few assets selected and secondly, when there are too many assets selected.

- If there are too few assets in the portfolio, we deduct a portion γ from the selected assets and assign them either randomly or uniformly distributed to the number of missing, i.e. previously not chosen assets, which will be selected randomly. If δ-normalization is used, one has to ensure that the additions are always above δ by choosing a correct level γ, and using a valid redistribution strategy.
- Even simpler, if too many assets are selected, then complete weights starting from the lowest positive weight will be reassigned, until the number of assets fulfills the constraint, without requiring additional validation.

Constraints of upper asset investment limits

Upper asset limits can be handled by redistributing the sum γ of excessive asset weight to other assets - either randomly, or by some special selection strategy, e.g. sorting the currently positive asset weights and redistribute γ according to the current non-excessive weights. Again, a valid redistribution scheme in accordance to other modification operators has to be set up in advance.

5.4 Implementation & Numerical Results

5.4.1 Implementation

The evolutionary algorithm was implemented in Matlab R2007a. The code is based on the evolutionary computation approach to scenario-based risk-return portfolio optimization for general risk measures as shown in (12), but has been extended to an object-oriented design, and can handle general constraints using a penalty-based approach, as well as specialized modification operators for constraints such as upper asset limits and cardinality constraints as shown in Section 5.3.3 above.[1]

The object oriented structure of the underlying evolutionary stochastic portfolio optimization problem is not the main focus of this chapter, and will be sketched in an informal way in the following. There are two main classes: `loss` and `population`.

[1] The code is available under an academic license on email request to the author.

The class `loss` handles the loss-distribution-based view on data and risk management. There are two groups of methods. The first group is used to fill an instance with data, i.e. `setScenario(matrix)`, and `setPortfolio(vector)` are used to define the scenario set, as well as one specific portfolio.

If non-equally weighted scenarios are used, the probabilities can be set with `setProbability(vector)`. The second group of methods is used to evaluate arbitrary functionals given the current data of the instance, e.g. `evalFunctional(string)`, or by using short-hand notation for specific functionals, e.g. `evalMean()`.

The class `population` handles the evolutionary structure, and contains evolutionary operators (mutation and crossover), auxiliary modification functions, as well as normalization methods.

5.4.2 Data

The scenario set used for the calculations below consists of weekly historical data of 14 selected Dow Jones Euro STOXX Supersector indices (Automobiles & Parts, Banks, Basic Resources, Chemicals, Construction & Materials, Financial Services, Food & Beverage, Health Care, Industrial Goods & Services, Insurance, Oil & Gas, Technology, Telecommunications, Utilities). In the results below, the numbers $1-14$ refer to the position in this list.

Two different time-horizons (both with a duration of 7 years) were selected, based on different financial market situations. The first time-frame T_1 is January 1991 to December 1997 (365 scenarios of weekly index changes), and the second one T_2 is January 1998 to December 2004 (366 scenarios of weekly index changes).

5.4.3 Naive portfolio selection

We will start our analysis by choosing three naive portfolio selection strategies:

- x_{eqw} is an equally weighted portfolio vector, i.e. $x_{eqw,a} = \frac{1}{a}, \forall a \in \mathscr{A}$,
- x_{hex} consists of n assets with the highest expected value given the respective scenario set. Let $x_{ex}^{[m]}$ be the asset with the m-highest expectation in this set, then each $x_{hex,a}$ is defined by

$$x_{hex,a} = \begin{cases} \frac{1}{n} & \forall x_{ex}^{[i]}, i = 1, \ldots, n \\ 0 & \text{otherwise} \end{cases}$$

- x_{lrv} consists of n assets with the lowest risk measured with the variance given the scenario set, and setting up the portfolio as shown by x_{hex} above.

Using $n = 3$ we obtain the following results for these three naive strategies using both data sets T_1 and T_2: The portfolios as well as histograms of the respective loss distributions are shown in fig. 5.2, fig. 5.3, and fig. 5.4 respectively.

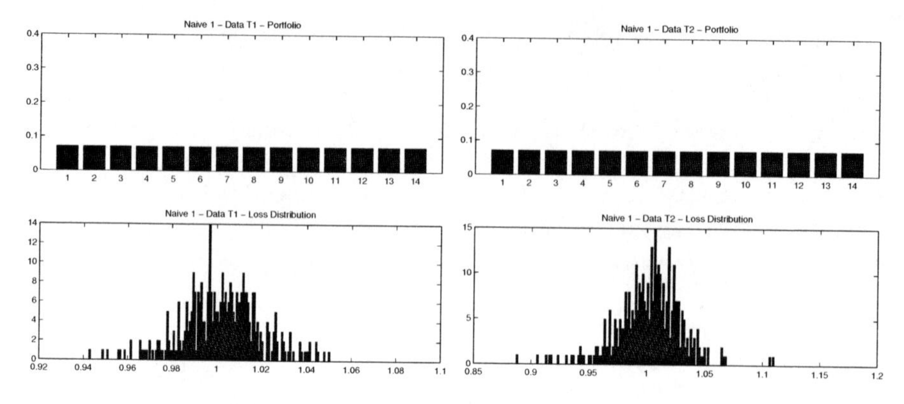

Fig. 5.2. Naive portfolio x_{eqw}. Data set T_1 (left) and T_2 (right)

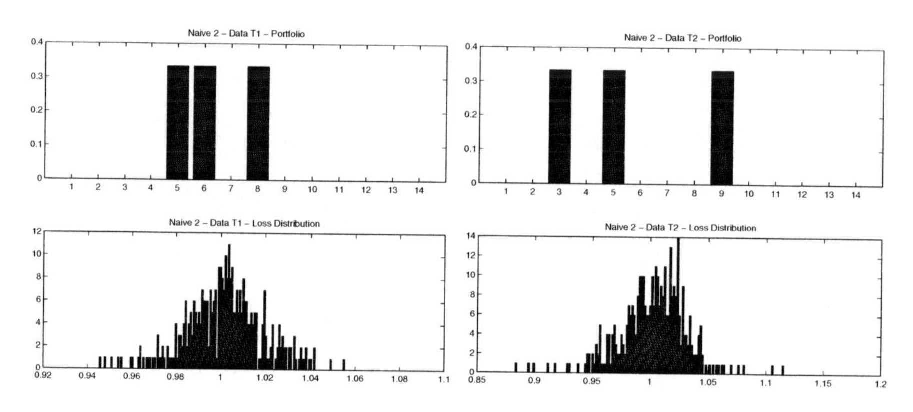

Fig. 5.3. Naive portfolio x_{hex}. Data set T_1 (left) and T_2 (right)

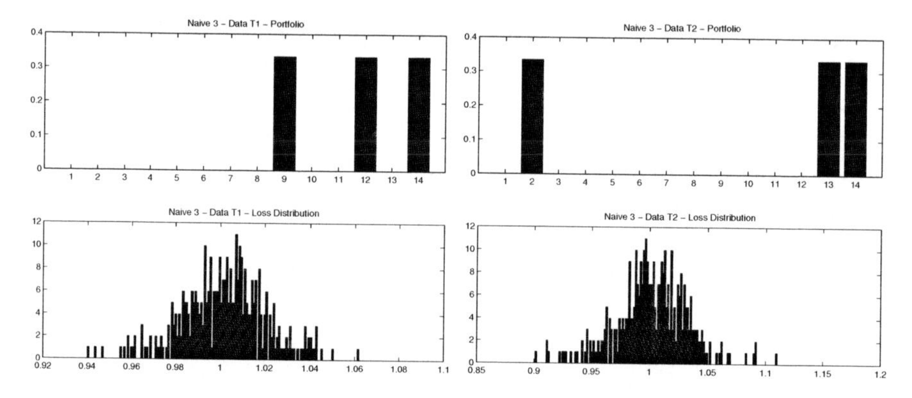

Fig. 5.4. Naive portfolio x_{lrs}. Data set T_1 (left) and T_2 (right)

5.4.4 Evolutionary stochastic portfolio optimization

In the following, we will apply the evolutionary stochastic portfolio optimization approach, which was discussed above to calculate optimal portfolios using different risk measures. The evolutionary parameters to compute the results are shown in table 5.3. The population size parameter c has been set to 30 in this concrete case. Whenever a random chromosome is a generated, the *realistic portfolio modifier* is applied, which sets a random number of assets (up to $a-2$) to 0.

Table 5.3. Evolutionary parameters

- Initial population size: 200 randomly created chromosomes (with realistic portfolio modifier).
- Maximum iterations: 50.
- Each new population consists of
  - c of the best chromosomes of the last population (elitist selection),
  - c random chromosomes (with realistic portfolio modifier),
  - $c/2$ with a randomly amount of genes mutated by factor 0.5 as well as $c/2$ mutated by factor 2,
  - $c/2$ 1-point crossover children,
  - $c \cdot 3$ intermediate/blend crossover children with random weights.

Unconstrained portfolio optimization

We may now calculate unconstrained portfolios given the four risk measures discussed in section 5.2.4 above. The following additional risk parameters have been used: For normalization purposes, a $\delta = 0.03$ has been selected, quantile-based risk measures are computed using an $\alpha = 0.1$-quantile, and the risk weighting has been set to $\kappa = 1$ (including the correct sign given the respective risk measure). The results are shown in Figs. 5.5, 5.6, 5.7, and 5.8.

The effects of market risk management using stochastic portfolio optimization with probabilistic risk measures can be seen very well in these figures. Compare the spread of the distributions to the naive strategies, especially in time-frame T_2. It can be observed, that bad expected return (less than 0.9) can be avoided without loosing expected return on the profit side.

Furthermore, differences in the structure and shape of the resulting loss distribution can also be clearly seen given the respective risk-measure class, i.e. deviation-based (Markowitz, MADSD), and quantile-based (VaR, ES). The rather sharp cut at the quantile is observable, when quantile-based measures are used.

Constrained portfolio optimization

The calculation of efficient frontiers for risk visualization purposes will conclude the section on numerical examples. We will add the minimum return constraint and

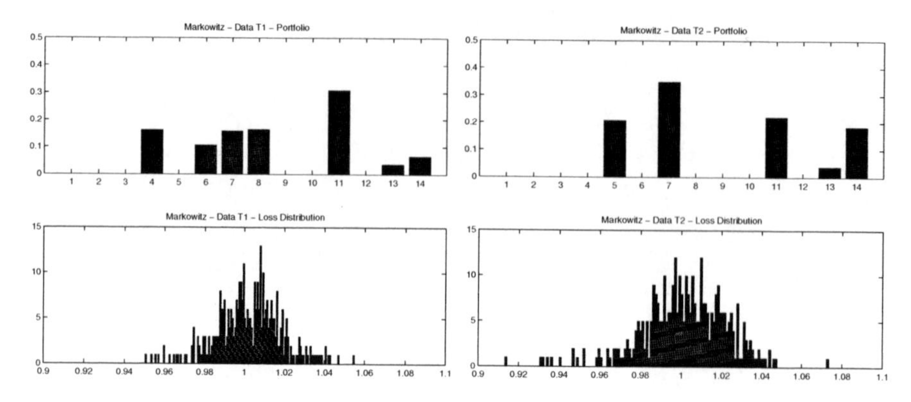

Fig. 5.5. Markowitz. Data set T_1 (left) and T_2 (right)

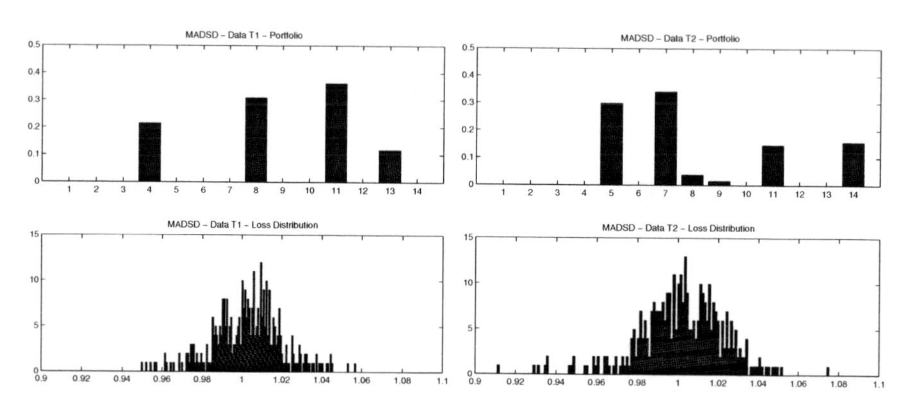

Fig. 5.6. Mean absolute downside semi-deviation. Data set T_1 (left) and T_2 (right)

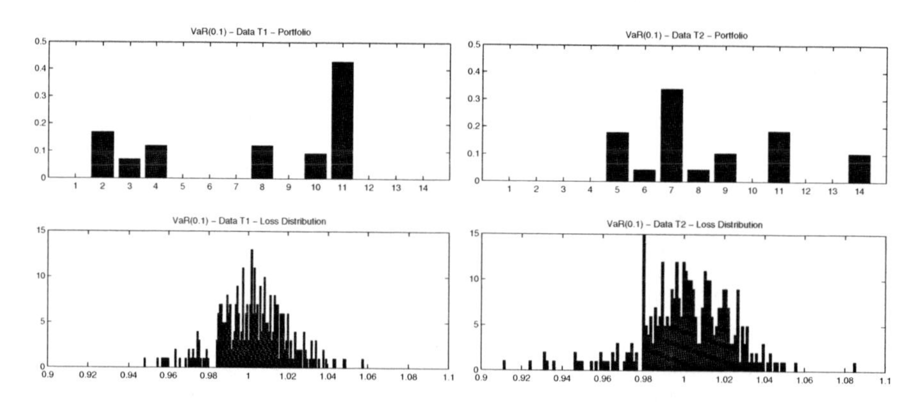

Fig. 5.7. Value-at-Risk ($\alpha = 0.9$). Data set T_1 (left) and T_2 (right)

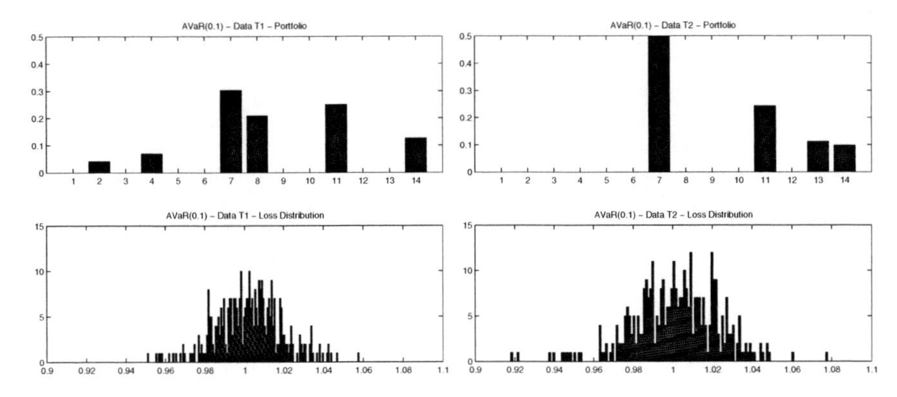

Fig. 5.8. Expected Shortfall ($\alpha = 0.9$). Data set T_1 (left) and T_2 (right)

compute a number of optimal portfolios by iteratively solving portfolio optimizations over a range of minimum mean requirements between a minimum required mean m_{min} and a maximum required mean m_{max}. Valid choices for m_{min} are e.g. either the mean of the portfolio given some unconstrained portfolio optimization or the mean of the scenario means. m_{max} is usually set to some β quantile of the scenario means.

If a chromosome is violating the constraint, its fitness is reduced to 90% of the original fitness value. The resulting efficient frontiers for all four risk measures for each time-frame are shown in fig. 5.9, fig. 5.10, fig. 5.11, and fig. 5.12 respectively. Each portfolio on the efficient frontier represents an optimal selection in the sense that at the same level of risk, no portfolio with higher expected return can be chosen, and likewise at the same level of expected return, no portfolio with less risk can be created. The investor may now choose an optimal portfolio either by selecting an appropriate level of risk or expected return.

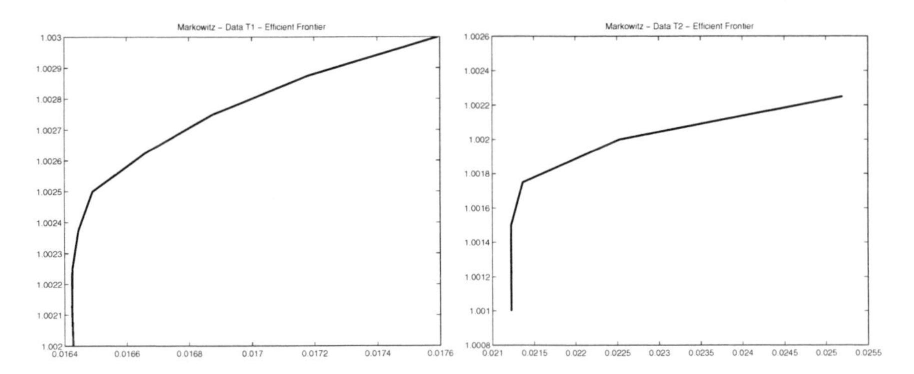

Fig. 5.9. Markowitz. Data set T_1 (left) and T_2 (right)

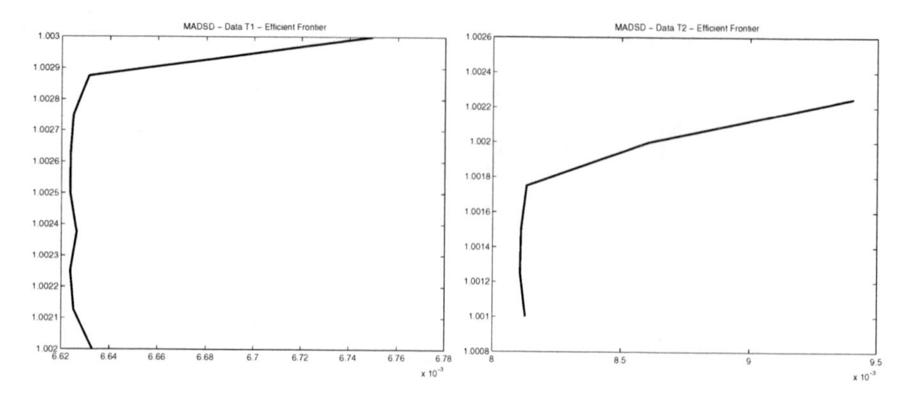

Fig. 5.10. Mean absolute downside semi-deviation. Data set T_1 (left) and T_2 (right)

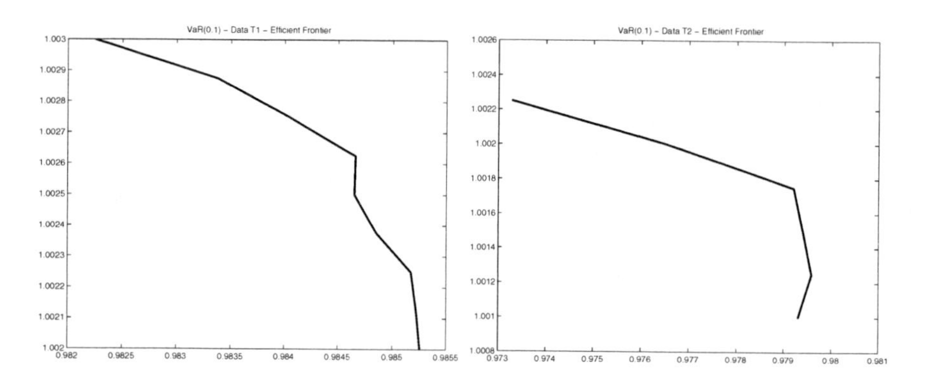

Fig. 5.11. Value-at-Risk ($\alpha = 0.9$). Data set T_1 (left) and T_2 (right)

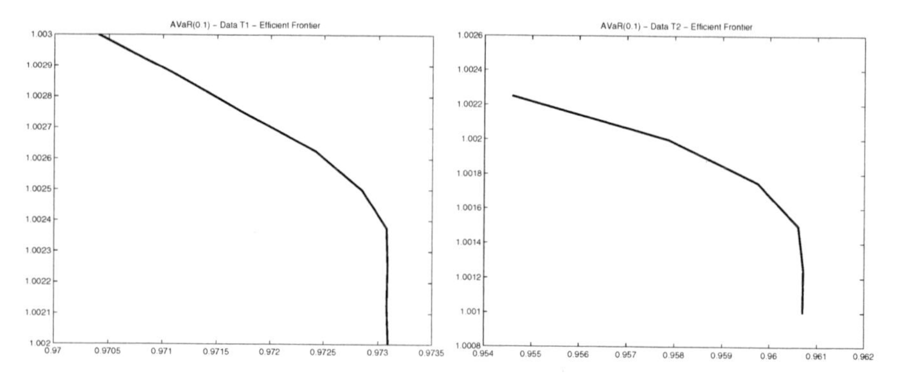

Fig. 5.12. Expected Shortfall ($\alpha = 0.9$). Data set T_1 (left) and T_2 (right)

5.5 Conclusion

In this chapter, the topic of evolutionary stochastic portfolio optimization has been discussed. Combining selected theory from the fields of Stochastic Programming, evolutionary computation, portfolio optimization, as well as financial risk management led to a generalized framework for computing optimal portfolios under uncertainty for various probabilistic risk measures. Both the unconstrained, as well as the constrained case have been implemented. Every real-world portfolio constraint can be integrated using the general penalty function approach. However, some for some selected constraints, special operations can be applied. This was exemplified with cardinality constraints and upper asset investment limits. In summary, this chapter provides further evidence, that evolutionary stochastic portfolio optimization is another successful application of bio-inspired natural computing algorithms for practical financial engineering.

Further research includes a systematic treatment of portfolio constraints for special sub-application areas, e.g. hedge fund management. An important constraint in this field is to minimize the absolute correlation of the portfolio to some reference market index. A further important addition is an extension to the multi-stage case. While the number of dimensions grows with respect to time-stages, the number of assets under consideration is usually smaller, as multi-stage optimization is used for strategic asset allocation in contrast to the single-stage case, which is used for tactical and even operational asset allocation. Thus, the underlying evolutionary concepts presented in this chapter can be applied for optimizing the multi-stage case too.

References

[1] Artzner P, Delbaen F, Eber JM, Heath D (1999) Coherent measures of risk. Mathematical Finance 9(3):203–228
[2] Beale EML (1955) On Minimizing a Convex Function Subject to Linear Inequalities. Journal of the Royal Statistical Society, Series B 17:173–184
[3] Blum C, Roli A (2003) Metaheuristics in combinatorial optimization: Overview and conceptual comparison. ACM Computing Surveys 35(3):268–308
[4] Brabazon A, O'Neill M (2006) Biologically inspired algorithms for financial modelling. Natural Computing Series, Springer-Verlag, Berlin
[5] Coello Coello CA (2002) Theoretical and numerical constraint-handling techniques used with evolutionary algorithms: a survey of the state of the art. Computer Methods in Applied Mechanics and Engineering 191(11-12):1245–1287
[6] Dantzig GB (1955) Linear Programming under Uncertainty. Management Science 1:197–206
[7] Ehrgott M, Klamroth K, Schwehm C (2004) An mcdm approach to portfolio optimization. European Journal of Operational Research 155(3):752–770
[8] Embrechts P, Lindskog F, McNeil AJ (2003) Modelling dependence with copulas and applications to risk management. In: Rachev S (ed) Handbook of heavy tailed distributions in finance, Elsevier, chap 8, pp 329–384

[9] Föllmer H, Schied A (2004) Stochastic finance, de Gruyter Studies in Mathematics, vol 27. Walter de Gruyter & Co., Berlin, an introduction in discrete time
[10] Gilli M, Këllezi E, Hysi H (2006) A data-driven optimization heuristic for downside risk minimization. The Journal of Risk 8(3):1–18
[11] Gomez MA, Flores CX, Osorio MA (2006) Hybrid search for cardinality constrained portfolio optimization. In: GECCO '06: Proceedings of the 8th annual conference on Genetic and evolutionary computation, ACM Press, pp 1865–1866
[12] Hochreiter R (2007) An evolutionary computation approach to scenario-based risk-return portfolio optimization for general risk measures. In: EvoWorkshops 2007, Springer, Lecture Notes in Computer Science, vol 4448, pp 199–207
[13] Hochreiter R, Pflug GC (2007) Financial scenario generation for stochastic multi-stage decision processes as facility location problems. Annals of Operations Research 152(1):257–272
[14] Krokhmal P, Palmquist J, Uryasev S (2002) Portfolio optimization with conditional value-at-risk objective and constraints. The Journal of Risk 4(2):11–27
[15] Lin D, Li X, Li M (2005) A genetic algorithm for solving portfolio optimization problems with transaction costs and minimum transaction lots. In: Wang L, Chen K, Ong YS (eds) Advances in Natural Computation, First International Conference, ICNC 2005, Changsha, China, August 27-29, 2005, Proceedings, Part III, Springer, Lecture Notes in Computer Science, vol 3612, pp 808–811
[16] Maringer D (2005) Portfolio Management with Heuristic Optimization, Advances in Computational Management Science, vol 8. Springer
[17] Markowitz HM (1952) Portfolio selection. The Journal of Finance 7(1):77–91
[18] McNeil AJ, Frey R, Embrechts P (2005) Quantitative risk management. Princeton Series in Finance, Princeton University Press
[19] Michalewicz Z, Schönauer M (1996) Evolutionary algorithms for constrained parameter optimization problems. Evolutionary Computation 4(1):1–32
[20] Miettinen K, Mäkelä MM, Toivanen J (2003) Numerical comparison of some penalty-based constraint handling techniques in genetic algorithms. Journal of Global Optimization 27(4):427–446
[21] Pennanen T, Koivu M (2005) Epi-convergent discretizations of stochastic programs via integration quadratures. Numerische Mathematik 100(1):141–163
[22] Pflug GC, Römisch W (2007) Modeling, Measuring and Managing Risk. World Scientific
[23] Rockafellar RT (2007) Coherent approaches to risk in optimization under uncertainty, tutorial Notes, Version May 14th, 2007
[24] Rockafellar RT, Uryasev S (2000) Optimization of Conditional Value-at-Risk. The Journal of Risk 2(3):21–41
[25] Rockafellar RT, Uryasev S, Zabarankin M (2006) Generalized deviations in risk analysis. Finance and Stochastics 10(1):51–74
[26] Ruszczyński A, Shapiro A (eds) (2003) Stochastic programming, Handbooks in Operations Research and Management Science, vol 10. Elsevier Science B.V., Amsterdam

[27] Schlottmann F, Mitschele A, Seese D (2005) A multi-objective approach to integrated risk management. In: Coello CAC, Aguirre AH, EZitzler (eds) Proceedings of the Evolutionary Multi-Criterion Optimization Conference (EMO 2005), Springer, Lecture Notes in Computer Science, vol 3410, pp 692–706

[28] Streichert F, Ulmer H, Zell A (2003) Evolutionary algorithms and the cardinality constrained portfolio selection problem. In: Selected Papers of the International Conference on Operations Research (OR 2003), Springer, pp 253–260

[29] Streichert F, Ulmer H, Zell A (2004) Comparing discrete and continuous genotypes on the constrained portfolio selection problem. In: et al KD (ed) Genetic and Evolutionary Computation (GECCO 2004) - Proceedings, Part II, Springer, Lecture Notes in Computer Science, vol 3103, pp 1239–1250

[30] Streichert F, Ulmer H, Zell A (2004) Evaluating a hybrid encoding and three crossover operators on the constrained portfolio selection problem. In: CEC2004. Congress on Evolutionary Computation, 2004, IEEE Press, vol 1, pp 932–939

[31] Subbu R, Bonissone P, Eklund N, Bollapragada S, Chalermkraivuth K (2005) Multiobjective financial portfolio design: a hybrid evolutionary approach. In: The 2005 IEEE Congress on Evolutionary Computation, IEEE Press, vol 2, pp 1722–1729

[32] Till J, Sand G, Urselmann M, Engell S (2007) A hybrid evolutionary algorithm for solving two-stage stochastic integer programs in chemical batch scheduling. Computers & Chemical Engineering 31(5-6):630–647

[33] Tokoro Ki (2001) A statistical selection mechanism of ga for stochastic programming problems. In: Proceedings of the 2001 Congress on Evolutionary Computation, vol 1, pp 487–492

[34] Urselmann M, Emmerich M, Till J, Sand G, Engell S (2007) Design of problem-specific evolutionary algorithm/mixed-integer programming hybrids: Two-stage stochastic integer programming applied to chemical batch scheduling. Engineering Optimization 39(5):529–549

[35] Uryasev S (2000) Conditional Value-at-Risk: Optimization algorithms and applications. Financial Engineering News 14:1–5

[36] Venkatraman S, Yen GG (2005) A generic framework for constrained optimization using genetic algorithms. IEEE Transactions on Evolutionary Computation 9(4):424–435

[37] Wallace SW, Ziemba WT (eds) (2005) Applications of stochastic programming, MPS/SIAM Series on Optimization, vol 5. Society for Industrial and Applied Mathematics (SIAM)

[38] Wang SM, Chen JC, Wee HM, Wang KJ (2006) Non-linear stochastic optimization using genetic algorithm for portfolio selection. International Journal of Operations Research 3(1):16–22

6

Non-linear Principal Component Analysis of the Implied Volatility Smile using a Quantum-inspired Evolutionary Algorithm

Kai Fan[1,2], Conall O'Sullivan[2], Anthony Brabazon[1], Michael O'Neill[1]

[1] Natural Computing Research and Applications Group, University College Dublin, Ireland.
kai.fan@ucd.ie; anthony.brabazon@ucd.ie; m.oneill@ucd.ie

[2] School of Business, University College Dublin, Ireland.
conall.osullivan@ucd.ie

Summary. Principal Component Analysis (PCA) is a standard statistical technique that is frequently employed in the analysis of large correlated data sets. We examine a technique for non-linear PCA by transferring the data from the non-linear space to linear space, where the weights on the non-linear functions are optimised using a Quantum-inspired Evolutionary Algorithm. This non-linear principal component analysis is used to examine the dynamics of the implied volatility smile derived from FTSE 100 stock index options over a sample period of 500 days.

6.1 Introduction

This chapter introduces a non-linear principal component analysis (NLPCA) methodology which uses a quantum-inspired evolutionary algorithm rather than neural network as in the traditional NLPCA approach. The NLPCA is used to determine the non-linear principal components that drive the variations in the implied volatility smile over time. The implied volatility smile (IVS) is how markets represent option prices. Option prices change from day to day to reflect changes in the asset price that the options are written on, and changes in market conditions, such as volatility and risk aversion and general economic trends. The pricing and hedging of assets depend on the evolution of the IVS.

6.1.1 Quantum-inspired Evolutionary Algorithm

Quantum mechanics is an extension of classical mechanics which models behaviours of natural systems that are observed particularly at very short time or distance scales. An example of such a system is a sub-atomic particle, such as a free electron. A complex-valued (deterministic) function of time and space co-ordinates, called

K. Fan et al.: *Non-linear Principal Component Analysis of the Implied Volatility Smile using a Quantum-inspired Evolutionary Algorithm*, Studies in Computational Intelligence (SCI) **100**, 89–107 (2008)
www.springerlink.com

the *wave-function*, is associated with the system: it describes the *quantum state* the system is in. The standard interpretation of quantum mechanics is that this abstract wave-function allows us to calculate probabilities of outcomes of concrete experiments. The squared modulus of the wave-function is a probability density function (PDF): it describes the probability that an observation of, for example, a particle will find the particle at a given time in a given region of space. The wave-function satisfies the *Schrödinger equation*. This equation can be thought of as describing the time evolution of the wave-function, and so the PDF, at each point in space. As time goes on, the PDF becomes more "spread out" over space, and our knowledge of the position of the particle becomes less precise, until an observation is carried out; then, according to the usual interpretation, the wave-function "collapses" to a particular classical state (or *eigenstate*), in this case a particular position, and the spreading out of the PDF starts all over again.

Before the observation we may regard the system as being in a linear combination of all possible classical states (this is called *superposition of states*); then the act of observation causes one such classical state to be chosen, with probability given by the PDF. Note that the wave function may interfere with itself (for example, if a barrier with slits is placed in the "path" of a particle) and this interference may be constructive or destructive, that is, the probability of detecting a particle in a given position may go up or go down.

More generally, we may seek to observe properties of quantum systems other than position, e.g., energy, momentum, or the quantum *spin* of an electron, photon or other particle. Such properties are called *observables*. Observables may be either continuous (e.g., position of a particle) or discrete (e.g., the energy of an electron in a bound state in an atom). Some observables may only take finitely many values, e.g., there are only two possible values for a given particle's spin: "up" or "down". This last is an example of a *two-state system*: in such a system the quantum state ψ is a linear superposition of just two eigenstates, say $|0\rangle$ and $|1\rangle$ in the standard Dirac bra-ket notation, that is,

$$\psi = \alpha|0\rangle + \beta|1\rangle, \tag{6.1}$$

where α and β are complex numbers with $|\alpha|^2 + |\beta|^2 = 1$. Here $|0\rangle$ and $|1\rangle$ are basis vectors for a 2-dimensional complex Hilbert space. A two-state system where the states are normalised and orthogonal, as here, may be regarded as a *quantum bit* or *qubit* (Geometrically, a qubit is a compact 2-dimensional complex manifold, called the Bloch sphere). It is thought of as being in eigenstates $|0\rangle$ and $|1\rangle$ simultaneously, until an observation is made and the quantum state collapses to $|0\rangle$ (with probability $|\alpha|^2$) or $|1\rangle$ (with probability $|\beta|^2$). The relation $|\alpha|^2 + |\beta|^2 = 1$ captures the fact that precisely one of $|0\rangle$, $|1\rangle$ must be observed, so their probabilities of observation must sum to 1.

A *quantum computer* is one which works with qubits instead of the (classical) bits used by usual computers. Benioff (1) first considered a Turing machine which used a tape containing what we would call qubits. Feynman (7) developed examples of physical computing systems not equivalent to the standard model of deterministic computation, the Turing machine. In recent years there has been a substantial interest in the theory and design of quantum computers, and the design of programs which could run on such computers, stimulated by Shor's discovery of a quantum factoring algorithm which would run faster than possible clasically. One interesting strand of research has been the use of natural computing (for example, genetic programming (GP)) to generate quantum circuits or programs (algorithms) for quantum computers (19). There has also been associated work in a reverse direction which draws inspiration from concepts in quantum mechanics in order to design novel natural computing algorithms. This is currently an area of active research interest. For example, quantum-inspired concepts have been applied to the domains of evolutionary algorithms (9, 10, 16, 21, 22), social computing (23), neuro-computing (8, 14, 20), and immuno-computing (12, 15). A claimed benefit of these algorithms is that because they use a quantum representation, they can maintain a good balance between exploration and exploitation. It is also suggested that they offer computational efficiencies as use of a quantum representation can allow the use of smaller population sizes than typical evolutionary algorithms.

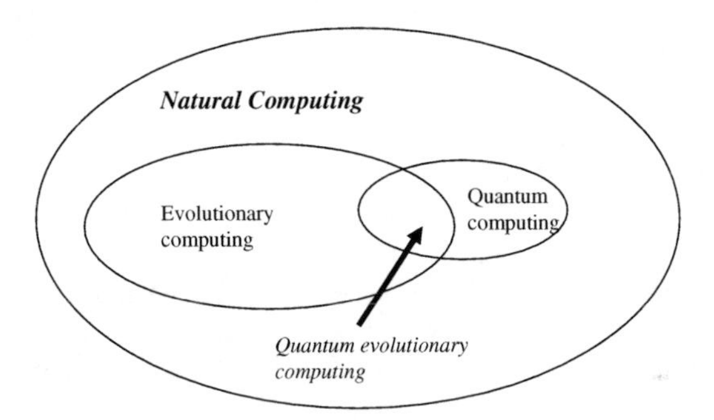

Fig. 6.1. Quantum-inspired evolutionary computing

Quantum-inspired evolutionary algorithms (QIEA) offer interesting potential. As yet, due to their novelty, only a small number of recent papers have implemented a QIEA, typically reporting good results (21, 22). Consequently, we have a limited understanding of the performance of these algorithms and further testing is required in order to determine both their effectiveness and their efficiency. It is also noted that although a wide-variety of biologically-inspired algorithms have been applied for financial modelling (2), only a few studies have yet applied the QIEA methodology in the finance domain (5). This chapter partly addresses both of these research gaps.

6.1.2 Structure of Chapter

The rest of this chapter is organised as follows. The next section provides a concise overview of QIEA, concentrating on the quantum-inspired genetic algorithm

(QIGA), and introduces NLPCA based on QIGA. We then outline the experimental methodology adopted. The remaining sections provide the results of these experiments followed by a number of conclusions.

6.2 The Quantum-inspired Genetic Algorithm

The best-known application of quantum-inspired concepts in evolutionary computing is the quantum-inspired genetic algorithm (QIGA) (9, 10, 16, 21, 22). The QIGA is based on the concepts of a qubit and the superposition of states. In essence, in QIGAs the traditional representations used in evolutionary algorithms (binary, numeric and symbolic) are extended to include a quantum representation.

A crucial difference between a qubit and a (classical) bit is that multiple qubits can exhibit quantum entanglement. Entanglement is when the wave function of a system composed of many particles cannot be separated into independent wave functions, one for each particle. A measurement made on one particle can produce, through the collapse of the total wave function, an instantaneous effect on other particles with which it is entangled, even if they are far apart. Entanglement is a nonlocal property that allows a set of qubits to be highly correlated. Entanglement also allows many states to be acted on simultaneously, unlike bits that can only have one value at a time. The use of entanglement in quantum computers is sometimes called *quantum parallelism*, and gives a possible explanation for the power of quantum computing: because the state of the quantum computer (i.e., the state of the system considered as a whole) can be in a quantum superposition of many different classical computational states, these classical computations can all be carried out at the same time.

The quantum equivalent of a classical operator on bits is an *evolution* (not to be confused with the concept of evolution in evolutionary algorithms). It transforms an input to an output, e.g., by rotation or a Hadamard gate, and operates without measuring the value of the qubit(s). Thus it effectively does a parallel computation on all the qubits at once and gives rise to a new superposition.

In the language of evolutionary computation a system of m qubits may be referred to as a *quantum chromosome* and can be written as a matrix with two rows

$$\begin{bmatrix} \alpha_1 \; \alpha_2 \; \dots \; \alpha_m \\ \beta_1 \; \beta_2 \; \dots \; \beta_m \end{bmatrix}. \tag{6.2}$$

A key point when considering quantum systems is that they can compactly convey information on a large number of possible system states. In classical bit strings, a string of length m can represent 2^m possible states. However, a quantum space of m qubits has 2^m *dimensions* (as a complex manifold).[1] Thus, a single qubit register of length m can simultaneously represent *all* possible bit strings of length 2^m, e.g., an 8 qubit system can simultaneously encode 256 distinct strings. This implies that it is possible to modify standard evolutionary algorithms to work with very few, or even

[1] It can be shown that, because of entanglement, an m-qubit physical system has $2^{m+1}-2$ degrees of freedom, much larger than the $2m$ degrees a classical version would have.

a single quantum individual, rather than having to use a large population of solution encodings. The qubit representation can also help to maintain diversity during the search process of an evolutionary algorithm, due to its capability to represent multiple system states simultaneously.

6.2.1 Representing a Quantum System

There are many ways that a quantum system could be defined in order to encode a set of binary (solution) strings. For example, in the following 3 qubit quantum system, the quantum chromosome is defined using the three pairs of amplitudes below

$$\begin{bmatrix} \frac{1}{\sqrt{2}} & \frac{\sqrt{3}}{2} & \frac{1}{2} \\ \frac{1}{\sqrt{2}} & \frac{1}{2} & \frac{\sqrt{3}}{2} \end{bmatrix} \tag{6.3}$$

These numbers are the probabilities that a qubit (unit of information) will be observed in a particular eigenstate rather than another. Taking the first qubit, the occurrence of either state 0 or 1 is equally likely as both α_1 and β_1 have the same amplitude. Following on from the definition of the 3 qubit system, the (quantum) state of the system is given by

$$\frac{\sqrt{3}}{4\sqrt{2}}|000\rangle+\frac{3}{4\sqrt{2}}|001\rangle+\frac{1}{4\sqrt{2}}|010\rangle+\frac{\sqrt{3}}{4\sqrt{2}}|011\rangle+\frac{\sqrt{3}}{4\sqrt{2}}|100\rangle+\frac{3}{4\sqrt{2}}|101\rangle+\frac{1}{4\sqrt{2}}|110\rangle+\frac{\sqrt{3}}{4\sqrt{2}}|111\rangle \tag{6.4}$$

To provide intuition on this point, consider the system state $|000\rangle$. The associated probability amplitude for this state is $\frac{\sqrt{3}}{4\sqrt{2}}$ and this is derived from the probability amplitudes of the 0 state for each of the three individual qubits ($\frac{1}{\sqrt{2}} * \frac{\sqrt{3}}{2} * \frac{1}{2} = 0.25$). The associated probabilities of each of the individual states ($|000\rangle, |001\rangle, |010\rangle, |011\rangle, |100\rangle, |101\rangle, |110\rangle, |111\rangle$) are $\frac{3}{32}, \frac{9}{32}, \frac{1}{32}, \frac{3}{32}, \frac{3}{32}, \frac{9}{32}, \frac{1}{32}, \frac{3}{32}$ respectively. Taking the first of these states as an example, $(\frac{\sqrt{3}}{4\sqrt{2}})^2 = \frac{3}{32}$.

6.2.2 Real-Valued Quantum-inspired Evolutionary Algorithms

In the initial literature which introduced the QIGA, a binary representation was adopted, wherein each quantum chromosome was restricted to consist of a series of 0s and 1s. The methodology was modified to include real-valued vectors by da Cruz et al., (4). As with binary-representation QIGA, real-valued QIGA maintains a distinction between a quantum population and an observed population of, in this case, real-valued solution vectors. However the quantum individuals have a different form to those in binary-representation QIGA. The quantum population $Q(t)$ is comprised of N quantum individuals ($q_i : i = 1,2,3,\ldots,N$), where each individual i is comprised of G genes ($g_{ij} : j = 1,2,3,\ldots,G$). Each of these genes consist of a pair of values $q_{ij} = (p_{ij}, \sigma_{ij})$ where $p_{ij}, \sigma_{ij} \in \Re$ represent the mean and the width of a square pulse. Representing a gene in this manner has a parallel with the quantum concept of superposition of states as a gene is specified by a range of possible values, rather than by a single unique value. The original QIGA algorithms, e.g., (9, 10) are based very closely on physical qubits, but the "quantum-inspired" algorithm of da Cruz et al. (4) used in this chapter draws less inspiration from quantum mechanics since it:

- does not use the idea of a quantum system (in particular, no qubits);
- only allows for constructive (not destructive) interference, and that interference is among "wave-functions" of *different* individuals;
- uses real numbers as weights, rather than the complex numbers which arise in superposition of states in physical systems;
- the PDFs used (uniform distributions) are not those arising in physical systems.

However, the da Cruz et al algorithm does periodically sample from a distribution to get a "classical" population, which can be regarded as a wave-function (quantum state) collapsing to a classical state upon observation.

Algorithm

The real-valued QIGA algorithm is as follows

```
Set t=0

Initialise Q(t) of N individuals with G genes

While (t < max t)
    Create the PDFs (and corresponding CDFs, which describe the probability
        distributions of real-valued random variables, see equation(6)) for
        each gene locus using the quantum individuals
    Create a temporary population, denoted E(T), of K real-valued solution
        vectors by observing Q(t) (via the CDFs)

    If (t=0) Then C(t)=E(t)
         (Note: the population C(T) is maintained between
             iterations of the algorithm)
    Else     E(t)=Outcome of crossover between E(t) and C(t)
             Evaluate E(t)
             C(t)= K best individuals from E(t) U C(t)
    End if

    With the N best individuals from C(t)
    Q(t+1)=Output of translate operation on Q(t)
    Q(t+1)=Output of resize operation on Q(t+1)
    t=t+1
Endwhile
```

Initialising the Quantum Population

A *quantum chromosome*, which is observed to give a specific solution vector of real-numbers, is made up of several quantum genes. The number of genes is determined by the required dimensionality of the solution vector. At the start of the algorithm, each quantum gene is initialised by randomly selecting a value from within the range of allowable values for that dimension. A gene's width value is set to the range of allowable values for the dimension. For example, if the known allowable values for dimension j are $[-75,75]$ then q_{ij} (dimension j in quantum chromosome i) is initially determined by randomly selecting a value from this range (say) -50. The corresponding width value will be 150. Hence, $q_{ij} = (-50,150)$. The square pulse need not be entirely within the allowable range for a dimension when it is initially created as the algorithm will automatically adjust for this as it executes. The height of the pulse arising from a gene j in chromosome i is calculated using

$$h_{ij} = \frac{1/\sigma_{ij}}{N} \tag{6.5}$$

where N is the number of individuals in the quantum population. This equation ensures that the probability density functions (PDFs) (see next subsection) used to generate the observed individual solution vectors will have a total area equal to one. Fig. 6.2 provides an illustration of a quantum gene where N=4.

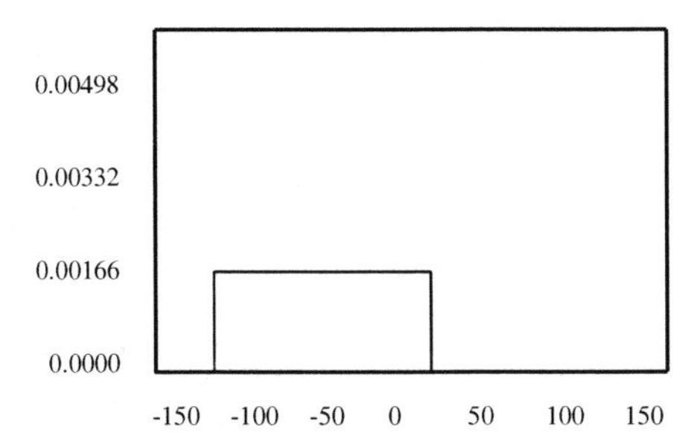

Fig. 6.2. A square pulse, representing a quantum gene, with a width of 150, centred on -50. The height of the gate is 0.005

Observing the Quantum Chromosomes

In order to generate a population of real-valued solution vectors, a series of observations must be undertaken using the population of quantum chromosomes (individuals). A pseudo-interference process between the quantum individuals is simulated by summing up the square pulses for each individual gene across all members of the quantum population. This generates a separate PDF (just the sum of the square pulses) for each gene and eq. 6.5 ensures that the area under this PDF is one. Hence, the PDF for gene j on iteration t is

$$PDF_j(t) = \sum_i^j g_{ij} \tag{6.6}$$

where g_{ij} is the square pulse of the jth gene of the ith quantum individual (of N). To use this information to obtain an observation, the PDF is first converted into its corresponding Cumulative Distribution Function (CDF)

$$CDF_j(x) = \int_{L_j}^{U_j} PDF_j(x)dx \tag{6.7}$$

where U_j and L_j are the upper and lower limits of the probability distribution. By generating a random number r from (0,1), the CDF can be used to obtain an observation of a real number x, where $x = CDF^{-1}(r)$. If the generated value x is outside the allowable real valued range for that dimension, the generated value is limited to its allowable boundary value. A separate PDF and CDF is calculated for each of the G gene positions. Once these have been calculated, the observation process is iterated to create a temporary population with K members, denoted E(t).

Crossover Mechanism

The crossover operation takes place between C(t) and the temporary population E(t). This step could be operationalised in a variety of ways with (4) choosing to adopt a variant of uniform crossover, without an explicit selection operator. After the K crossover operations have been performed, with the resulting children being copied into E(t), the best K individuals $\in C(t) \cup E(t)$ are copied into C(t).

Updating the Quantum Chromosomes

The N quantum chromosomes are updated using the N best individuals from C(t) after performing the crossover step. Each quantum gene's mean value is altered using

$$p_{ij} = c_{ij} \tag{6.8}$$

so that the mean value of the jth gene of the ith quantum chromosome is given by the corresponding j^{th} value of the ith ranked individual in $C(t)$.

The next step is to update the corresponding width value of the jth gene. The objective of this process is to vary the exploration / exploitation characteristics of the search algorithm, depending on the feedback from previous iterations. If the search process is continuing to uncover many new better solutions, then the exploration phase should be continued by keeping the widths relatively broad. However, if the search process is not uncovering many new better solutions, the widths are reduced in order to encourage finer-grained search around already discovered good regions of the solution space. There are multiple ways this general approach could be operationalised. For example, (4) suggests use of the 1/5th mutation rule from Evolutionary Strategies (17) whereby

$$if\ \phi < 1/5 \text{ then } \sigma_{ij} = \sigma_{ij} g \tag{6.9}$$

$$if\ \phi > 1/5 \text{ then } \sigma_{ij} = \sigma_{ij} / g$$

$$if\ \phi = 1/5 \text{ then } \sigma_{ij} = \sigma_{ij}$$

where σ_{ij} is the width of the ith quantum chromosome's j^{th} gene, g is a constant in the range $[0,1]$ and ϕ is the proportion of individuals in the new population that have improved their fitness. In this study we update the width of the ith quantum chromosome's j^{th} gene by comparing each successive generation's best fitness function. If the best fitness function has improved (disimproved) we shrink (enlarge) the width in order to improve the local (global) search.

QIGA vs Canonical Genetic Algorithm

A number of distinctions between the QIGA above and the canonical GA (CGA) can be noted. In the CGA, the population of solutions persists from generation to generation, albeit in a changing form. In contrast, in QIGA, the population of solutions in $P(t)$ are discarded at the end of each loop. The described QIGA, unlike CGA, does not have explicit concepts of crossover or mutation. However, the adaptation of the quantum chromosomes in each iteration does embed implicit selection as the best solution is selected and is used to adapt the quantum chromosome(s). The crossover and mutation steps are also implicitly present, as the adaptation of the quantum chromosome in effect creates diversity, as it makes different states of the system more or less likely over time. Another distinction between the QIGA and the CGA is that the CGA operates directly on representations of the solution (the members of the current population of solutions), whereas in QIGA the update step is performed on the probability amplitudes of the ground states for each qubit making up the quantum chromosome(s).

In the next section we use non-linear principal component analysis (NLPCA) to decompose the variation of the implied volatility smile into a number of non-linear principal components. To run NLPCA a set of weights on a number of non-linear mapping functions must be determined by optimising the proportion of variation explained by the principal components. Given the non-linearities inherent in options prices, and in the NLPCA method, QIEA is used to determine these weights in case the optimisation problem is not convex.

6.3 Non-Linear Principal Component Analysis

Suppose $X \in M^{m,n}$ is a panel data set that contains correlated data points along the columns, evaluated at different points in time along the rows. Given that X consists of correlated data points, the variation in X can be decomposed into a small number r of orthogonal principal components with $r < n$, resulting in a reduction of the dimension of the problem with only a small loss in information. The principal components from standard PCA are linear combinations (along the rows) of the original data set. If it is suspected that the data set contains non-linearities, a common procedure is to "linearise" the data set using suitable transformations prior to analysis. This approach has the advantage that it retains the simplicity of the underlying principal component analysis (PCA) whilst gaining the ability to cope with non-linear data. To do this we construct a modified data set X_{NL} from the original data set X:

$$X_{NL} = G(X), \tag{6.10}$$

where G is a function consisting of n individual mapping functions from linear to non-linear space:

$$G = w_1 g_1(X) + w_2 g_2(X) + \cdots + w_n g_n(X), \tag{6.11}$$

and where $g_i(X)$ is an individual non-linear mapping function of X and w_i is the weight on the function g_i. There are an infinite number of mapping functions $g_i(X)$ to choose from and in this paper we consider a small number of mapping functions we think are important given the domain knowledge of the problem under consideration (see next section). There are a total of four functions chosen in this study and they are given as follows:

- *Logistic mapping:*

$$g_1(X) = 4X \circ (1 - X), \tag{6.12}$$

  where $\circ$ denotes element by element matrix multiplication.
- *Exponential mapping:*

$$g_2(X) = \exp(X), \tag{6.13}$$

  where the exponential function is applied on an element by element basis.
- *Hénon mapping:*

$$g_3(X(t)) = 1 - 1.4(X(t))^2 + 0.3X(t-1), \tag{6.14}$$

  where $X(t)$ is a single row of the data set X.
- *Auto regressive process:*

$$g_4(X(t)) = 0.25X(t-1) + \varepsilon(t), \tag{6.15}$$

  where $X(t)$ is as above and $\varepsilon(t)$ is a standard normal random variable.

The objective of this data mapping is to compensate for any non-linearities within X. That is to linearise the data before implementing PCA. Provided this is performed as an integral part of the analysis, a non-linear variant of PCA will result. The method is described as follows: we perform standard PCA on the non-linear transformation of the original data set and optimise the weights on the different mapping functions with the objective of maximising the proportion of variation explained by the first principal component from standard PCA. A quantum-inspired evolutionary algorithm is used to find the weights on the non-linear mapping functions $g_i \in G$ given the potential for local minima. Future work will increase the number of functions considered and optimise not only the weights but the various parameters of the functions $g_i \in G$.

6.4 Implied Volatility Smile (IVS)

In this section we explain the meaning of the implied volatility smile. A European call (put) option on an asset S_t with maturity date T and strike price K is defined as a contingent claim with payoff at time T given by $\max[S_T - K, 0]$ ($\max[K - S_T, 0]$). The well known Black-Scholes (BS) formula for the price of a call option on an underlying asset S_t is given by

$$C_{BS}(S_t, K, r, q, \tau; \sigma) = S_t e^{-q\tau} N(d_1) - K e^{-r\tau} N(d_1) \tag{6.16}$$

$$d_1 = \frac{-\ln m + \left(r - q + \frac{1}{2}\sigma^2\right)\tau}{\sigma\sqrt{\tau}} \qquad d_2 = d_1 - \sigma\sqrt{\tau} \tag{6.17}$$

where $\tau = T - t$ is the time-to-maturity, t is the current time, $m = K/S$ is the moneyness of the option, r and q are the continuously compounded risk-free rate and dividend yield and $N(\cdot)$ is the cumulative normal distribution function. Suppose a market option price, denoted by $C_M(S_t, K)$, is observed. The Black-Scholes implied volatility for this option price is that value of volatility which equates the BS model price to the market option price as follows

$$\sigma_{BS}(S_t, K) > 0 \tag{6.18}$$

$$C_{BS}(S_t, K, r, q, \tau; \sigma_{BS}(S_t, K)) = C_M(S_t, K) \tag{6.19}$$

If the assumptions underlying the BS option pricing model were correct, the BS implied volatilities for options on the same underlying asset would be constant for different strike prices and maturities. However in reality the BS implied volatilities are varying over strike price and maturity. The variation of implied volatilities over strike price for a fixed maturity is known as the implied volatility smile. Given that the options are written on a single underlying asset this result seems at first paradoxical, i.e. we have a number of different implied volatilities for a single asset which should only have one measure for its volatility. The assumptions in the BS model can be relaxed, such as allowing the underlying asset price to follow a more complex data generating process than the log normal stochastic process (as assumed by BS), or allowing the underlying asset price to experience sudden discontinuous jumps etc. When the resulting complications of these more general assumptions are taken into account, the implied volatility smile begins to make sense and is simply highlighting the erroneous assumptions that underpin the BS model.

Implied volatilities are frequently used in the financial markets to quote the prices of options. The participants in the options markets do not believe that the BS model is correct, but use the model as a convenient way of quoting option prices. The reason is that implied volatilities usually have to be updated less frequently than option prices themselves and implied volatilities vary less dramatically than option prices with strike price and maturity. Option traders and brokers monitor movements in volatility smiles closely. As option prices change over time the implied volatility smile (for various maturities) also changes.

If we stack the implied volatility smile (for one particular maturity) according to the time the IVS data was recorded, what results is a time series of panel data with highly correlated entries. Implied volatilities at different strikes are highly correlated because as the volatility of the asset rises all implied volatilities rise yet some may rise more than others. However the economic forces of no-arbitrage (no free-lunch) ensures that the implied volatilities cannot get too detached from one another because if they did this represents a riskless trading opportunity for savvy investors, who sell the more expensive option (with the higher implied volatility) and hedge it with cheaper options (with lower implied volatilities).

PCA is an ideal tool to reduce the complexity of such a data set by explaining the variation of the IVS over time in terms of a small number of orthogonal principal factors. The approach has been applied and the dynamical properties of the implied volatility smile has been studied in recent years using increasingly advanced principal component approaches. See Heynen, Kemma and Vorst (11) and Fengler, Härdle and Schmidt (6) for PCA applied to the term structure of at-the-money implied volatilities (implied volatilities for different maturities and a fixed strike price) and see Skiadopoulos, Hodges and Clewlow (18) for PCA applied to implied volatility smiles. The evidence suggests that changes in the implied volatility smile are driven predominantly by three factors. The first factor is a level factor which controls the overall height of the implied volatility smile. The second and third factors are slope and curvature factors across the strike price axis. However options and the implied volatilities associated with options are multi-dimensional non-linear instruments and standard PCA may neglect some of non-linear subtleties inherent in option implied volatilities. This is the reason that NLPCA is applied to the IVS in this paper.

6.5 Results

6.5.1 Data

The data used in this study is option implied volatilities across 11 different strikes and a number of different maturities on the FTSE 100 index. The data consists of end-of-day settlement option implied volatilities from the 26th of March 2004 to the 17th of March 2006 consisting of 500 trading days. FTSE 100 index options are European style options and the underlying asset is the FTSE 100 performance index. To price options on this index one must adjust the index by extracting the present value of future cash dividend payments before each option's expiration date. The annualised dividend yield of the FTSE 100 index was downloaded from Datastream. The one-month LIBOR (London inter-bank offered rate) rate was used as the risk-free rate where the LIBOR rate was converted into a continuously compounded rate. The forward price used in the option calculations is then $F_t = S_0 e^{(r-q)t}$ where S_0 is the current index price level, F_t is the price for the forward contract maturing at time t, r is the continuously compounded risk-free rate and q is the continuously compounded dividend yield. Settlement prices of call and put option are calculated from the implied volatilities using the Black-Scholes formula.

As calendar time passes, the option contracts wind down towards maturity and the observed implied volatility surface (implied volatilities plotted across strike price and maturity) is constantly changing in terms of its moneyness and maturity values, see fig. 6.3. To obtain implied volatilities on a fixed grid of moneyness and maturity the market implied volatilities were interpolated using a non-parametric Nadaraya-Watson estimator, see Cont and da Fonseca (3), so that we have interpolated implied volatilities on a fixed grid of moneyness and maturity for all the days in the data sample. On day t an interpolated estimate for implied volatility at a moneyness m and a time to maturity τ is given by

$$I_t(m,\tau) = \frac{\sum_i^{n_m} \sum_j^{n_\tau} I_t(m_i,\tau_j) f(m-m_i,\tau-\tau_j)}{\sum_i^{n_m} \sum_j^{n_\tau} f(m-m_i,\tau-\tau_j)}, \tag{6.20}$$

$$f(x,y) = (2\pi)^{-1} \exp\left(-x^2/h_1\right) \exp\left(-y^2/h_2\right), \tag{6.21}$$

where n_m and n_τ are the number of different option moneyness levels and maturities available on a particular day in the sample, m_i is the moneyness and τ_j is the maturity of the $(i,j)^{th}$ observed option and h_1 and h_2 are the bandwidths of the estimator across moneyness and maturity. The bandwidths for the estimator were chosen using cross validation, where one implied volatility is removed and is then interpolated from all the other available implied volatilities on that date. The difference between the interpolated and the observed implied volatility is the cross validation error. This is calculated for all implied volatilities available on a particular day and this error is miminised by optimising over h_1 and h_2. For each day t in the sample we define the implied volatility smile at a fixed maturity τ_j by

$$IVS(t) = \left\{I_t(1,\tau_j), \ldots, I_t(n_m,\tau_j)\right\}. \tag{6.22}$$

We then stack these implied volatility smiles over time to form the data matrix $X = \{IVS(1), \ldots, IVS(500)\}'$. Non-linear principal component analysis (NLPCA) is conducted on the implied volatility smile and logarithm of the implied volatility smile for maturities ranging from 2 to 6 months.

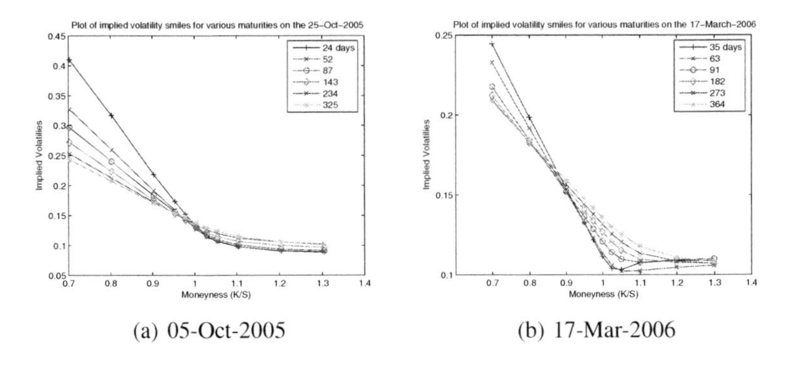

(a) 05-Oct-2005 (b) 17-Mar-2006

Fig. 6.3. Implied Volatility Smiles on two different dates.

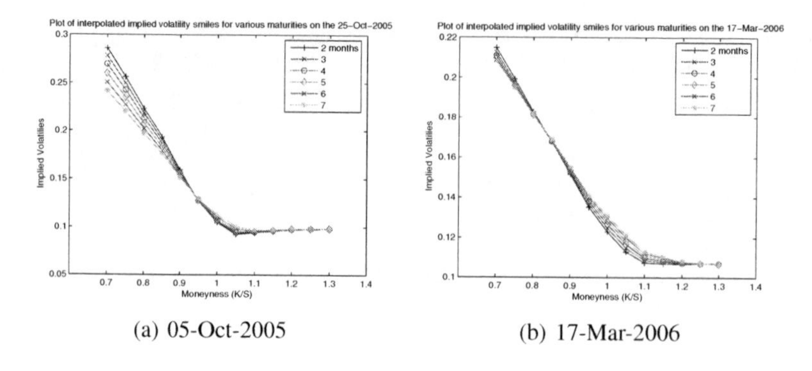

Fig. 6.4. Interpolated Implied Volatility Smiles on the same dates as above.

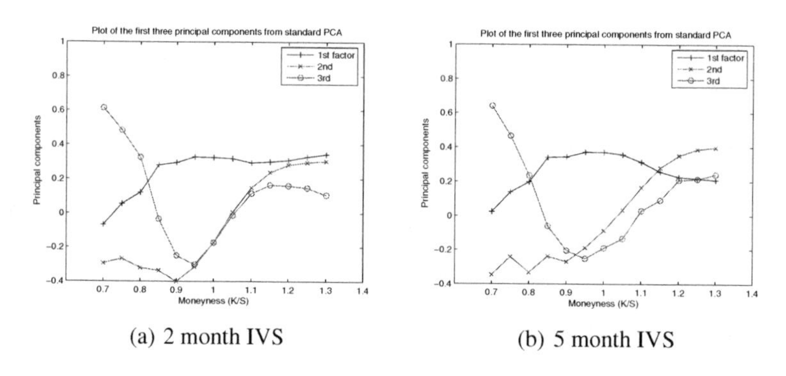

Fig. 6.5. Three linear principal components for an IVS with a maturity of 2 and 5 months.

6.5.2 Analysis

The first three principal components from linear PCA explain up to approximately 96% of the variation in the level of the implied volatility smile, depending on the maturity of the IVS considered. As 96% in PCA analysis may be overfitting, we would rather target the first principal component than the first three components and this why the objective function in the NLPCA was chosen to be proportion of variation explained by the first principal component.

The analysis of the eigenfactors from standard PCA for the implied volatility smiles of each maturity shows that the first factor has a positive effect on all implied volatilities. This eigenfactor can be interpreted as a level or a volatility factor. An increase in this factor causes the whole IVS to increase and causes all options to become more expensive since options are increasing functions of volatility. The second factor has a negative effect for implied volatilities with $K < S$, e.g. out-of-the-money puts, and a positive effect for implied volatilities with $K > S$, e.g. out-of-the-money calls. This factor can be interpreted as a skew factor and increase in this factor causes out-of-the money calls to become more expensive relative to out-of-the-money puts.

The third factor has a positive effect for implied volatilities with $K < S$ and $K > S$ e.g. out-of-the-money calls and puts, and a negative effect for implied volatilities that are close to the money with $K \approx S$. This factor can be interpreted as a curvature factor an an increase in this factor causes out-of-the money calls and puts to become more expensive relative to near-the-money calls and puts.

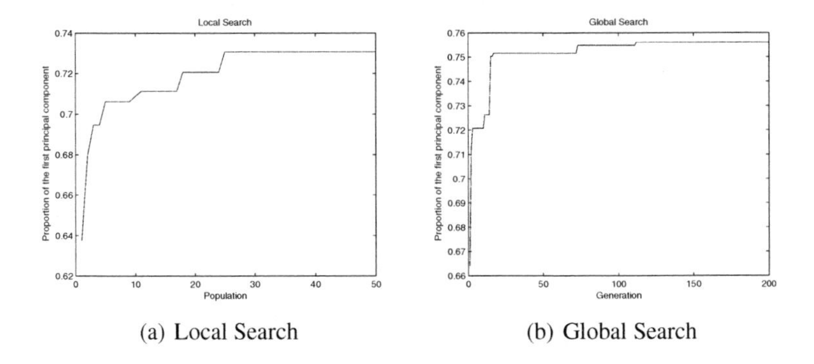

(a) Local Search

(b) Global Search

Fig. 6.6. Local and Global search

Table 6.1. Parameters setting in Quantum-inspired Genetic Algorithm

Population size	Generation Number	Crossover rate	Shrinkage	Enlargement
200	50	0.7	0.8	1.2

Table 6.2. Results of QIGA. The proportion explained by the first principal component (PC) from the last generation are averaged over 30 runs and compared with the parameter values from 30 runs of a Matlab optimiser.

Maturity	Linear PCA (%)	Non-linear PCA (%)	Standard deviation (%)	Matlab optimiser (%)
2 months	64.15	80.47	0.04	80.97
3 months	69.57	80.27	0.04	81.17
4 months	72.90	80.87	0.03	81.30
5 months	77.01	81.67	0.01	82.04
6 months	80.27	84.35	0.01	84.38

In our NLPCA-QIEA analysis, the weights on the mapping functions are optimised by using a quantum-inspired genetic algorithm to maximise the objective function which is the proportion of variation in the data explained by the first principal

Table 6.3. Results of QIGA. The optimal and average weights on mapping functions from the last generation are averaged over 30 runs.

Maturity(months)	Logistic	Exp	Hénon	AR
2	0.000	0.3652	0.9814	0.3669
3	0.000	0.3634	0.9791	0.3769
4	0.000	0.3614	0.9800	0.3885
5	0.000	0.3704	0.9800	0.3542
6	0.000	0.3653	0.9815	0.3798

component. The weights are also optimised using the Matlab function *fminsearch*. *Fminsearch* uses the simplex search method of Lagarias et al (13). This is a direct search method that does not use numerical or analytic gradients. Fig. 5 depicts the evolution of the objective function versus the generation number. The parameter settings in the QIEA are given in Table 1. NLPCA is more efficient than linear PCA especially for the options with shorter times-to-maturity. For example, for the 2 month IVS the 1st principal component from NLPCA explains approximately 80% of the variation of the data versus only 64% for standard PCA. However the outperformance of NLPCA is to be expected given the extra degrees of freedom involved since it uses four non-linear functions that first operate on the data before PCA is applied. It is interesting to note that for the two month IVS the first component from NLPCA with four non-linear functions explains 80% of the variation whilst the first *three* components from linear PCA explain up to 96% of the variation in the data. Although 96% is a higher level of explanatory power this is more than likely overfitting historical data at the expense of poor out-of-sample performance. If we forecast the evolution of the IVS out-of-sample using the techniques in this paper, a parsimonious procedure would be to include a more general set of time series models in the set of non-linear functions and use these to forecast the first factor from NLPCA and reconstruct future IVSs from the weights derived from historical analysis. This would be more parsimonious than fitting a separate time series model to three linear principal components and then reconstructing the future IVS as would have to be done in linear PCA. Thus, at least for shorter term options, the NLPCA method can explain 80% of the variation in the data with one linear combination of non-linear functions of the data versus approximately 64% for linear PCA. Thus rather than increasing the number of principal components in the analysis we have shown that another route is to use non-linear principal components to achieve a statistical significant increase in explanatory power.

It is interesting to note the weights on the various functions that were derived in the NLPCA. The weight on the logistic function are always very close to zero thus this mapping had nothing to contribute to the NLPCA. The weights on the exponential function and the autoregressive function were approximately the same at a little over $\frac{1}{3}$ and the weights on the Hénon function were close to one. The exponential function is capturing the skew effect mentioned earlier. The auto regressive function is capturing serial correlation in the daily movements of the IVS (something that cannot be done under linear PCA). The fixed value on the AR function was taken to be 0.25 thus when we multiply this by the weight of approximately 0.35 this results in serial correlation coefficient of approximately 0.088. Thus there is positive serial correlation in the data and this represents a possible trading strategy. The Hénon function is capturing a combination of a curvature effect (mentioned earlier), due to the squaring of the data, and a time series effect due to the dependence on past values. The weight on this function is close to one meaning this function is relatively important in the analysis because the average values of the entries in the vector $1 - 1.4X(t)^2 + 0.3X(t-1)$ are close to one and the weight multiplied by this average value are relatively large compared to the values from the other functions. This implies that the function depending on the curvature of the current IVS and the past level of the IVS is very important for explaining the variation in the IVS over time.

6.6 Conclusions

A non-linear principal component analysis was conducted on the implied volatility smile derived from FTSE 100 stock index options. It was shown, at least for shorter term options, that the NLPCA method can explain 80% of the variation in the data with one non-linear principal component versus approximately 64% for one linear principal component in linear PCA. Thus the non-linear functions used in the NLPCA captured some of the higher order non-linear factors that affect the data and effectively increased the explanatory power of the method.

The weights on these non-linear functions were optimised using a quantum-inspired evolutionary algorithm (QIEA). Although the problem considered in this paper was not high-dimensional, it has potential to be a highly non-linear non-convex optimisation problem due to the fact that the options data analysed are highly non-linear and method used to describe the variation in the options data is a non-linear method. Thus it was thought that this was a reasonable problem to test out the QIEA with a view to using it for more extensive analysis in future work. This future work consists of expanding the number of non-linear functions being considered with a focus on including a larger number of time series models. This would be very useful in predicting the IVS out-of-sample and in constructing options trading strategies. Future work could also look at multi-objective NLPCA where the proportion of variation explained by the first factor is maximised followed by the proportion of variation explained by the second factor, etc. Also it would be useful to relax the restriction on the parameters of the non-linear functions used in NLPCA and allow the QIEA to find optimal values for these parameters. All of these extensions will result in very high-dimensional optimisation problems where the use of evolutionary algorithms such as the QIEA may be essential.

References

[1] Benioff P (1980) The Computer as a Physical System: A Microscopic Quantum Mechanical Hamiltonian Model of Computers as Represented by Turing Machines. Journal of Statistical Physics 22: 563–591

[2] Brabazon A and O'Neill M (2006) Biologically-inspired Algorithms for Financial Modelling. Berlin Springer.

[3] Cont R and da Fonseca J (2002) Dynamics of implied volatility surfaces. Journal of Quantitative finance 2: 45–60

[4] da Cruz A, Vellasco M and Pacheco M (2006) Quantum-inspired evolutionary algorithm for numerical optimization. Proceedings of the 2006 IEEE Congress on Evolutionary Computation (CEC 2006), 16-21 July, Vancouver, 9180–9187, IEEE Press

[5] Fan K, Brabazon A, O'Sullivan C and O'Neill M (2007) Quantum Inspired Evolutionary Algorithms for Calibration of the VG Option Pricing Model. Proceedings of the Genetic and Evolutionary Computation Conference (GECCO 2007), London, July 7-11, 2007, 1983–1990, New York: ACM Press

[6] Fengler M, Härdle W and Schmidt P (2002) Common factors governing VDAX movements and the maximum loss. Jounal of Financial Markets and Portfolio Management 1:16-19

[7] Feynman R (1982) Simulating Physics with Computers. International Journal of Theoretical Physics 21(6&7):467–488

[8] Garavaglia S (2002) A quantum-inspired self-organizing map (QISOM) Proceedings of 2002 International Joint Conference on Neural Networks (IJCNN 2002), 12-17 May 2002, 1779–1784, IEEE Press

[9] Han K-H and Kim J-H (2002) Quantum-inspired evolutionary algorithm for a class of combinatorial optimization. IEEE Transactions on Evolutionary Computation 6(6):580–593

[10] Han K-H and Kim J-H (2002) Quantum-inspired evolutionary algorithms with a new termination criterion, H_ε gate and two-phase scheme. IEEE Transactions on Evolutionary Computation 8(3):156–169

[11] Heynen R and Kemma K and Vorst T (1994) Analysis of the term structure of implied volatilities. Journal of Financial and Quantitative Analysis 29: 31–56

[12] Jiao L and Li Y (2005) Quantum-inspired immune clonal optimization. Proceedings of 2005 International Conference on Neural Networks and Brain (ICNN&B 2005) 13-15 Oct. 2005, 461–466, IEEE Press

[13] Lagarias J C, Reeds J A, Wright M H and Wright P E (1998) Convergence Properties of the Nelder-Mead Simplex Method in Low Dimensions. SIAM Journal of Optimization 9(1):112–147

[14] Lee C-D, Chen Y-J, Huang H-C, Hwang R-C and Yu G-R (2004) The nonstationary signal prediction by using quantum NN. Proceedings of 2004 IEEE International Conference on Systems, Man and Cybernetics, 10-13 Oct. 2002, 3291–3295, IEEE Press

[15] Li Y, Zhang Y, Zhao R and Jiao L (2004) The immune quantum-inspired evolutionary algorithm. Proceedings of 2004 IEEE International Conference on Systems, Man and Cybernetics, 10-13 Oct. 2002, 3301–3305, IEEE Press

[16] Narayanan A and Moore M (1996) Quantum-inspired genetic algorithms. Proceedings of IEEE International Conference on Evolutionary Computation, May 1996, 61–66, IEEE Press

[17] Rechenberg I (1973) Evolutionsstrategie: Optimierung Technisher Systeme nach Prinzipien der Biologischen Evolution. Fromman-Holzboog Verlag, Stuggart

[18] Skiadopoulos G, Hodges S and Clewlow L (1999) The Dynamics of the S&P 500 Implied Volatility Surface. Review of Derivatives Research 3:263–282

[19] Spector L (2004) Automatic Quantum Computer Programming: A Genetic Programming Approach. Kluwer Academic Publishers, Boston, MA

[20] Tsai X-Y, Chen Y-J, Huang H-C, Chuang S-J and Hwang R-C (2005) Quantum NN vs NN in Signal Recognition. Proceedings of the Third International Conference on Information Technology and Applications (ICITA 05), 4-7 July 2005, 308–312, IEEE Press

[21] Yang S, Wang M and Jiao L (2004) A genetic algorithm based on quantum chromosome. Proceedings of IEEE International Conference on Signal Processing (ICSP 04), 31 Aug- 4 Sept. 2004, 1622–1625, IEEE Press

[22] Yang S, Wang M and Jiao L (2004) A novel quantum evolutionary algorithm and its application. Proceedings of IEEE Congress on Evolutionary Computation 2004 (CEC 2004), 19-23 June 2004, 820–826, IEEE Press

[23] Yang S, Wang M and Jiao L (2004) A Quantum Particle Swarm Optimization. Proceedings of the Congress on Evolutionary Computation 2004 1:320–324, New Jersey: IEEE Press

7

Estimation of an *EGARCH* Volatility Option Pricing Model using a Bacteria Foraging Optimisation Algorithm

Jing Dang[1,2], Anthony Brabazon[1], Michael O'Neill[1], and David Edelman[2]

[1] Natural Computing Research and Applications Group,
University College Dublin, Ireland.
jing.dang1@ucdconnect.ie; anthony.brabazon@ucd.ie;
m.oneill@ucd.ie

[2] School of Business, University College Dublin, Ireland.
davide@ucd.ie

Summary. The bacterial foraging optimisation algorithm is a novel natural computing algorithm which is based on mimicking the foraging behavior of E.coli bacteria. This chapter illustrates how a bacteria foraging optimisation algorithm (BFOA) can be constructed. The utility of this algorithm is tested by comparing its performance on a series of benchmark functions against that of the canonical genetic algorithm (GA). Following this, the algorithm's performance is further assessed by applying it to estimate parameters for an EGARCH model which can then be applied for pricing volatility options. The results suggest that the BFOA can be used as a complementary technique to conventional statistical computing techniques in parameter estimation for financial models.

7.1 Introduction

This chapter introduces a novel natural computing algorithm, the bacteria foraging optimisation (BFOA) algorithm, which draws metaphorical inspiration from the foraging behavior of *Escherichia coli* bacteria in order to design an optimisation algorithm. This algorithm was introduced by Passino (25) in 2002 and has been applied to solve a range of real-world problems. Mishra (20) illustrates an application of the BFOA to deal with harmonic estimation for a signal distorted with additive noise. This system performed very well when compared with conventional discrete Fourier transform and genetic algorithm methods. Kim used a hybrid system based on the genetic algorithm and the BFOA (17, 18) to tune a PID Controller for an automatic voltage regulator (AVR) system. Ulagammai et al trained a wavelet neural network (WNN) using a BFOA for load forecasting (LF) in an electric power system (31). In this chapter, we further examine the ability of BFOA to solve optimization problems, applying it to optimise nonlinear financial problems. We consider a version of the BFOA proposed by Passino (25). A number of alternative formulations of the

J. Dang et al.: *Estimation of an* EGARCH *Volatility Option Pricing Model using a Bacteria Foraging Optimisation Algorithm*, Studies in Computational Intelligence (SCI) **100**, 109–127 (2008)
www.springerlink.com

BOFA have been developed and the interested reader is referred to recent work by (19, 28–30). Initially, we assess the performance of the BFOA on six benchmark problems, comparing the BOFA with the canonical GA. We then apply the algorithm to estimate the parameters of an EGARCH model which in turn can be used to approximate a volatility option pricing model.

7.1.1 Structure of Chapter

The rest of this chapter is organized as follows. The next section provides a concise overview of the BFOA. The following section illustrates the comparative study of the BFOA and GA. We then outline the experimental methodology adopted to estimate the option pricing model. The remaining sections provide the results of these experiments followed by a number of conclusions.

7.2 The Bacteria Foraging Optimisation Algorithm (BFOA)

Natural selection tends to eliminate animals with poor foraging strategies (methods for locating, handling and ingesting food) and favor the propagation of genes of those animals that have successful foraging strategies, since they are more likely to enjoy reproductive success. After many generations, poor foraging strategies tend to disappear being morphed into better ones. This observation led to the idea of using foraging strategies as a source of inspiration for the design of optimisation algorithms. In forgaging, animals seek 'value for money', in other words they implicitly seek to maximise the energy obtained per unit of time (cost) spent foraging, in the face of constraints presented by its own physiology (e.g., sensing and cognitive capabilities) and environment (e.g., density of prey, risks from predators, physical characteristics of the search area). Although we usually do not think of bacteria of being complex creatures, they are in fact capable of surprisingly sophisticated behaviours. One of these behaviours is their foraging strategy for nutrients. In this chapter we focus on the forgaging behaviour of E. coli bacteria. During the lifetime of E. coli, they undertake multiple activities including chemotaxis and reproduction. They are also subject to events such as environmental dispersal.

Chemotaxis is the ability of a bacterium to move toward distant sources of nutrients. In this stage, an E.coli bacterium alternates between swimming (moving in a relatively straight line) and tumbling (changing direction). In bacterial reproduction, healthy bacteria split into two 'child' bacteria. In dispersal, a bacterium can be dispersed to a random location, for example, by being carried elsewhere by the wind. A broad outline of the BFOA is presented in Algorithm 1.

Algorithm 1. Canonical BFOA

```
Randomly distribute initial values for θ^i, i = 1,2,...,S across the optimisation
domain

for Elimination-dispersal loop do
    for Reproduction loop do
        for Chemotaxis loop do
            for Bacterium i do
                Compute the value of cost function J(i,j)
                Let J_last = J(i,j)
                Tumble: Generate a random variable φ(j) as a unit length
                random direction
                Move: let θ^i(j+1) = θ^i(j) + C(i) * φ(j)
                Compute J(i,j+1) with θ^i(j+1)
                Swim: let m=0
                while m < N_s do
                    let m=m+1
                    if J(i,j+1) < J_last then
                        let J_last = J(i,j+1)
                        let θ^i(j+1) = θ^i(j) + C(i) * φ(j) and compute the
                        new J(i,j+1)
                    else
                        let m = N_s
                    end
                end
            end
        end
        Let J^i_health = Σ_{j=1}^{N_c+1} J(i,j) be the health of bacterium i.
        Sort bacteria in order of ascending cost J_health
        The S_r = S/2 bacteria with the highest J_health value die and other
        S_r bacteria with the best value split
    end
    Eliminate and disperse the bacteria to random locations on the
    optimisation domain with probability p_ed
end
```

As depicted in Algorithm 1, the BFOA considered in this chapter contains three main steps, namely, chemotaxis, reproduction, and elimination-dispersal. A description of each of these steps is as follows.

A. Chemotaxis

Microbiological studies show that E.coli bacteria move by rotating flagella which are attached to their cell body. When all the flagella rotate counterclockwise, the E.coli bacteria move forward, when all the flagella rotate clockwise, the E. coli bacteria slow down and tumble in its place. The foraging of E. coli bacteria is accompanied by the alternation of the two modes of operation during its entire lifetime. As a result bacteria are physically able to move towards nutrients and away from noxious substances. The chemotactic step is achieved through tumbling and swimming via flagella (see fig. 7.1).

In the BFOA, a *tumble* is represented by a unit walk with random direction, and a *swim* is indicated as a unit walk with the same direction in the last step. The position update of a bacterium can be described as

$$\theta^i(j+1) = \theta^i(j) + C(i) * \phi(j) \tag{7.1}$$

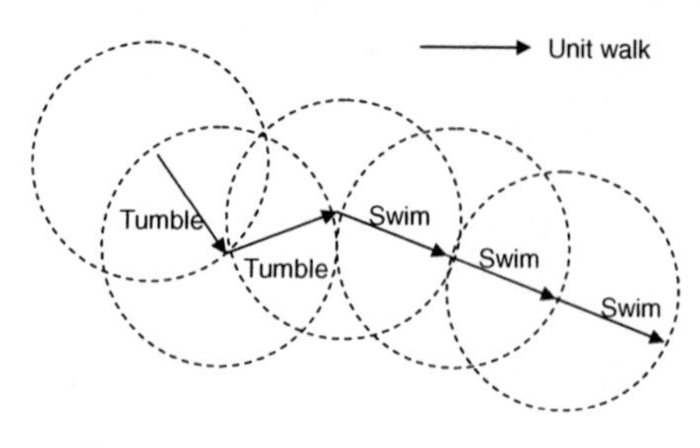

Fig. 7.1. Chemotactic step

where $\theta^i(j)$ indicates the position of the ith bacterium at the jth chemotactic step. $C(i)$ is the step size taken in the random direction, which is specified by $\phi(j)$, a unit length random direction. Let $J(i,j)$ denote the cost at the position of the ith bacterium $\theta^i(j)$. If at $\theta^i(j+1)$, the cost $J(i,j+1)$ is better (in other words lower, here we are assuming that the objective is cost minimisation) than the cost at $\theta^i(j)$, another swim step is taken, and is continued as long as it continues to reduce the cost, but only up to a maximum number of steps, N_s. This means that the bacterium will tend to keep moving if it is heading towards a better region of the fitness landscape.

B. Reproduction After N_c chemotactic steps, a reproduction operation is performed. Let N_{re} be the number of reproduction operations taken during the algorithm. The accumulated cost of each bacterium is calculated as the sum of the cost during its life, i.e., $\sum_{j=1}^{N_c+1} J(i,j)$. All bacteria are sorted in order of ascending accumulated cost (higher accumulated cost indicates that a bacterium has lower fitness value, which implies an unsuccessful life of foraging). In the reproduction step, only the first half of population survive and a surviving bacterium is cloned into two identical ones, which are placed at the same location. Thus, the population size of bacteria is kept constant.

C. Elimination - dispersal The elimination-dispersal step happens after a certain number of reproduction steps. Bacteria are selected stochastically from the population (an individual bacterium is chosen with probability p_{ed}), to be dispersed and moved randomly to a new position within the optimisation domain.

Looking at the elements of the algorithm, we can consider that the chemotactic step is primarily (although not necessarily exclusively) a local search operator, the reproduction step assists in convergence of the population, and the elimination-dispersal step helps prevent premature convergence. A variety of BOFAs exist. In this chapter we do not consider the social swarming effect observed in some bacterial colonies, where bacteria can signal, and influence the behaviour of, their peers by releasing chemical molecules (see Passino (25)). Future work will investigate this mechanism. In this chapter we focus on a simpler model where the chemotactic step size parameter $C(i)$ is adapted to control the convergence speed, i.e., $C(i)$ starts from

$\frac{Range\ of\ search\ domain}{100}$, and shrinks after each reproduction step. The basic BFOA bears some similarities to the canonical genetic algorithm (GA) and both can be used to solve real-valued optimisation problems. In the next section, we compare the BFOA and the canonical GA on some benchmark problems.

7.3 Comparative Study with Genetic Algorithm

The comparison with the genetic algorithm is undertaken to understand the relative performance characteristics of the BFOA. The BFOA and GA are population-based search algorithms. As shown from table 7.1, the nutrient concentration function and the fitness function used in the BFOA and GA respectively both define an environment. In the BFOA, bacteria in the most favorable environments gain a selective advantage for reproduction, which is similar to the selection process in GA. In the BFOA, the bacteria with higher fitness (lower cost) split in a cloning operation into two children at the same location. In GA, crossover generally acts to create children in the neighbourhood of their parents on the fitness landscape. In the BFOA, elimination-dispersal results in physical dispersion of bacteria. Mutation in GA results in a similar process. Both mechanisms allow the population to escape from local optima.

Table 7.1. Comparison of BFOA and canonical GA

BFO	GA
Nutrient concentration function	Fitness function
Bacterial reproduction	Selection and crossover
Elimination and dispersal	Mutation

Benchmark Function Tests

Six benchmark functions (8, 24) are chosen in order to test the performance of the BFOA. In order to get a better understanding of the algorithm, the results are compared to those from a canonical GA. Details of the mathematical form of (and illustrative graphs of) the benchmark functions the are shown in table 7.2 and fig's. 7.2 and 7.3. The benchmark functions are as follows.

f_1: *Sphere function* (also known as De Jong's function 1) is the simplest test function, which is continuous, convex and unimodal.

f_2: *Schwefel's function* produces rotated hyper-ellipsoids with respect to the coordinate axes. It is continuous, convex and uni-modal.

f_3: *Rosenbrock's function* (also known as De Jong's function 2, or Banana function) is a classic optimization problem. The global optimum is inside a long, narrow,

Table 7.2. Benchmark functions

f	Function	Mathematical representation	Range	$f(x_i^*)$	x_i^*
f_1	Sphere	$f(x) = \sum_{i=1}^{p} x_i^2$	$-5.12 \leq x_i \leq 5.12$	0	0
f_2	Schwefel	$f(x) = \sum_{i=1}^{p} (\sum_{j=1}^{i} x_j)^2$	$-65.536 \leq x_i \leq 65.536$	0	0
f_3	Rosenbrock	$f(x) = \sum_{i=1}^{p-1} 100(x_{i+1} - x_i^2)^2 + (1 - x_i)^2$	$-2.048 \leq x_i \leq 2.048$	0	1
f_4	Rastrigin	$f(x) = 10p + \sum_{i=1}^{p} (x_i^2 - 10cos(2\pi x_i))$	$-5.12 \leq x_i \leq 5.12$	0	0
f_5	Ackley	$f(x) = 20 + e - 20exp\left(-0.2\sqrt{\frac{1}{p}\sum_{i=1}^{p} x_i^2}\right) - exp\left(\frac{1}{p}\sum_{i=1}^{p} \cos(2\pi x_i)\right)$	$-30 \leq x_i \leq 30$	0	0
f_6	Griewangk	$f(x) = \sum_{i=1}^{p} \frac{x_i^2}{4000} - \prod_{i=1}^{p} \cos\left(\frac{x_i}{\sqrt{i}}\right) + 1$	$-600 \leq x_i \leq 600$	0	0

parabolic shaped flat valley. To find the valley is trivial, however convergence to the global optimum is more difficult.
f_4: *Rastrigin's function* is based on function 1 with the addition of cosine modulation to produce many local minima. Thus, the test function is highly multi-modal. However, the location of the minima are regularly distributed.
f_5: *Ackley's function* is a widely-used multi-modal test function.
f_6: *Griewangk's function* is similar to Rastrigin's function. It has many widespread local minima. However, the location of the minima are regularly distributed.

The parameters used for BFOA are shown in table 7.3. The values were chosen after some initial trial and error experimentation, and were selected in order to balance search speed vs accuracy. The chemotactic step size $C(i)$ starts from $\frac{Range\ of\ search\ domain}{100}$, and shrinks after each reproduction step. The larger $C(i)$ values promote more exploration in the earlier phases of the algorithm with the smaller values encouraging more fine-grained search later.

Table 7.3. Initializing the BFOA's Parameters

Parameters	Definition
$D = 5$	Dimension of the search space
$S = 50$	Number of bacteria (population size)
$N_c = 20$	Maximum number of chemotactic steps
$N_s = 4$	Maximum number of swimming steps
$N_{re} = 2$	Maximum number of reproduction steps
$S_r = S/2$	Number of bacteria for reproduction / cloning
$N_{ed} = 2$	Maximum number of elimination-dispersal steps
$p_{ed} = 0.25$	The probability that each bacterium will be eliminated / dispersed
$C(i)$	Chemotactic step size for bacterium i

In order to have a reasonably fair comparison between the BFOA and the GA results, we use same population size of 50 and the same number of fitness function evaluations in each. Whereas in GA, the number of fitness function evaluations is equivalent to the number of generations times the population size, in BFOA number of fitness function evaluations is the count of total chemotactic steps taken for each

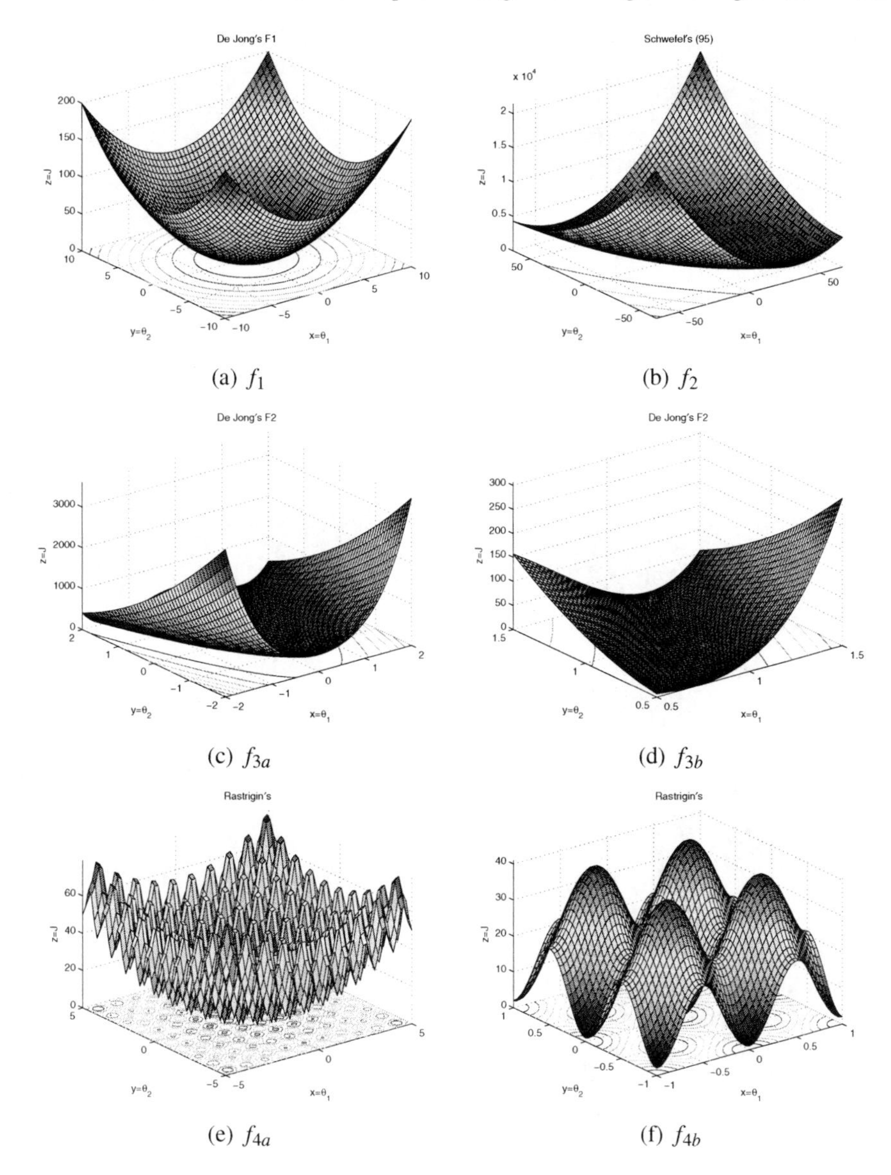

(a) f_1 (b) f_2

(c) f_{3a} (d) f_{3b}

(e) f_{4a} (f) f_{4b}

Fig. 7.2. Two dimensional visualization of benchmark functions f_1, f_2, f_3 and f_4.

bacterium during the algorithm. In the GA, we set the crossover rate at 0.7 and the mutation rate at 0.05. The results are shown in table 7.4, averaged over 30 runs of each algorithm. The first column lists the minimal (optimal) objective value found during the 30 runs within the whole population. The second and third columns list

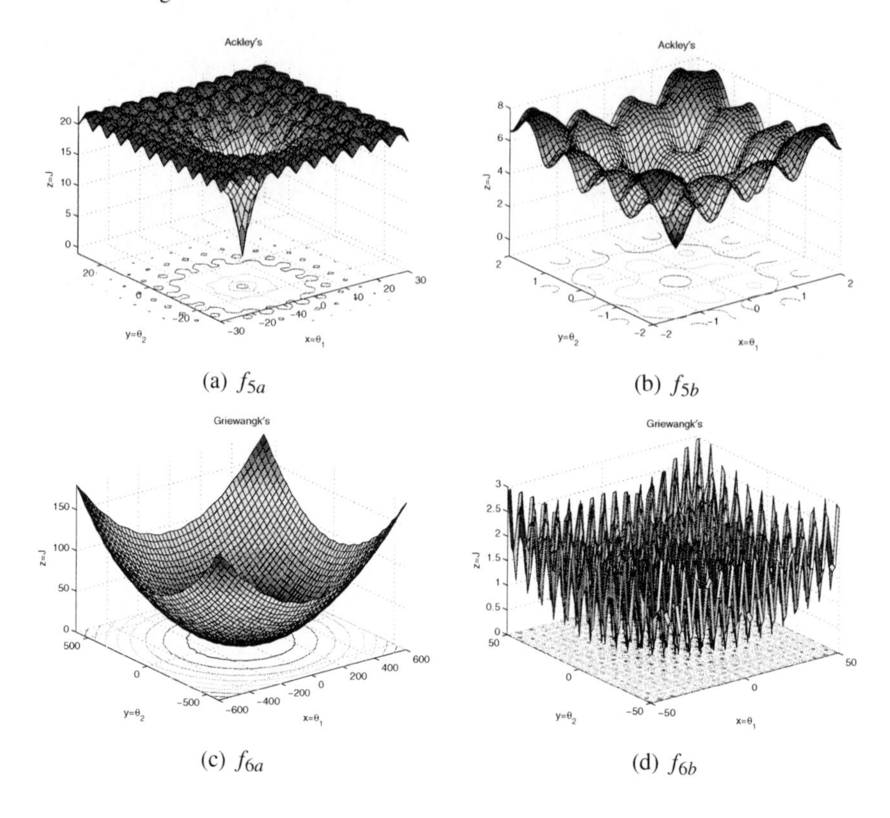

Fig. 7.3. Two dimensional visualization of benchmark functions f_5 and f_6

the mean and standard deviation for the minimal value over the 30 runs. The *Time* column shows the average processing time taken for each run.
As seen in table 7.4, the BFOA finds better results than GA except for f_4 - Rastrigin's function. The average run time for the BFOA is shorter than that of the GA. It is also noted that although the BFOA generally found good 'best-of-run' values, it performed less well it terms of some of the mean and standard deviation values, particularly in the case of functions f_2, f_4 and f_5. The optimisation process for f_3 using the BFOA and GA is illustrated in fig. 7.4. From the graph it can be seen that GA converges earlier than the BFOA, however, the BFOA finds a better result over time. In summary, the results suggest that even a canonical form of the BOFA is capable of acting as a reasonably efficient and generally, quite effective, optimiser. However, it is also noted that the mean and standard deviation values for the BOFA can be variable. Of course, it would be possible to reformulate the BOFA to incorporate additional mechanisms such as swarming in order to improve its results further. This is left for future work.

Following the above benchmarking, it appears reasonable to apply the BFOA to solve real financial problems, such as parameter estimation. Traditionally, statisti-

Table 7.4. Results of BFO and GA with 30 runs for benchmark functions testing

Algorithm	Best	Mean	S.D.	Time(s)
$f_1 : Sphere\ function$				
BFO	0.000376	0.00194	0.00103	0.047
GA	0.002016	0.00202	1.32e-018	1.07
$f_2 : Schwefel's\ function$				
BFO	0.1717	6.55	9.508	0.049
GA	0.1982	0.55	0.614	1.21
$f_3 : Rosenbrock's\ function$				
BFO	0.03989	0.578	0.73	0.050
GA	0.06818	2.46	1.43	1.17
$f_4 : Rastrigin's\ function$				
BFO	2.032	10.4	3.84	0.058
GA	0.399	0.7841	0.69	1.06
$f_5 : Ackley's\ function$				
BFO	0.0361	3.11	4.14	0.085
GA	1.0895	1.0895	9.03e-016	1.14
$f_6 : Griewangk's\ function$				
BFO	0.3271	0.687	0.17	0.113
GA	0.7067	0.722	0.015	1.20

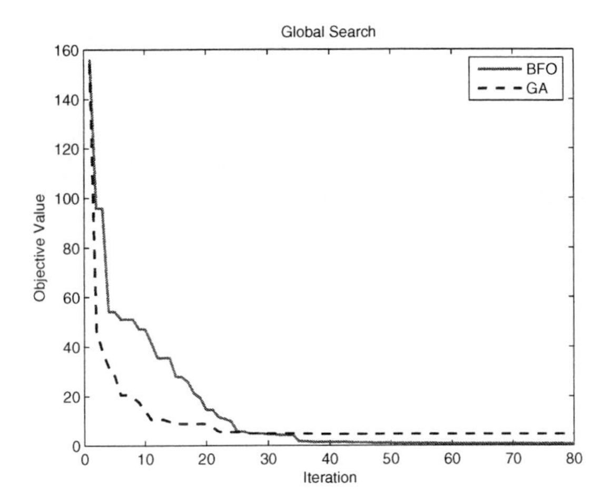

Fig. 7.4. Objective Value vs. Number of Generations. The GA converges earlier than the BFOA, however, the BFOA finds a better solution.

cal computing methods are usually employed in finance for parameter estimation. These usually require gradient information about the objective function and often also require initial estimates of the parameters which are being optimised (i.e. to provide good starting conditions). However, a strength of the BFOA, and other natural computing algorithms, is that they are not confined by these limitations. In the following section, we illustrate the application of the BFOA by using it to estimate parameters of a EGARCH model for the purpose of volatility option pricing.

7.4 Volatility Option Pricing Model

7.4.1 Volatility Option

Volatility is a measure of how much a stock can move over a specified period of time. The more variability there is in the price changes of a stock or index, the higher its volatility. Volatility is defined as the standard deviation of daily percentage changes of the stock price. Options are financial instruments that convey the right, but not the obligation, to engage in a future transaction on some underlying security. For example, the buyer of a European call option has the right, but not the obligation to buy an agreed quantity of a particular security (the underlying instrument) from the seller of the option at a certain time (the expiration date) for a certain price (the strike price).

In February 2006, options on the S&P 500 volatility index (VIX Options) began trading on the Chicago Board of Exchange (CBOE). This was the first product on market volatility to be traded on a regulated securities exchange. The S&P 500 Volatility Index (VIX) was created in 1993 as the first measure of volatility in the overall market. The VIX is designed to reflect investors' consensus view of expected stock market volatility over the next 30 days. The index is commonly referred to as the market's "fear gauge" and serves as a proxy for investor sentiment, rising when investors are anxious or uncertain about the market and falling during times of confidence or complacency. VIX options offer investors the ability to make trades based on their view of the future direction or movement of the VIX. VIX options also offer the opportunity to hedge the volatility (as distinct from the price) risk of an investment portfolio.

A growing literature on volatility options emerged after the 1987 stock market crash. Brenner and Galai (3, 4) first suggested the idea of options written on a volatility index that would serve as the underlying asset. Towards this end, Whaley (32) constructed the VIX (currently termed VXO), a volatility index based on the S&P 100 option's implied volatilities.[1] Since then, a variety of other implied volatility indices have been developed (e.g., VDAX in Germany, VXN in CBOE, VX1 and VX6 in France).

Various models to price volatility options written on the instantaneous volatility have also been developed (e.g., Whaley (32), Grunbichler and Longstaff (11), and Detemple and Osakwe (7)). These models differ in the specification of the assumed stochastic process and in respect of the assumptions made about the volatility risk premium. For example, Grunbichler and Longstaff (11) specify a mean-reverting

[1] Implied volatility is the volatility that makes the theoretical value of an option equal to the market price of an option.

square root diffusion process for volatility. Their framework is similar to that of Hull and White (15), Stein and Stein (27) and others. These models assumed that the premium for volatility risk was proportional to the level of volatility. This approach is in the spirit of the equilibrium approach of Cox, Ingersoll and Ross (6). A more recent paper by Detemple and Osakwe (7) also uses a general equilibrium framework to price European and American style volatility options, emphasising mean-reversion in a log volatility model.

The literature on option pricing under stochastic volatility can be grouped into two categories, the bivariate diffusion, and the GARCH (generalized autoregressive conditional heteroskedasticity) approaches.[2] The former strand approaches option pricing with stochastic volatility in a diffusion framework, assuming that asset prices and their volatility follow a stochastic process. The latter develops the option pricing model in a GARCH framework. GARCH models are popular econometric modelling methods, having been firstly specified by Engle (10) and Bollerslev (1), they are specifically designed to model and forecast changes in variance, or volatility per se. These two strands of option pricing models are unified by a convergence result that the GARCH option pricing model weakly converged to a bivariate diffusion option pricing model (9, 22).

Fig. 7.5 provides some empirical data about the volatility of S&P 500 index, based on sample period from 02/01/1990 to 30/12/2006, the data source is CBOE. Fig. 5(a) shows the daily closing values of the S&P 500 equity index in the sample period. There appears no long-run average level about which the series evolves. This is evidence of a nonstationary time series. Fig. 5(b) illustrates the continuously compounded returns (the log returns)[3] associated with the price series in fig. Fig. 7.5a. In contrast to the price series in 5(a), the log returns appear to be quite stable over time, and the transformation from prices to returns has produced a stationary time series. Fig. 5(c) shows the closing level of the S&P 500 Volatility Index (VIX) during the sample period. We could intuitively find the volatility clustering effect, where large volatility movements are more likely to be succeeded by further large volatility movements of either sign than by small movements. 5(d) gives an example of the probability density function of VIX, the dots represents the frequency of VIX occurred within range of values in the x-axis, and the bell curve line represents the probability of ln(VIX) for a normal distribution (with mean μ = 2.89, and standard deviation σ=0.32). The illustration shows that VIX tends to follow a lognormal distribution.

We can see from fig. 7.5, that the volatility rates of S&P 500 are higher over certain periods and lower in others, and that periods of high volatility tend to cluster together. Therefore, we would expect the volatilities to be correlated to some extent. Also it

---

[2] The term "heteroskedasticity" refers to a condition which exists when the differences between actual and forecast values do not have a constant variance across an entire range of time-series observations.

[3] Denoting the successive price observations made at times $t-1$ and t as P_{t-1} and P_t respectively, then we could obtain the continuously compounded returns as $R_t = \log\frac{P_t}{P_{t-1}} = \log P_t - \log P_{t-1}$, this is the preferred method for most financial calculations since the log returns are more stationary and are continuously distributed.

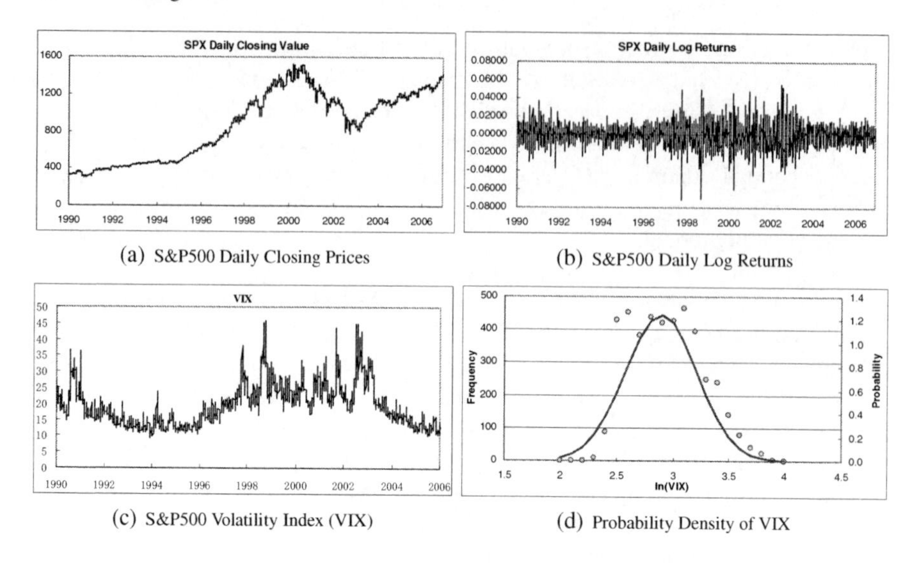

(a) S&P500 Daily Closing Prices

(b) S&P500 Daily Log Returns

(c) S&P500 Volatility Index (VIX)

(d) Probability Density of VIX

Fig. 7.5. Volatility of S&P500

is noticeable that volatility tends to mean-revert. In this chapter, we consider the mean-reverting log process (MRLP) option pricing model, proposed by Detemple and Osakwe (7). The relevance of this model is motivated by (i) substantial empirical evidence supporting the EGARCH (Exponential GARCH) model of Nelson (23) and (ii) the fact that EGARCH converges to a Gaussian process that is mean reverting in the log, thus matching our MRLP specification.

7.4.2 EGARCH Pricing Model

With the existence of too much noise in the newly traded volatility option data, we calibrate the MRLP option pricing model by estimating the corresponding EGARCH model and then taking the limit. The exponential GARCH (EGARCH) model[4] is an asymmetric model designed to capture the leverage effect, or negative correlation, between asset returns and volatility.

The EGARCH[5] (1,1) model considered in this chapter is set up as follows:

The conditional mean model:

$$y_t = C - \frac{1}{2}\sigma_t^2 + \varepsilon_t \tag{7.2}$$

[4] The EGARCH model was proposed by Nelson (23) , the nonnegativity constraints as in the linear GARCH model are taken out and so there are no restriction on the parameters in this model.

[5] The EGARCH model specified here is often referred to as the EGARCH in Mean (EGARCH-M) model, since the conditional variance term σ^2 in the variance equation also appears in the mean equation

$$\text{where } \varepsilon_t = \sigma_t z_t, \text{ and } z_t \sim N(0,1)$$
$$y_t = \log(\frac{S_t}{S_{t-1}}) \text{ , (the log returns of S\&P)}$$

The conditional variance model:

$$\log\sigma_t^2 = K + G_1 \log\sigma_{t-1}^2 + A_1[|z_{t-1}| - E(|z_{t-1}|)] + L_1 z_{t-1} \tag{7.3}$$

$$\text{where } z_{t-1} = \frac{|\varepsilon_{t-1}|}{\sigma_{t-1}}$$
$$E(|z_{t-1}|) = \sqrt{2/\pi}, \text{ if } z_t \sim \text{Gaussian}$$

Duan (9) shows that under the locally risk-neutralized probability measure Q, the asset return dynamic takes the form in eq. 7.2. There are five parameters to be estimated using the BFOA, namely, C, K, G_1, A_1, and L_1. C is the conditional mean constant, K is the conditional variance constant, G_1 (GARCH term) is the coefficient related to lagged conditional variances, A_1 (ARCH term) is the coefficient related to lagged innovations, L_1 is the leverage coefficient for asymmetric EGARCH-M(1,1) model. The coefficient of σ^2 (GARCH in Mean term) is fixed at $-\frac{1}{2}$, and hence it is not being estimated here.

The left-hand side of eq. 7.3 is the log value of the conditional variance. This implies that the leverage effect is exponential, rather than quadratic, and the forecasts of the conditional variance are guaranteed to be nonnegative. The presence of leverage effects can be tested by the hypothesis that $\alpha_3 < 0$. The impact is asymmetric if $\alpha_3 \neq 0$. The weak limit of this model converges to the unique strong solution of the MRLP (mean-reverting log process) stochastic volatility diffusion model. The limiting process is:

$$d\, ln(S_t) = (r - \delta - \frac{1}{2}V^2)dt + V_t(\rho dZ_{1t} + \sqrt{1-\rho^2}dZ_{2t}) \tag{7.4}$$

$$d\, ln(V_t) = (\alpha - \lambda ln(V_t))dt + \sigma dZ_{1t} \tag{7.5}$$

Detemple and Osakwe (7) derived analytic pricing formulae for European volatility options as a functions of parameters α, λ, σ and ρ, based on the MRLP volatility diffusion model. Where α/λ denotes a long run mean for log (V), $exp\left((\alpha + \frac{1}{4}\sigma^2)/\lambda\right)$ $\sqrt{285}$ denoting a long run mean annualized volatility (based on 285 days), and ρ represents the correlation between Z_1 and Z_2. These parameters for the option pricing model can be calculated as below (9):

$$\begin{aligned} \alpha &= \frac{K}{2} + \frac{A_1}{\sqrt{2\pi}} \\ \lambda &= 1 - G_1 \\ \sigma &= \frac{1}{2}\sqrt{{L_1}^2 + (\frac{\pi-2}{\pi}){A_1}^2} \\ \rho &= \frac{L1}{2\sigma} \end{aligned} \tag{7.6}$$

We employ the BFOA to optimize the EGARCH model parameters: C, K, G_1, A_1 and L_1.

7.4.3 EGARCH Parameter Estimation using the BFOA

The EGARCH model can be estimated by maximum likelihood estimation (MLE). The idea behind maximum likelihood parameter estimation is to determine the parameters that maximize the probability (likelihood) of the available sample data. From a statistical point of view, the method of maximum likelihood is considered to be robust and it yields estimators with good statistical properties. Although the methodology for maximum likelihood estimation is simple, the implementation is computationally intensive. For the EGARCH models specified in eqs. 7.2 and 7.3, the objective is to maximise the log likelihood function (LLF) as follows:

$$LLF = -\frac{1}{2}\sum_{t=1}^{T}[\log(2\pi\sigma_t^2) + \frac{\varepsilon_t^2}{\sigma_t^2}] \tag{7.7}$$

Given the observed log return series, the current parameter values, and the starting value of $z_1 \sim N(0,1)$, $\sigma_1^2 = exp(K)$, the σ_t^2 and ε_t are inferred by recursive substitution based on the conditional mean/variance equation (equation 7.2 and 7.3):

$$\begin{aligned}
\sigma_t^2 &= exp(K + G_1 \log\sigma_{t-1}^2 + A_1[|z_{t-1}| - E(|z_{t-1}|)] + L_1 z_{t-1}) \\
z(t) &= (-C + y_t + \frac{1}{2}\sigma_t^2)/\sigma_t \\
\varepsilon_t &= \sigma_t z_t
\end{aligned} \tag{7.8}$$

The log-likelihood function then uses the inferred residuals ε_t and conditional variances σ_t^2 to evaluate the appropriate log-likelihood objective function in eq. 7.7. We employ the BFOA as an optimisation tool searching for the optimal parameters and maximising the log-likelihood objective function. Since minimising the negative log-likelihood ($-LLF$) is the same as maximising the log-likelihood (LLF), we use $-LLF$ as our nutrient function (the objective function). The goal is to minimize the $-LLF$ value, by finding optimal values of the parameters C, K, G_1, A_1, L_1 within the search domain.

7.5 Results

The EGARCH model is fitted to the return series of S&P 500 daily index using the BFOA. The S&P 500 (Ticker SPX) equity index is obtained from the CBOE, with the sample period from 02/01/1990 to 30/12/2006, for a total of 4084 daily observations. The parameters used in the BFOA are listed in table 7.5. They are chosen based on trial and error for this particular problem. $C(i)$ is varied from 0.01 to 0.1 in step size of 0.01, and running the BFOA for 10 trials in each case respectively.
Fig. 7.6 depicts the evolution of the objective function, measured using negative maximum likelihood ($-LLF$), as a function of the iteration number for a single run of the algorithm. Figs. 7(a), 7(b), 7(c), 7(d) and 7(e) depict the evolution of the parameters C, K, G_1, A_1, and L_1 as a function of the iteration number for a single run

Table 7.5. BFOA's Parameters

Dimension of the search space: $D = 5$
Population size: $S = 50$
Chemotactic steps: $N_c = 20$
Swimming steps: $N_s = 4$
Reproduction steps: $N_{re} = 4$
Number of bacteria for reproduction/splitting: $S_r = S/2$
Elimination-dispersal steps: $N_{ed} = 2$
Probability that each bacterium will be eliminated/dispersed $p_{ed} = 0.25$
Chemotactic step size for bacterium i: $C(i) = 0.08$

of the algorithm. In the early generations, the BFOA mainly performs global search for the optimum value, with quicker convergence than in latter generations, where the focus is on more local search. From the 40th iteration, the optimal objective value becomes worse and the effect lasting for a few generations, this is due to the elimination-dispersal step conducted in iteration 40, by allowing the optimal value to be worse, we can jump out of a local minimum, and move towards global minimum.

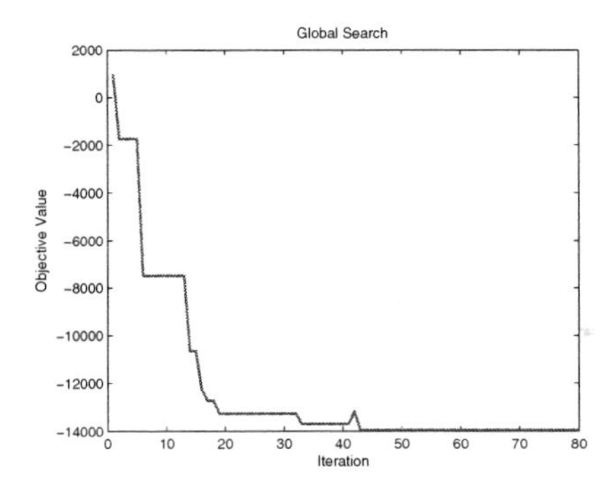

Fig. 7.6. Objective Value vs. Iteration

The best results over 30 runs are reported in the second column of table 7.6. The best results averaged over 30 runs are reported in the third column. The standard deviation of the best results over 30 runs are reported in the fourth column. In order to provide a benchmark for the results obtained by the BFOA, a Matlab optimising function *fmincon* was used. The function *fmincon* uses sequential quadratic programming (SQP) methods, which closely mimic Newton's method for constrained optimization. It requires information about the gradient of the objective function and initial estimates of the optimising parameters, while the BFOA does not require these. Running the BFOA over 30 trials, we obtain the results shown in table 7.6.
From table 7.6, we obtain the the optimal objective (the minimal $-LLF$) value of -14180.98, which is slightly lower than -14244.13 obtained in Matlab using the

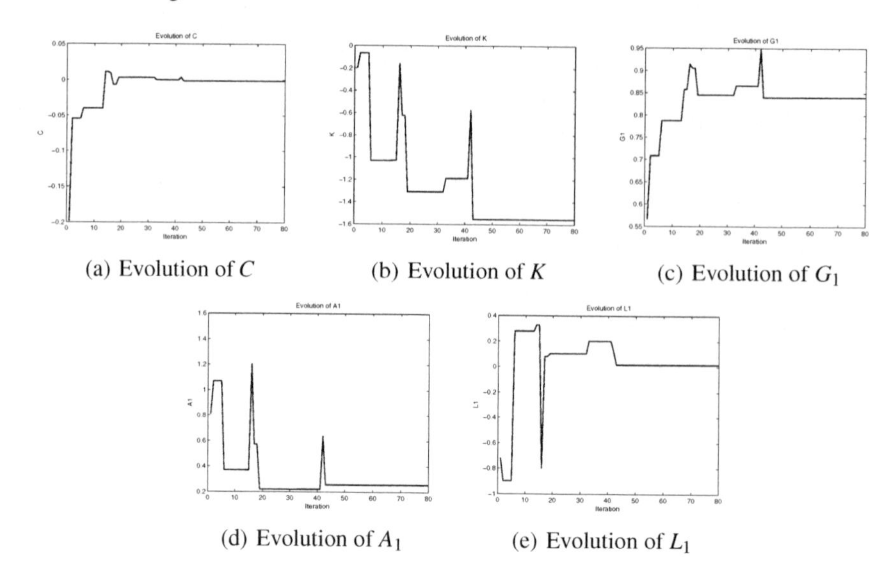

(a) Evolution of C (b) Evolution of K (c) Evolution of G_1

(d) Evolution of A_1 (e) Evolution of L_1

Fig. 7.7. Evolution of parameters over generations

Table 7.6. Results of BFOA over 30 runs

Parameter	Optima	Mean	Standard Deviation	Matlab optimisation
$-LLF$ (Objective)	-14180.98	-14099.87	32.3997	-14244.13
C	0.0005	0.0003	0.00067	0.0002
K	-0.3588	-0.301	0.0478	-0.3643
G_1	0.9107	0.904	0.0056	0.9610
A_1	0.1634	0.235	0.0489	0.1782
L_1	-0.1197	-0.0495	0.0473	-0.1184

default *fmincon* function. The result is reasonably acceptable and the standard deviation is relatively small, indicating the stability of BFOA. The estimated optimal parameters value are: $C = 0.0005$, $K = -0.3588$, $G_1 = 0.9107$, $A_1 = 0.1634$, $L_1 = -0.1197$. The leverage effect term L_1 is negative and statistically different from zero, indicating the existence of the leverage effect in future stock returns during the sample period. With the flexibility of the BFOA, it is believed that by further evolving the BFOA's parameters such as chemotactic step size $C(i)$, number of chemotactic steps N_c etc, we can improve the accuracy of the results, however, there is always trade off between accuracy (achieved by adding complexity to the algorithm) and convergence speed.

Based on the above results and eq. 7.6, the resulting stochastic volatility option pricing model parameters are: $\alpha = -0.1142, \lambda = 0.0893, \sigma = 0.0775$ and $\rho = -0.7722$. The negative correlation ρ corresponds to the asymmetric relationship between returns and changes in volatility, i.e., the leverage effect. The negative α implied mean reversion with a long run mean for log (V) of $\alpha/\lambda = -1.2790$, and a long run mean annualised volatility (based on 285 days) of $exp\left((\alpha+\frac{1}{4}\sigma^2)/\lambda\right)\sqrt{285} =$ 4.7783 percent. The speed of reversion λ, is small, indicating strong autocorrelation in volatility which in turn implies volatility clustering. These are consistent with the empirical results noted in fig. 7.5. Furthermore, based on the estimated parameters of the volatility option pricing model, hedgers can manage the risk/volatility in their existing investment/portfolio. Traders can also use the generated theoretical volatility options prices as a trading guide to make arbitrage or speculative profits.

7.6 Conclusion

In this chapter, we introduced and assessed the recently developed bacteria foraging optimisation algorithm (BFOA). The BOFA is a novel natural computing algorithm which is loosely based on the foraging behavior of E.coli bacteria. This chapter illustrated how a bacteria foraging optimisation algorithm (BFOA) can be constructed and tested the utility of the algorithm by comparing its performance on a series of benchmark functions against that of the canonical genetic algorithm (GA). Following this, the algorithm was applied in a proof of concept study to estimate parameters for an EGARCH model which could be applied for pricing volatility options. The results suggest that BFOA can be used as a complementary technique to conventional statistical computing techniques in parameter estimation for financial models. Future work will concentrate on applying more realistic models of bacteria foraging in the design of the BFOA. In particular, the algorithm has been extended to include a swarming effect. This bears parallel with financial markets where individual can share information and where entire markets can sometimes display herding effects. It is also noted that recent work in bacteria foraging (19, 28–30) has extended the BOFA into dynamic environments, and this also offers opportunities for financial application.

References

[1] Bollerslev T (1986) Generalized Autoregressive Conditional Heteroskedasticity. Journal of Econometrics 31:307–327
[2] Brabazon A and O'Neil M (2006) Biologically Inspired Algorithms for Financial Modelling. Berlin Springer
[3] Brenner M and Galai D (1989) New Financial Instruments for Hedging Changes in Volatility. Financial Analyst Journal, Jul/Aug 61–65
[4] Brenner M and Galai D (1993) Hedging Volatility in Foreign Currencies. Journal of Derivatives 1:53–59.
[5] Cox J C and Ross S A (1976) The Valuation of Options for Alternative Stochastic Processes. Journal of Financial Economics 3:145–166
[6] Cox J C, Ingersoll J E and Ross S A (1985) An Intertemporal General Equilibrium Model of Asset Prices. Econometrica 53:363–384

[7] Detemple J B and Osakwe C (2000) The Valuation of Volatility Options. European Finance Review. 4(1):21–50
[8] Digalakis J G, Margaritis K G (2000) An experimental study of benchmarking functions for Genetic Algorithms. Proceedings of IEEE Conference on Transactions, Systems, Man and Cybernetics 5:3810–3815
[9] Duan J C (1997) Augmented GARCH (p,q) Process and its Diffusion Limit. Journal of Econometrics 79:97–127
[10] Engle R F (1982) Autoregressive conditional heteroskedasticity with estimates of the variance of U.K. inflation. Econometrica 50:987–1008
[11] Grunbichler A and Longstaff F (1996) Valuing Futures and Options on Volatility. Journal of Banking and Finance 20:985–1001
[12] Heston S L (1993) A Closed-Form Solution for Options with Stochastic Volatility with Applications to Bond and Currency Options. Review of Financial Studies 6(2):237–343
[13] Heston S L and Nandi S (1997) A Closed-Form GARCH Option Pricing Model. Working Paper 97-9, Federal Reserve Bank of Atlanta
[14] Hentschel L (1995) All in the Family: Nesting Symmetric and Asymmetric GARCH Models. Journal of Financial Economics 39:71–104
[15] Hull J and White A (1987) The Pricing of Options on Assets with Stochastic Volatility. Journal of Finance 42:281–300
[16] Hull J (2006) Options, Futures, and Other Derivatives (Sixth Ed.), Pearson Prentice Hall
[17] Kim D H and Cho J H (2005) Intelligent Control of AVR System Using GA-BF. Springer LNAI 3684, 854–859
[18] Kim D H, Abraham A and Cho J H (2007) A Hybrid Genetic Algorithm and Bacterial Foraging Approach for Global Optimization. Information Sciences 177:3918–3937
[19] Li M S, Tang W J, Wu Q H and Saunders J R (2007) Bacterial Foraging Algorithm with Varying Population for Optimal Power Flow. Springer (EvoWorkshops 2007), LNCS 4448, 32–41
[20] Mishra S (2005) A Hybrid Least Square-Fuzzy Bacterial Foraging Strategy for Harmonic Estimation. IEEE Transactions on Evolutionary Computation 9(1):61–73
[21] Mishra S and Bhende C N (2007) Bacterial Foraging Technique-Based Optimized Active Power Filter for Load Compensation. IEEE Transactions on Power Delivery 22(1):457–465
[22] Nelson D B (1990) ARCH Models as Diffusion Approximations. Journal of Econometrics 45:7–39
[23] Nelson D B (1991) Conditional Heteroskedasticity in Asset Returns: A New Approach. Econometrica 59(2):347–370
[24] Ortiz-Boyer D, Hervás-Martínez C and García-Pedrajas N (2005) CIXL2: A Crossover Operator for Evolutionary Algorithms Based on Population Features. Journal of Artificial Intelligence Research 24:1–48
[25] Passino K M (2002) Biomimicry of bacterial foraging for distributed optimization and control, IEEE Control Systems Magazine 2(3):52–67

[26] Ramos V, Fernandes C and Rosa A C (2005) On Ants, Bacteria and Dynamic Environments. Natural Computing and Applications Workshop (NCA 2005), IEEE Computer Press, Sep. 2005, 25–29

[27] Stein E M and Stein J C (1991) Stock Price Distributions with Stochastic Volatility: An Analytical Approach. Review of Financial Studies 4:727–752

[28] Tang W J, Wu Q H and Saunders J R (2006) A Novel Model for Bacterial Foraging in Varying Environments. Springer (ICCSA 2006), LNCS 3980, 556–565

[29] Tang W J, Wu Q H and Saunders J R (2006) Bacterial Foraging Algorithm for Dynamic Environments, Proceedings of IEEE Congress on Evolutionary Computation (CEC 2006), 1324–1330

[30] Tang W J, Wu Q H and Saunders J R (2007) Individual- Based Modeling of Bacterial Foraging with Quorum Sensing in a Time-Varying Environment. Springer (EvoBIO 2007), LNCS 4447, 280–290

[31] Ulagammai M, Venkatesh P, Kannan P S and Padhy N P (2007) Application of Bacterial Foraging Technique Trained Artificial and Wavelet Neural Networks in Load Forecasting. Neurocomputing 70(16-18):2659–2667

[32] Whaley R E (1993) Derivatives on Market Volatility: Hedging Tools Long Overdue. Journal of Derivatives 1:71–84

[33] Whaley R E (2000) The Investor Fear Gauge, Journal of Portfolio Management 26(3):12–17

Part II

Model Induction

8

Fuzzy-Evolutionary Modeling for Single-Position Day Trading

Célia da Costa Pereira and Andrea G. B. Tettamanzi

Università Degli Studi di Milano
Dipartimento di Tecnologie dell'Informazione, Milan,
Italy.
celia.pereira@unimi.it, andrea.tettamanzi@unimi.it

Summary. This chapter illustrates a data-mining approach to single-position day trading which uses an evolutionary algorithm to construct a fuzzy predictive model of a financial instrument. The model is expressed as a set of fuzzy IF-THEN rules. The model takes as inputs the open, high, low, and close prices, as well as the values of a number of popular technical indicators on day t and produces a *go short*, *do nothing*, *go long* trading signal for day $t+1$ based on a dataset of past observations of which actions would have been most profitable. The approach has been applied to trading several financial instruments (large-cap stocks and indices): the experimental results are presented and discussed. A method to enhance the performance of trading rules based on the approach by using ensembles of fuzzy models is finally illustrated. The results clearly indicate that, despite its simplicity, the approach may yield significant returns, outperforming a buy-and-hold strategy.

8.1 Introduction

Single-position automated day-trading problems (ADTPs) involve finding an automated trading rule for opening and closing a single position[1] within a trading day. They are a neglected subclass of the more general automated intraday trading problems, which involve finding profitable automated technical trading rules that open and close positions within a trading day. An important distinction that may be drawn is the one between static and dynamic trading problems. A *static* problem is when the entry and exit strategies are decided before or on market open and do not change thereafter. A *dynamic* problem allows making entry and exit decisions as market action unfolds.

Dynamic problems have been the object of much research, and different flavors of evolutionary algorithms have been applied to the discovery and/or the optimization of dynamic trading rules (1, 8, 12, 13, 16, 19, 22, 23, 28, 34). Static problems

[1] A *position* is a non-zero balance for a financial instrument in a trader's account. Therefore, a long (short) position is opened by buying (short-selling) some quantity of a given financial instrument, and is closed by selling (buying back) an equal quantity of the same financial instrument.

C. da Costa Pereira and A.G.B. Tettamanzi: *Fuzzy-Evolutionary Modeling for Single-Position Day Trading*, Studies in Computational Intelligence (SCI) **100**, 131–159 (2008)
www.springerlink.com

are technically easier to approach, as the only information that has to be taken into account is information available before market open. This does not mean, however, that they are easier *to solve* than their dynamic counterparts.

This chapter will focus on solving a class of static single-position automated day-trading problems by means of a data-mining approach which uses an evolutionary algorithm to construct a fuzzy predictive model of a financial instrument. The model takes as inputs the open, high, low, and close prices, as well as the values of a number of popular technical indicators on day t and produces a *go short*, *do nothing*, *go long* trading signal for day $t+1$ based on a dataset of past observation of which actions would have been most profitable.

The chapter is organized as follows: Section 8.2 discusses how trading rules are evaluated and compared by investors; Section 8.3 states the problem addressed by this chapter, namely the static single-position day trading problem, and situates it within the context of trading problems. Next, the two basic tools used for approaching such problem are introduced: Section 8.4 provides a gentle introduction to fuzzy logic, with a particular reference to fuzzy rule-based systems, and Section 8.5 introduces evolutionary computation in general and distributed evolutionary algorithms in particular. Section 8.6 describes a fuzzy-evolutionary modeling approach to single-position day trading and the data used for modeling; in particular, several technical analysis indicators used by the approach are defined and briefly discussed. Section 8.7 reports the protocol and the results of experiments carried out to assess the approach and Section 8.8 discusses an ensemble technique to improve the performance of trading rules discovered by the approach. Section 8.9 concludes.

8.2 Evaluating Trading Rules

Informally, we may think of a trading rule R as some sort of decision rule which, given a time series $X = \{x_t\}_{t=1,2,\ldots,N}$ of prices of a given financial instrument, for each time period t returns some sort of trading signal or instruction. The reason one might want to write such a trading rule is to consistently operate a strategy on a market, with the intent of gaining a profit. Instead of considering absolute profit, which depends on the quantities traded, it is a good idea to focus on returns.

8.2.1 Measures of Profit

For mathematical reasons, it is convenient to use log-returns instead of the usual returns, because they are additive under compounding. The average log-return of rule R when applied to time series X of length N is

$$r(R;X) = \frac{Y}{N}\sum_{t=1}^{N} r(R;X,t), \tag{8.1}$$

where $r(R;X,t)$ is the return generated by rule R in the tth day of time series X, and Y is the number of market days in a year. T his is the most obvious performance index for a trading rule. However, as a performance measure, average log-return completely overlooks the risk of a trading rule.

8.2.2 Measures of Risk-Adjusted Return

Following the financial literature on investment evaluation (7), the criteria for evaluating the performance of trading rules, no matter for what type of trading problem, should be measures of risk-adjusted investment returns. The reason these are good metrics is that, in addition to the profits, consistency is rewarded, while volatile patterns are not. Common measures within this class are the Sharpe ratio and its variations, the Treynor ratio, Jensen's performance index, and the upside-potential ratio. The main ingredients of all these measures are:

- the risk-free log-return r_f (in practice, one can use the log-return of short-dated government bonds of the currency in question);
- the average log-return of a rule, defined in Equation 8.1;
- the average log-return of time series X,

$$r(X) = \frac{Y}{N-1} \sum_{t=2}^{N} r(X,t) = \frac{Y}{N-1} \sum_{t=2}^{N} \ln \frac{x_t^C}{x_{t-1}^C}, \tag{8.2}$$

  where x_t^C is the closing price of the tth day;
- the standard deviation of the log-returns (i.e., the risk) of rule R on X,

$$\sigma(R;X) = \sqrt{\frac{Y}{N} \sum_{t=1}^{N} [r(R;X,t) - r(R;X)]^2}; \tag{8.3}$$

- the downside risk (14, 27) of rule R on X,

$$\mathrm{DSR}_\theta(R;X) = \sqrt{\frac{Y}{N} \sum_{t=1}^{N} [r(R;X,t) < \theta][\theta - r(R;X,t)]^2}; \tag{8.4}$$

Sharpe Ratio and Variations

The Sharpe ratio (26) is probably the most popular measure of risk-adjusted returns for mutual funds and other types of investments; in the case of trading rules, the Sharpe ratio of rule R on time series X is

$$\mathrm{SR}(R;X) = \frac{r(R;X) - r_f}{\sigma(R;X)}, \tag{8.5}$$

where we are making the assumption that the risk-free rate r_f is constant during the timespan covered by X. A critique of this measure is that it treats positive excess returns, i.e., windfall profits, exactly the same way as it treats negative excess return; in fact, traders, just like investors, do not regard windfall profits as something to avoid as unexpected losses. A variation of the Sharpe ratio which acknowledges this fact is Sortino ratio (27), which in our case is

$$\mathrm{SR}_d(R;X) = \frac{r(R;X) - r_f}{\mathrm{DSR}_{r_f}(R;X)}. \quad (8.6)$$

Unlike the Sharpe ratio, the Sortino ratio adjusts the expected return for the risk of falling short of the risk-free return; positive deviations from the mean are not taken into account to calculate risk.

8.3 The Trading Problem

Static ADTPs can be classified according to the type of positions allowed, entry strategy, profit-taking strategy, and stop-loss or exit strategy. By combining these options (and perhaps more), one can name all different types of trading problems. In the rest of this chapter, we will focus on a particular class of static ADTP, namely the class of problems whereby the trading strategy allows taking both long and short positions at market opening, a position is closed as soon as a pre-defined profit has been reached, or otherwise at market close as a means of preventing losses beyond the daily volatility of an instrument.

Such problems make up the simplest class of problems when it comes to rule evaluation: all that is required is open, high, low, and close quotes for each day, since a position is opened at market open, if the rule so commands, and closed either with a fixed profit or at market close.

8.3.1 The BOFC Problem

Static ADTPs can be classified according to

- the type of positions allowed;
- the entry strategy;
- the profit-taking strategy;
- the stop-loss or exit strategy.

This chapter focuses on a particular ADTP which can be described as follows: take a long or short position on open, take profit if a fixed return is achieved, otherwise close the position at the end of the day. We will refer to this problem as the BOFC problem, because:

- both (B) long and short positions are allowed;
- the entry is on open (O) at market price;
- the profit-taking strategy is on a fixed (F) return;
- the exit strategy is on close (C) at market price.

8.3.2 Trading Rules for Static ADTPs

In its most general form, a trading rule for a static ADTP should specify:

1. a sign (long or short) for the position;
2. a limit price for entry, including market price;
3. a take-profit level;
4. a stop-loss level.

All prices should be expressed relative to some other price: the limit price for entry might be expressed relative to the previous close, the take-profit and stop-loss levels might be expressed relative to the price at which the position is opened. Formally, we can think of a trading rule R as a function of a day in a time series whose value is a 4-tuple

$$R(X,t) = \langle s, r_L, r_{TP}, r_{SL} \rangle, \tag{8.7}$$

where

1. $s \in \{-1, 1\}$ is the sign of the position to be opened: -1 for a short position, 1 for a long position;
2. $r_L \in \mathbb{R}$ is used to calculate the limit price for entry according to the formula $p_L = x_{t-1}^C e^{r_L}$;
3. $r_{TP} > 0$ is the log-return target from which a take-profit limit of $xe^{r_{TP}}$ is calculated, where x is the price at which the position is opened;
4. $r_{SL} > 0$ is the stop-loss log-return; the stop-loss level is $xe^{-r_{SL}}$.

Although this general framework is capable of accommodating all static ADTPs, for the BOFC problem, $s \in \{-1, 1\}$, $r_L \in \{-\infty, +\infty\}$ (i.e., a binary entry condition: $+\infty$ = enter, $-\infty$ = do not enter), $r_{TP} =$ constant, given by the user, $r_{SL} = +\infty$, meaning a stop-loss price level of 0, which is never triggered. Given these constraints, the result of a trading rule is just a ternary decision: *go short* ($s = -1$, $r_L = -\infty$), *do nothing* ($s \cdot r_L = -\infty$), or *go long* ($s = +1$, $r_L = +\infty$).

8.3.3 Rule Evaluation Requirements of the BOFC Problem

While for the purpose of designing a profitable trading rule R, i.e., solving any static ADTP, the more information is available the better, whatever the trading problem addressed, for the purpose of evaluating a given trading rule the quantity and granularity of quote information required varies depending on the problem.

In order to evaluate the performance of rules for the BOFC problem, the open, high, low, and close quotes for each day are needed. The required time series is $X = \{x_t^O, x_t^H, x_t^L, x_t^C\}_{t=1,\ldots,N}$. The log-return generated by rule R in the tth day of time series X is

$$r(R;X,t) = \begin{cases} r_{TP} & \text{if } s \cdot r_L = +\infty \text{ and } \bar{r} > r_{TP}, \\ s \ln \frac{x_t^C}{x_t^O} & \text{if } s \cdot r_L = +\infty \text{ and } \bar{r} \leq r_{TP}, \\ 0 & \text{otherwise}, \end{cases} \tag{8.8}$$

where

$$\bar{r} = \begin{cases} \ln \frac{x_t^H}{x_t^O} & \text{if } s > 0, \\ \ln \frac{x_t^O}{x_t^L} & \text{if } s < 0. \end{cases} \tag{8.9}$$

The BOFC problem is therefore among the most complex single-position day-trading problems whose solutions one can evaluate when disposing only of open, high, low, and close quotes for each day. The reason we chose to focus on such problem is indeed that while such kind of quotes are freely available on the Internet for a wide variety of securities and indices, more detailed data can in general only be obtained for a fee.

We approach this problem by evolving trading rules that incorporate fuzzy logic. The adoption of fuzzy logic is useful in two respects: first of all, by recognizing that concept definitions may not always be crisp, it allows the rules to have what is called an *interpolative behavior*, i.e., gradual transitions between decisions and their conditions; secondly, fuzzy logic provides for linguistic variables and values, which make rules more natural to understand for an expert. The next section introduces fuzzy logic, with a particular emphasis on the concepts that are relevant to our approach.

8.4 Fuzzy Logic

Fuzzy logic was initiated by Lotfi Zadeh with his seminal work on fuzzy sets (35). Fuzzy set theory provides a mathematical framework for representing and treating vagueness, imprecision, lack of information, and partial truth.

Very often, we lack complete information in solving real world problems. This can be due to several causes. First of all, human expertise is of a qualitative type, hard to translate into exact numbers and formulas. Our understanding of any process is largely based on imprecise, "approximate" reasoning. However, imprecision does not prevent us from performing successfully very hard tasks, such as driving cars, improvising on a chord progression, or trading financial instruments. Furthermore, the main vehicle of human expertise is natural language, which is in its own right ambiguous and vague, while at the same time being the most powerful communication tool ever invented.

8.4.1 Fuzzy Sets

Fuzzy sets are a generalization of classical sets obtained by replacing the characteristic function of a set A, χ_A which takes up values in $\{0,1\}$ ($\chi_A(x) = 1$ iff $x \in A$, $\chi_A(x) = 0$ otherwise) with a *membership function* μ_A, which can take up any value in $[0,1]$. The value $\mu_A(x)$ is the membership degree of element x in A, i.e., the degree to which x belongs in A.

A fuzzy set is completely defined by its membership function. Therefore, it is useful to define a few terms describing various features of this function, summarized in fig. 8.1. Given a fuzzy set A, its *core* is the (conventional) set of all elements x such that $\mu_A(x) = 1$; its *support* is the set of all x such that $\mu_A(x) > 0$. A fuzzy set is *normal* if its core is nonempty. The set of all elements x of A such that $\mu_A(x) \geq \alpha$, for a given $\alpha \in (0,1]$, is called the α-cut of A, denoted A_α.

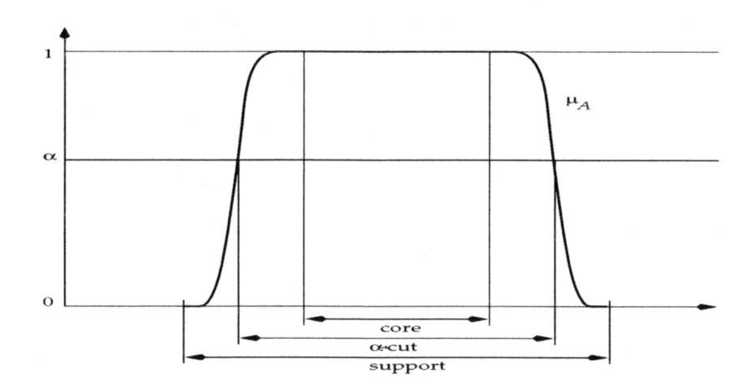

Fig. 8.1. Core, support, and α-cuts of a set A of the real line, having membership function μ_A.

If a fuzzy set is completely defined by its membership function, the question arises of how the shape of this function is determined. From an engineering point of view, the definition of the ranges, quantities, and entities relevant to a system is a crucial design step. In fuzzy systems all entities that come into play are defined in terms of fuzzy sets, that is, of their membership functions. The determination of membership functions is then correctly viewed as a problem of design. As such, it can be left to the sensibility of a human expert or more objective techniques can be employed. Alternatively, optimal membership function assignment, of course relative to a number of design goals that have to be clearly stated, such as robustness, system performance, etc., can be estimated by means of a machine learning or optimization method. In particular, evolutionary algorithms have been employed with success to this aim. This is the approach we follow in this chapter.

8.4.2 Operations on Fuzzy Sets

The usual set-theoretic operations of union, intersection, and complement can be defined as a generalization of their counterparts on classical sets by introducing two families of operators, called triangular norms and triangular co-norms. In practice, it is usual to employ the min norm for intersection and the max co-norm for union. Given two fuzzy sets A and B, and an element x,

$$\mu_{A\cup B}(x) = \max\{\mu_A(x), \mu_B(x)\}; \tag{8.10}$$

$$\mu_{A\cap B}(x) = \min\{\mu_A(x), \mu_B(x)\}; \tag{8.11}$$

$$\mu_{\bar{A}}(x) = 1 - \mu_A(x). \tag{8.12}$$

8.4.3 Fuzzy Propositions and Predicates

In classical logic, a given proposition can fall in either of two sets: the set of all true propositions and the set of all false propositions, which is the complement of the former. In fuzzy logic, the set of true proposition and its complement, the set of false propositions, are fuzzy. The degree to which a given proposition P belongs to the set of true propositions is its degree of truth, $T(P)$.

The logical connectives of negation, disjunction, and conjunction can be defined for fuzzy logic based on its set-theoretic foundation, as follows:

$$\text{Negation} \quad T(\neg P) = 1 - T(P); \tag{8.13}$$

$$\text{Disjunction} \quad T(P \vee Q) = \max\{T(P), T(Q)\}; \tag{8.14}$$

$$\text{Conjunction} \quad T(P \wedge Q) = \min\{T(P), T(Q)\}. \tag{8.15}$$

Much in the same way, a one-to-one mapping can be established as well between fuzzy sets and fuzzy predicates. In classical logic, a predicate of an element of the universe of discourse defines the set of elements for which that predicate is true and its complement, the set of elements for which that predicate is not true. Once again, in fuzzy logic, these sets are fuzzy and the degree of truth of a predicate of an element is given by the degree to which that element is in the set associated with that predicate.

8.4.4 Fuzzy Rule-Based Systems

A prominent role in the application of fuzzy logic to real-world problems is played by fuzzy rule-based systems. Fuzzy rule-based systems are systems of fuzzy rules that embody expert knowledge about a problem, and can be used to solve it by performing fuzzy inferences. The ingredients of a fuzzy rule-based systems are *linguistic variables*, *fuzzy rules*, and *defuzzification* methods.

Linguistic Variables

A linguistic variable (36) is defined on a numerical interval and has linguistic values, whose semantics is defined by their membership function. For example, a linguistic variable *temperature* might be defined over the interval $[-20°C, 50°C]$; it could have linguistic values like *cold, warm,* and *hot,* whose meanings would be defined by appropriate membership functions.

Fuzzy Rules

A fuzzy rule is a syntactic structure of the form

$$\text{IF } \textit{antecedent} \text{ THEN } \textit{consequent}, \tag{8.16}$$

where each *antecedent* and *consequent* are formulas in fuzzy logic. Fuzzy rules provide an alternative, compact, and powerful way of expressing functional dependencies between various elements of a system in a modular and, most importantly, intuitive fashion. As such, they have found broad application in practice, for example in the field of control and diagnostic systems (25).

Inference in Fuzzy Rule-Based Systems

The semantics of a fuzzy rule-based system is governed by the calculus of fuzzy rules (37). In summary, all rules in a fuzzy rule base take part simultaneously in the inference process, each to an extent proportionate to the truth value associated with its antecedent. The result of an inference is represented by a fuzzy set for each of the dependent variables. The degree of membership for a value of a dependent variable in the associated fuzzy set gives a measure of its compatibility with the observed values of the independent variables. Given a system with n independent variables $x_1,\ldots,x_n$ and m dependent variables $y_1,\ldots,y_m$, let R be a base of r fuzzy rules

$$\begin{array}{ll} \text{IF } P_1(x_1,\ldots,x_n) & \text{THEN } Q_1(y_1,\ldots,y_m), \\ \vdots & \vdots \\ \text{IF } P_r(x_1,\ldots,x_n) & \text{THEN } Q_r(y_1,\ldots,y_m), \end{array} \tag{8.17}$$

where $P_1,\ldots,P_r$ and $Q_1,\ldots Q_r$ represent fuzzy predicates respectively on independent and dependent variables, and let τ_P denote the truth value of predicate P. Then the membership function describing the fuzzy set of values taken up by dependent variables $y_1,\ldots,y_m$ of system R is given by

$$\begin{aligned} &\tau_R(y_1,\ldots,y_m;x_1,\ldots,x_n) \\ &\quad = \sup_{1\leq i\leq r} \min\{\tau_{Q_i}(y_1,\ldots,y_m), \tau_{P_i}(x_1,\ldots,x_n)\}. \end{aligned} \tag{8.18}$$

The Mamdani Model

The type of fuzzy rule-based system just described, making use of the min and max as the triangular norm and co-norm, is called the Mamdani model. A Mamdani system (17) has rules of the form

$$\text{IF } x_1 \text{ is } A_1 \text{ AND } \ldots \text{ AND } x_n \text{ is } A_n \text{ THEN } y \text{ is } B, \tag{8.19}$$

where the A_is and B are linguistic values (i.e., fuzzy sets) and each clause of the form "x is A" has the meaning that the value of variable x is in fuzzy set A.

Defuzzification Methods

There may be situations in which the output of a fuzzy inference needs to be a crisp number y^* instead of a fuzzy set R. Defuzzification is the conversion of a fuzzy quantity into a precise quantity.

At least seven methods in the literature are popular for defuzzifying fuzzy outputs (15), which are appropriate for different application contexts. The *centroid method* is the most prominent and physically appealing of all the defuzzification methods. It results in a crisp value

$$y^* = \frac{\int y\mu_R(y)dy}{\int \mu_R(y)dy}, \tag{8.20}$$

where the integration can be replaced by summation in discrete cases. The next section introduces evolutionary algorithms, a biologically inspired technique which we use to learn and optimize fuzzy rule bases.

8.5 Distributed Evolutionary Algorithms

Evolutionary algorithms (EAs) are a broad class of stochastic optimization algorithms, inspired by biology and in particular by those biological processes that allow populations of organisms to adapt to their surrounding environment: genetic inheritance and survival of the fittest.

An evolutionary algorithm (EA) maintains a population of candidate solutions for the problem at hand, and makes it evolve by iteratively applying a (usually quite small) set of stochastic operators, known as *mutation*, *recombination*, and *selection*.

Mutation randomly perturbs a candidate solution; recombination decomposes two distinct solutions and then randomly mixes their parts to form novel solutions; and selection replicates the most successful solutions found in a population at a rate proportional to their relative quality.

The initial population may be either a random sample of the solution space or may be seeded with solutions found by simple local search procedures, if these are available.

The resulting process tends to find, given enough time, globally optimal solutions to the problem much in the same way as in nature populations of organisms tend to adapt to their surrounding environment. Books of reference and synthesis in the field of EC are (4, 5, 11); recent advances are surveyed in (33).

Evolutionary algorithms have enjoyed an increasing popularity as reliable stochastic optimization, search and rule-discovering methods in the last few years. The original formulation by Holland and others in the seventies was a sequential one. That approach made it easier to reason about mathematical properties of the algorithms and was justified at the time by the lack of adequate software and hardware. However, it is clear that EAs offer many natural opportunities for parallel implementation (20). There are several possible parallel EA models, the most popular being the fine-grained or *grid* (18), the coarse-grain or *island* (31), and the master-slave or *fitness parallelization* (10) models. In the grid model, large populations of individuals are spatially distributed on a low-dimensional grid and individuals interact locally within a small neighborhood. In the master-slave model, a sequential EA is executed on what is called the *master* computer. The master is connected to several *slave* computers to which it sends individuals when they require evaluation. The slaves evaluate the individuals (fitness evaluation makes up most of the computing time of an EA) and send the result back to the master.

In the island model, the population is divided into smaller subpopulations which evolve independently and simultaneously according to a sequential EA. Periodic migrations of some selected individuals between islands allow to inject new diversity into converging subpopulations. Microprocessor-based multicomputers and workstation clusters are well suited for the implementation of this kind of parallel EA. Being coarse-grained, the island model is less demanding in terms of communication speed and bandwidth, which makes it a good candidate for a cluster implementation.

8.6 The Approach

Data mining is a process aimed at discovering meaningful correlations, patterns, and trends between large amounts of data collected in a dataset. A model is determined by observing past behavior of a financial instrument and extracting the relevant variables and correlations between the data and the dependent variable. We describe below a data-mining approach based on the use of evolutionary algorithms, which recognize patterns within a dataset, by learning models represented by sets of fuzzy rules.

8.6.1 Fuzzy Models

A fuzzy model is described through a set of fuzzy rules. A rule is made by one or more antecedent clauses ("IF ...") and a consequent clause ("THEN ..."). Clauses are represented by a pair of indices referring respectively to a variable and to one of its fuzzy sub-domains, i.e., a membership function.

Using fuzzy rules makes it possible to get homogenous predictions for different clusters without imposing a traditional partition based on crisp thresholds, that often do not fit the data, particularly in financial applications. Fuzzy decision rules are useful in approximating non-linear functions because they have a good interpolative power and are intuitive and easily intelligible at the same time. Their characteristics allow the model to give an effective representation of the reality and simultaneously avoid the "black-box" effect of, e.g., neural networks.

The output of the evolutionary algorithm is a set of rules written in plain consequential sentences. The intelligibility of the model and the high explanatory power of the obtained rules are useful for a trader, because the rules are easy to interpret and explain. An easy understanding of a forecasting method is a fundamental characteristic, since otherwise a trader would be reluctant to use forecasts.

8.6.2 The Evolutionary Algorithm

The described approach incorporates an EA for the design and optimization of fuzzy rule-based systems that was originally developed to automatically learn fuzzy controllers (24, 29), then was adapted for data mining, (6) and is at the basis of MOLE, a general-purpose distributed engine for modeling and data mining based on EAs and fuzzy logic (30).

A MOLE classifier is a rule base, whose rules comprise up to four antecedent and one consequent clause each. Input and output variables are partitioned into up to 16 distinct linguistic values each, described by as many membership functions. Membership functions for input variables are trapezoidal, while membership functions for the output variable are triangular. Classifiers are encoded in three main blocks:

1. a set of trapezoidal membership functions for each input variable; a trapezoid is represented by four fixed-point numbers, each fitting into a byte;
2. a set of symmetric triangular membership functions, represented as an area-center pair, for the output variable;

3. a set of rules, where a rule is represented as a list of up to four antecedent clauses (the IF part) and one consequent clause (the THEN part); a clause is represented by a pair of indices, referring respectively to a variable and to one of its membership functions.

An island-based distributed EA is used to evolve classifiers. The sequential algorithm executed on every island is a standard generational replacement, elitist EA. Crossover and mutation are never applied to the best individual in the population.

The recombination operator is designed to preserve the syntactic legality of classifiers. A new classifier is obtained by combining the pieces of two parent classifiers. Each rule of the offspring classifier can be inherited from one of the parent programs with probability $1/2$. When inherited, a rule takes with it to the offspring classifier all the referred domains with their membership functions. Other domains can be inherited from the parents, even if they are not used in the rule set of the child classifier, to increase the size of the offspring so that their size is roughly the average of its parents' sizes. Like recombination, mutation produces only legal models, by applying small changes to the various syntactic parts of a fuzzy rulebase.

Migration is responsible for the diffusion of genetic material between populations residing on different islands. At each generation, with a small probability (the migration rate), a copy of the best individual of an island is sent to all connected islands and as many of the worst individuals as the number of connected islands are replaced with an equal number of immigrants. A detailed description of the evolutionary algorithm and of its genetic operators can be found in (24).

8.6.3 The Data

In principle, the modeling problem we want to solve requires finding a function which, for a given day t, takes the past history of time series X up to t and produces a trading signal *go short*, *do nothing*, or *go long*, for the next day. One could therefore try to approach this problem directly, by evolving trading rules in the most general form by means of genetic programming (19, 22, 23, 34). The search space for such trading rules is, clearly, incredibly huge. However, there exists an impressive body of expertise used everyday by practitioners in the financial markets about summarizing all important information of the past history of a financial time series into few relevant statistics. That body of expertise is called *technical analysis*.

> Technical analysis is the study of past financial market data, primarily through the use of charts, to forecast price trends and make investment decisions. In its purest form, technical analysis considers only the actual price behavior of the market or instrument, based on the premise that price reflects all relevant factors before an investor becomes aware of them through other channels (32).

The idea is then to reduce the dimensionality of the search space by limiting the inputs of the models we look for to a collection of the most popular and time-honored technical analysis statistics and indicators.

A technical indicator is a short-term statistics of a time series used by technical analysts to predict future price movements of financial instruments. Price data include any combination of the open, high, low, and close, plus volume information, over a (generally short) period of time. As we will see in the next sections, some indicators use only the closing prices, while others incorporate other information like volume and opening prices into their formulas. Some others consist of a combination of two or more different indicators.

The advantage of using different technical indicators, is that we can dispose of different ways to analyze price movements. Some indicators, like moving averages, are defined by using simple formulas and their mechanics are thus relatively easy to understand. Others, like stochastic oscillators, have more complex definitions. Regardless of the complexity of their definition, technical indicators can provide valid elements for predicting the direction of market price movements.

Popular Statistics and Technical Indicators

In this section, we provide definitions and a basic discussion of several popular technical indicators that have been adopted as inputs to the data mining process and to the models such process looks for.

A *moving average* is the average of the closing values of a financial instrument over a given time period. Generally speaking, moving averages tend to smooth out the short-term oscillations of a time series and identify longer-term trends. The two most popular types of moving average indicators are the *simple* moving average (SMA) and the *exponential* moving average (EMA). The difference between these indicators is uniquely on the weight each of them carry to recent prices compared to old prices. SMA considers equally important all the prices in a considered period, whereas EMA considers that recent prices are more important than the old ones. In general, a buy signal is generated when the instrument's price rises above a moving average and a sell signal is generated when the instrument's price falls below a moving average.

Simple and exponential moving averages are used in our approach by combining them with the closing price value at current day t. For both forms, we consider the moving average of closing values over the n days before t, with $n \in \{5, 10, 20, 50, 100, 200\}$ noted $\mathrm{SMA}_n(t)$ and $\mathrm{EMA}_n(t)$ respectively, and we then consider the position of the closing price x_t^C with respect to the simple (exponential) moving average, noted $x_t^C : \mathrm{SMA}_n(t)$ ($x_t^C : \mathrm{EMA}_n(t)$), as follows:

$$x_t^C : \mathrm{SMA}_n(t) = \ln \frac{x_t^C}{\mathrm{SMA}_n(t)} \tag{8.21}$$

and

$$x_t^C : \mathrm{EMA}_n(t) = \ln \frac{x_t^C}{\mathrm{EMA}_n(t)}. \tag{8.22}$$

If the above logarithms are negative, i.e., the closing price at t is lower than the moving average/exponential moving average value for the previous n days, this means that the price tends to decrease. Otherwise, it means the price tends to increase. In order to get further information about the market trend, we analyze also how the above result evolves from day $t-1$ to t. The formulas for these differences are:

$$\Delta(x_t^C : \mathrm{SMA}_n(t)) = x_t^C : \mathrm{SMA}_n(t) - x_{t-1}^C : \mathrm{SMA}_n(t-1). \tag{8.23}$$

$$\Delta(x_t^C : \mathrm{EMA}_n(t)) = x_t^C : \mathrm{EMA}_n(t) - x_{t-1}^C : \mathrm{EMA}_n(t-1). \tag{8.24}$$

The *Moving Average Convergence/Divergence* (MACD) is a technical indicator which shows the difference between a fast and a slow EMA of closing prices. The simplest version of this indicator is composed of two lines: the MACD line, which is the difference between the two EMAs, and a *signal line*, which is an EMA of the MACD line itself. We use the standard periods recommended back in the 1960s by Gerald Appel, which are 12 and 26 days (21):

$$\mathrm{MACD}(t) = \mathrm{EMA}_{26}(t) - \mathrm{EMA}_{12}(t) \tag{8.25}$$

The signal line we have used is also standard. It consists of an EMA_9 of the MACD line. This allows to further smooth the trend of market price values:

$$\mathrm{signal}(t) = \mathrm{EMA}_9(t; \mathrm{MACD}) \tag{8.26}$$

We also consider Thomas Aspray's difference between the MACD and the signal line, which is often plotted as a solid block *histogram* style:

$$\mathrm{histogram}(t) = \mathrm{MACD}(t) - \mathrm{signal}(t). \tag{8.27}$$

Its change from the previous day is given by:

$$\Delta(\mathrm{MACD}(t) - \mathrm{signal}(t)) = \mathrm{histogram}(t) - \mathrm{histogram}(t-1). \tag{8.28}$$

The *rate of change* (ROC) ia a simple technical indicator showing the difference between the closing price on a current day t and the corresponding price n days before day t. Because this indicator is measured in terms of the old closing price, it represents the increase as a fraction,

$$\mathrm{ROC}_n(t) = \frac{x_t^C - x_{t-n}^C}{x_{t-n}^C}, \tag{8.29}$$

which is positive when the price value is increasing over the last n days, negative otherwise. We use the popular version with $n = 12$.

Developed by George C. Lane in the late 1950s, the *Stochastic Oscillator* is a momentum indicator that shows the location of the current close relative to the high/low range over a set number of periods. Closing levels that are consistently near the top of the range indicate accumulation (buying pressure) and those near the bottom of the range indicate distribution (selling pressure). The fast stochastic oscillator, $\%K_n$ calculates the ratio of two closing price statistics — the difference between the latest closing price and the lowest closing price in the last n days over the difference between the highest and lowest closing prices in the last n days:

$$\%K_n(t) = \frac{x_t^C - \min_{i=t-n}^t \{x_i^C\}}{\max_{i=t-n}^t \{x_i^C\} - \min_{i=t-n}^t \{x_i^C\}} \cdot 100. \tag{8.30}$$

We use the popular time period n of 14 days. $\%K_n = 0$ when the current closing price is a low for the last n days; $\%K_n = 100$ when the current closing price is a high for the last n days.

The slow stochastic oscillator $\%D_n$ calculates the simple moving average of $\%K_n$ across a period of k days:

$$\%D_n(t) = \mathrm{SMA}_k(t; \%K_n). \tag{8.31}$$

We use the popular value of $k = 3$ days.

The *relative strength index* (RSI) is another oscillator which determines the strength of a financial instrument by comparing upward and downward close-to-close movements in a period of n days. For each day t, the amounts of upward change, $U(t) > 0$, or downward change, $D(t) > 0$, are calculated as follows. On an "up" day t,

$$U(t) = x_t^C - x_{t-1}^C,$$
$$D(t) = 0.$$

Conversely, on a "down" day t,

$$U(t) = 0,$$
$$D(t) = x_t^C - x_{t-1}^C.$$

If the closing price in $t-1$ is the same as the closing price in t, both $U(t)$ and $D(t)$ are zero. An exponential moving average of up days in the period of n days, U, is calculated. Similarly, an exponential moving average of down days in the period of n days, D, is calculated. The ratio of those averages is the relative strength,

$$\mathrm{RS}(t) = \frac{\mathrm{EMA}_n(t;U)}{\mathrm{EMA}_n(t;D)} \tag{8.32}$$

This is converted to a relative strength index between 0 and 100:

$$\mathrm{RSI}(t) = 100 - 100 \cdot \frac{1}{1 + \mathrm{RS}(t)}. \tag{8.33}$$

This can be rewritten as follows to emphasize the way RSI expresses the up as a proportion of the total up and down (averages in each case),

$$\mathrm{RSI}(t) = 100 \cdot \frac{\mathrm{EMA}_n(t;U)}{\mathrm{EMA}_n(t;U) + \mathrm{EMA}_n(t;D)}. \tag{8.34}$$

The basic idea behind the use of RSI is that when there is a high proportion of daily movements in one direction, this suggests an extreme, and prices are likely to "change direction". More precisely, a financial instrument is considered overbought if RSI ≥ 70, meaning that a speculator should consider the option of selling. Conversely, a speculator should consider buying if if RSI ≤ 30 (oversold).

The *money flow index* indicator (MFI) indicates the balance between money flowing into and out of a financial instrument. The construction and interpretation of this indicator is similar to RSI, with the only difference that volume is also taken into account. The *typical price* $\bar{x}_t$ is the average of high, low and close values in day t:

$$\bar{x}_t = \frac{x_t^H + x_t^L + x_t^C}{3}. \tag{8.35}$$

Money flow is the product of typical price on a day t and the volume on that day:

$$\mathrm{MF}(t) = \bar{x}_t \cdot x_t^V. \tag{8.36}$$

Totals of the money flow amounts over the given n days are then formed. *Positive* money flow is the total for those days where the typical price is higher than the previous day's typical price, and *negative* money flow is the total of those days where the typical price is below the previous day's typical price. A money ratio is then formed:

$$\mathrm{MR}(t) = \frac{\mathrm{MF}^+(t)}{\mathrm{MF}^-(t)}, \tag{8.37}$$

from which a *money flow index* ranging from 0 to 100 is formed,

$$\mathrm{MFI}(t) = 100 - \frac{100}{1 + \mathrm{MR}(t)}. \tag{8.38}$$

When analyzing the money flow index, one needs to take into consideration the following points:

- divergences between the indicator and price movement — if prices grow while MFI falls (or *vice versa*), there is a great probability of a price turn;
- MFI over 80 or under 20 signals a potential peak or bottom of the market.

The *accumulation/distribution* indicator (AccDist) is a cumulative total volume technical analysis indicator created by Marc Chaikin, which adds or subtracts each day's volume in proportion to where the close is between the day's high and low. For example, many up days occurring with high volume in a down trend could signal that the demand for the financial instrument is starting to increase. In practice this indicator is used to find situations where the indicator is heading in the opposite directions than the price. Once this divergence has been identified, traders will wait to confirm the reversal and make their transaction decisions using other technical indicators.
A *close location value* is defined as:

$$\mathrm{CLV}(t) = \frac{(x_t^C - x_t^L) - (x_t^H - x_t^C)}{x_t^H - x_t^L}. \tag{8.39}$$

This ranges from -1, when the close is the low of the day, to $+1$ when it is the high. For instance if the close is $3/4$ the way up the range, then CLV is $+0.5$. The accumulation/distribution index adds up volume multiplied by the CLV factor, ie.

$$\text{AccDist}(t) = \text{AccDist}(t-1) + x_t^V \cdot \text{CLV}(t). \tag{8.40}$$

Finally, the *on-balance volume* indicator (OBV) is intended to relate price and volume in the stock market. OBV is based on a cumulative total volume. Volume on an up day (i.e., close value at day t higher than close value at day $t-1$) is added and volume on a down day is subtracted. The formula is

$$\text{OBV}(t) = \text{OBV}(t-1) + \begin{cases} x_t^V & \text{if } x_t^C > x_{t-1}^C, \\ 0 & \text{if } x_t^C = x_{t-1}^C, \\ -x_t^V & \text{if } x_t^C < x_{t-1}^C. \end{cases} \tag{8.41}$$

The starting point for an OBV total is arbitrary. Only the shape of the resulting indicator is used, not the actual level of the total.

Combining Statistics and Technical Indicators

After a careful scrutiny of the most popular technical indicators outlined in the previous subsection, we concluded that more data were needed if we wanted an evolutionary algorithm to discover meaningful models expressed in the form of fuzzy IF-THEN rules. Combinations of statistics and technical indicators are required that mimic the reasonings analysts and traders carry out when they are looking at a technical chart, comparing indicators with current price, checking for crossings of different graphs, and so on.

Combinations may take the form of differences between indicators that are pure numbers or that have a fixed range, or of ratios of indicators such as prices and moving average, that are expressed in the unit of measure of a currency. Following the use of economists, we consider the natural logarithm of such ratios, and we define the following notation: given two prices x and y, we define

$$x : y \equiv \ln \frac{x}{y}. \tag{8.42}$$

Eventually, we came up with the following combinations:

- all possible combinations of the Open (O), High (H), Low (L), Close (C), and previous-day Close (P) prices: $O:P, H:P, L:P, C:P, H:O, C:O, O:L, H:L, H:C, C:L$;
- close price compared to simple and exponential moving averages, $C:\text{SMA}_n$, $C:\text{EMA}_n$, $n \in \{5, 10, 20, 50, 100, 200\}$;
- the daily changes of the close price compared to simple and exponential moving averages, $\Delta(C:\text{SMA}_n)$, $\Delta(C:\text{EMA}_n)$, where $\Delta(x) \equiv x(t) - x(t-1)$;
- the MACD histogram, i.e., MACD – signal, and the daily change thereof, $\Delta(\text{Histogram})$;
- Fast stochastic oscillator minus slow stochastic oscillator, $\%K - \%D$, and the daily change thereof, $\Delta(\%K - \%D)$.

The full list of the statistics, technical indicators, and their combinations used as model inputs is given in table 8.1.

Table 8.1. The independent variables of the dataset.

Name	Formula	Explanation
Open	x_t^O	the opening price on day t
High	x_t^H	the highest price on day t
Low	x_t^L	the lowest price on day t
Close	x_t^C	the closing price on day t
Volume	x_t^V	the volume traded on day t
O:P	$x_t^O : x_{t-1}^C$	opening price on day t vs. previous-day closing price
H:P	$x_t^H : x_{t-1}^C$	high on day t vs. previous-day closing price
L:P	$x_t^L : x_{t-1}^C$	low on day t vs. previous-day closing price
C:P	$x_t^C : x_{t-1}^C$	close on day t vs. previous-day closing price
H:O	$x_t^H : x_t^O$	high on day t vs. same-day opening price
C:O	$x_t^C : x_t^O$	closing on day t vs. same-day opening price
O:L	$x_t^O : x_t^L$	opening price on day t vs. same-day lowest price
H:L	$x_t^H : x_t^L$	high on day t vs. same-day low
H:C	$x_t^H : x_t^C$	high on day t vs. same-day closing price
C:L	$x_t^C : x_t^L$	closing price on day t vs. same-day low
dVolume	$x_t^V : x_{t-1}^V$	change in volume traded on day t
C:MAn	$x_t^C : \mathrm{SMA}_n(t)$	n-day simple moving averages, for $n \in \{5, 10, 20, 50, 100, 200\}$.
dC:MAn	$\Delta(x_t^C : \mathrm{SMA}_n(t))$	daily change of the above
C:EMAn	$x_t^C : \mathrm{EMA}_n(t)$	n-day exponential moving averages, for $n \in \{5, 10, 20, 50, 100, 200\}$.
dC:EMAn	$\Delta(x_t^C : \mathrm{EMA}_n(t))$	daily change of the above
MACD	$\mathrm{MACD}(t)$	Moving average convergence/divergence on day t
Signal	$\mathrm{signal}(t)$	MACD signal line on day t
Histogram	$\mathrm{MACD}(t) - \mathrm{signal}(t)$	MACD histogram on day t
dHistogram	$\Delta(\mathrm{MACD}(t) - \mathrm{signal}(t))$	daily change of the above
ROC	$\mathrm{ROC}_{12}(t)$	rate of change on day t
K	$\%K_{14}(t)$	fast stochastic oscillator on day t
D	$\%D_{14}(t)$	slow stochastic oscillator on day t
K:D	$\%K_{14}(t) - \%D_{14}(t)$	fast vs. slow stochastic oscillator
dK:D	$\Delta(\%K_{14}(t) - \%D_{14}(t))$	daily change of the above
RSI	$\mathrm{RSI}_{14}(t)$	relative strength index on day t
MFI	$\mathrm{MFI}_{14}(t)$	money-flow index on day t
AccDist	$\Delta(\mathrm{AccDist}(t))$	The change of the accumulation/distribution index on day t
OBV	$\Delta(\mathrm{OBV}(t))$	The change of on-balance volume on day t
PrevClose	x_{t-1}^C	closing price on day $t-1$

8.6.4 Fitness

Modeling can be thought of as an optimization problem, where we wish to find the model M^* which maximizes some criterion which measures its accuracy in predicting $y_i = x_{im}$ for all records $i = 1, \ldots, N$ in the training dataset. The most natural criteria for measuring model accuracy are the mean absolute error and the mean square error.

One big problem with using such criteria is that the dataset must be *balanced*, i.e., an equal number of representatives for each possible value of the predictive attribute y_i must be present, otherwise the under-represented classes will end up being modeled with lesser accuracy. In other words, the optimal model would be very good at predicting representatives of highly represented classes, and quite poor at predicting individuals from other classes. To solve this problem, MOLE divides the range $[y_{\min}, y_{\max}]$ of the predictive variable into 256 bins. The bth bin, X_b, contains all the indices i such that

$$1 + \lfloor 255 \frac{y_i - y_{\min}}{y_{\max} - y_{\min}} \rfloor = b. \tag{8.43}$$

For each bin $b = 1, \ldots, 256$, it computes the mean absolute error for that bin

$$\text{err}_b(M) = \frac{1}{\|X_b\|} \sum_{i \in X_b} |y_i - M(x_{i1}, \ldots, x_{i,m-1})|, \tag{8.44}$$

then the total absolute error (TAE) as an integral of the histogram of the absolute errors for all the bins, $\text{tae}(M) = \sum_{b:\|X_b\| \neq 0} \text{err}_b(M)$. Now, the mean absolute error for every bin in the above summation counts just the same no matter how many records in the dataset belong to that bin. In other words, the level of representation of each bin (which, roughly speaking, corresponds to a class) has been factored out by the calculation of $\text{err}_b(M)$. What we want from a model is that it is accurate in predicting all classes, independently of their cardinality.The fitness used by the EA is given by $f(M) = \frac{1}{\text{tae}(M)+1}$, in such a way that a greater fitness corresponds to a more accurate model.

8.7 Experiments

The approach described above has been applied to trading several financial instruments. This section reports some of the experiments that have been carried out and their results.

8.7.1 Aims of the Experimental Study

There would be many interesting questions to ask of a data-mining approach applied to single-position day trading like the one described above. Here, we address the following:

1. What are the generalization capabilities of the models obtained? This question has to do with being able to correctly model the behavior of the financial instrument used for learning for a timespan into the future: here the basic question to ask is, how well can we expect the model perform tomorrow with today's data, if we have used data up to yesterday to learn it? However, another relevant question is, how does the model's performance decay as we move into the future, i.e., the day after tomorrow, and then in three days, and so forth?
2. How much historical data is needed to obtain a reliable model? This is a critical issue, as for many financial instruments the history available may be limited: think for example of the stock of a new company resulting from the merger of two pre-existing companies, or of the stock of an old company which starts being traded as a result of an initial public offering. Furthermore, it could been argued that markets evolve as the context in which they function changes; therefore, very old data about a given financial instrument might not be of any help for learning a model that reflects current (and possibly future) behavior of that instrument. The issue has thus two faces, which are distinct but related:
   a) ideally, for a data-mining approach, the more data is available the better; however, finance is a domain where only historical data is available; there is no way of obtaining more data by performing experiments: so the question is, how much data is required for learning a meaningful model?
   b) if too little data exists to allow a meaningful model to be learned, too much data (especially, data that goes too deep into the past) might lead to a model which does not correctly reflect current behavior: how old data can be before it is useless or counterproductive for learning a reliable model?

8.7.2 Experimental Protocol

The following financial instruments have been used for the experiments:

- the Dow Jones Industrial Average index (DJI);
- the Nikkei 225 index (N225);
- the common stock of Italian oil company ENI, listed since June 18, 2001 on the Milan stock exchange;
- the common stock of world's leading logistics group Deutsche Post World Net (DPW), listed since November 20, 2000 on the XETRA stock exchange;
- the common stock of Intel Co. (INTC), listed on the NASDAQ.

For all the instruments considered, three datasets of different length have been generated, in an attempt to gain some clues on how much historical data is needed to obtain a reliable model:

- a "long-term" dataset, generated from the historical series of prices since January 1, 2002 till December 31, 2006, consisting of 1,064 records, of which 958 are used for training and the most recent 106 are used for testing;
- a "medium-term" dataset, generated from the historical series of prices since January 1, 2004 till December 31, 2006, consisting of 561 records, of which 505 are used for training and the most recent 56 are used for testing;

- a "short-term" dataset, generated from the historical series of prices since January 1, 2005 till December 31, 2006, consisting of 304 records, of which 274 are used for training and the most recent 30 are used for testing;

The validation dataset, in all cases, consists of records corresponding to the first half of 2007, which require a historical series starting from March 17, 2006 (200 market days before January 2, 2007) to be generated, due to the 200-day moving averages and their changes that need to be computed. Having six months of validation data allows us to evaluate the rate of decay of model performance as it "gets stale".

8.7.3 Results

For each combination of instrument and dataset, ten runs of the MOLE data mining engine with four islands of size 100 connected according to a ring topology and with a standard parameter setting have been performed. Each run lasted as many generations as required to reach convergence, defined as no improvement for 100 consecutive generations. The results are summarized in table 8.2. Fig. 8.2 shows how the automated trading strategy based on the models found by each run fared when applied to the validation set.

To correctly interpret the fitness values appearing in table 8.2, the following considerations may be of interest:

- a hypothetical model M_{rnd} that returns a completely random (i.e., uniformly distributed) value for the Action variable in the interval $[-1,1]$ will have an expected mean absolute error of 1 in the -1 bin, of $\frac{2}{3}$ in the 0 bin, and of 1 in the $+1$ bin, according to Equation 8.44, and thus an expected total average error of 2 and $\frac{2}{3}$, corresponding to an expected fitness $E[f(M_{\text{rnd}})] = \frac{3}{11} = 0.2727\ldots$ — model M_{rnd} is representative of a situation of complete ignorance;
- a "constant" model M_0 that always returns zero for the Action variable, which is equivalent to a *do nothing* trading signal, will have an expected mean absolute error of 1 in the -1 bin, of 0 in the 0 bin, and of 1 in the $+1$ bin, according to Equation 8.44, and thus an expected total average error of 2, corresponding to an expected fitness $E[f(M_0)] = \frac{1}{3} = 0.3333\ldots$;
- the other two "constant" models M_{-1} and M_{+1}, instead, will have an expected total average error of 3, corresponding to an expected fitness $E[f(M_{-1})] = E[f(M_{+1})] = \frac{1}{4} = 0.25$.

For the calculation of the Sortino ratio, we made the simplifying assumption of a constant risk-free rate of 3.925% (the average three-month Euribor rate in the first half of year 2007[2]) for the instruments denominated in euros, of 5.175% (the average discount rate of 13-week US Treasury Bills in the first half of 2007[3]) for those denominated in dollars, and of 0.650% (the yield of 3-month Japanese Government Bills in June 2007[4]) for those denominated in yen.

[2] Source: Euribor, URL `http://www.euribor.org/`.

[3] Source: TreasuryDirect, URL `http://www.treasurydirect.gov`.

[4] Source: Bloomberg

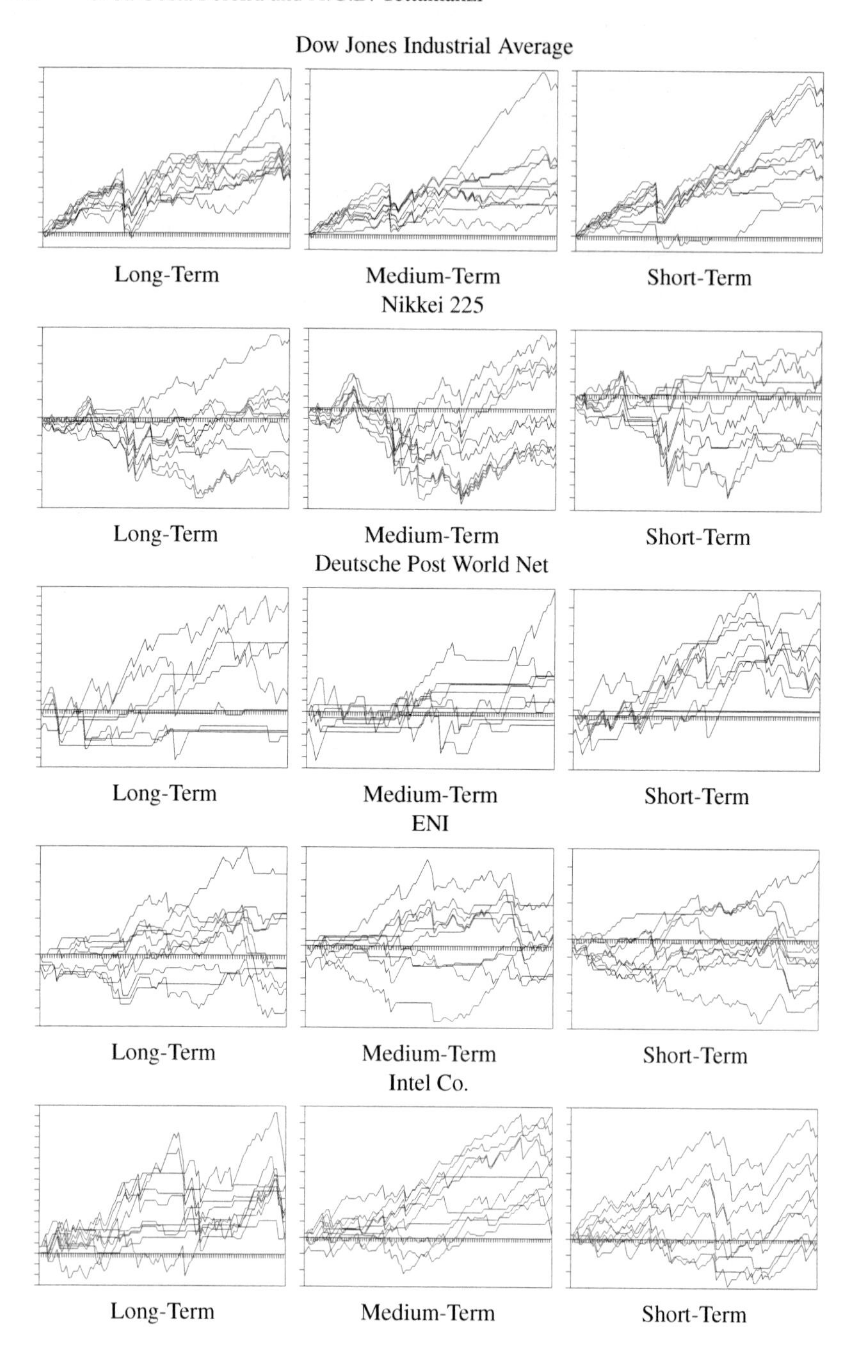

Fig. 8.2. Graphs of the cumulative returns obtained by the automated trading strategy based on the models generated by ten independent runs, for each instrument and training dataset.

Table 8.2. Summary of experimental results. Minimum, average, and maximum values are over the best models produced in ten independent runs of the island-based evolutionary algorithm, when applied to the corresponding validation set. When infinite Sharpe ratios occur, the average is computed over the finite values only.

Performance Measure	**Dataset**								
	Long-Term			Medium-Term			Short-Term		
	min	avg	max	min	avg	max	min	avg	max
Dow Jones Industrial Average									
Fitness	0.3297	0.3394	0.3484	0.3188	0.3327	0.3457	0.3183	0.3398	0.3671
Return*	0.1618	0.2303	0.4017	0.1062	0.2280	0.5503	0.0996	0.3225	0.5416
Sharpe Ratio	1.2250	2.0073	3.6700	0.6123	2.0901	4.9866	0.6064	3.0687	5.1994
Sortino Ratio	1.5380	2.5572	4.7616	0.7642	2.7949	6.5557	0.7215	4.0799	6.9968
Nikkei 225									
Fitness	0.3211	0.3414	0.3651	0.3241	0.3418	0.3575	0.3205	0.3351	0.3529
Return*	−0.1467	−0.0006	0.2119	−0.1118	0.0006	0.1436	−0.1063	−0.0161	0.1040
Sharpe Ratio	−1.6352	−0.0192	2.8201	−1.2708	−0.0238	1.5947	−1.6385	−0.1828	2.0360
Sortino Ratio	−1.9181	0.0782	4.1485	−1.5070	0.0253	2.1311	−1.9135	−0.1033	3.2197
ENI Stock									
Fitness	0.2459	0.3268	0.3500	0.2475	0.2907	0.3425	0.2402	0.2949	0.3277
Return*	−0.1389	0.0122	0.2120	−0.0856	0.0248	0.1547	−0.1936	−0.0372	0.2643
Sharpe Ratio	−2.0204	−0.2971	2.2852	−2.1096	−0.1935	1.7443	−2.5211	−0.8458	2.4507
Sortino Ratio	−2.3274	−0.2751	3.0867	−2.4578	−0.1799	2.4460	−2.8959	−0.9655	3.2188
Deutsche Post World Net Stock									
Fitness	0.3182	0.3306	0.3451	0.3200	0.3342	0.3506	0.3118	0.3299	0.3403
Return*	−0.0607	0.0476	0.2646	−0.0246	0.0547	0.2480	0.0117	0.1169	0.2820
Sharpe Ratio	$-\infty$	0.1380	3.7798	$-\infty$	0.2476	2.0083	−2.7351	0.5796	3.2487
Sortino Ratio	−15.8114	−2.3809	10.5500	−15.8114	−0.1780	12.7425	−10.2067	0.0920	4.6700
Intel Co. Stock									
Fitness	0.2490	0.3050	0.3443	0.2433	0.2838	0.3658	0.2394	0.2665	0.3333
Return*	0.0247	0.1015	0.1669	0.0131	0.2254	0.4292	−0.0244	0.1252	0.3632
Sharpe Ratio	−0.2128	0.6425	2.0494	−0.3872	2.0846	4.1456	$-\infty$	0.7836	2.7926
Sortino Ratio	−0.2467	0.8624	3.2520	−0.4569	2.9042	6.1129	−15.8114	−0.7107	3.4903

*) Annualized logarithmic return.

8.7.4 Discussion of Results

In order to have a better idea of the quality of the results, one can compare them to a buy-and-hold strategy over the same time period, summarized in table 8.3.

Table 8.3. Annualized logarithmic return and Sharpe ratio of a buy-and-hold strategy over the validation period for the instruments considered.

Instrument	**Log-Return**	**Sharpe Ratio**
Dow Jones Industrial Average	0.1461	1.4218
Nikkei 225	0.1111	0.7258
ENI	0.1168	0.8074
Deutsche Post World Net	0.1320	0.5509
Intel Co.	0.2365	1.0696

Besides evaluating the results on a quantitative basis, it is also of interest to perform a qualitative assessment of the model obtained, to see if some useful insight is captured by the data mining approach. Fig. 8.3 shows the best model of the Dow Jones Industrial Average obtained from the short-term dataset in Run #1. The model uses in its rules 13 independent variables only, out of the 54 shown in table 8.1, namely: Close, O:P, dVolume, C:MA5, C:MA10, dC:MA20, C:MA200, C:EMA10, K, dK:D, MFI, and OBV. Most rules make sense to an expert: for example, the second rule states that a closing price below the 200-day moving average speaks in favor going long, as for a mean-reverting series it can be expected that the price will go back up. Similar *a posteriori* explanations have been given by expert traders for these and other rules on which they were asked for an opinion. Overall, a tendency towards using volume information can be observed, both directly (dVolume) and through indicators, like MFI and OBV, whose definitions depend on volume.

IF MFI **is** high **AND** OBV **is** positive **AND** dVolume **is** slightlyNegative **AND**
dK:D **is** positive
THEN Action **is** Short

IF C:MA200 **is** negative **THEN** Action **is** Long

IF MFI **is** high **AND** dVolume **is** slightlyNegative **AND** C:MA5 **is** aroundZero
THEN Action **is** Short

IF MFI **is** between60and70 **AND** dC:MA20 **is** positive **THEN** Action **is** Short

IF dK:D **is** positive **AND** K **is** between50and70 **AND** dC:MA20 **is** smallPositive
AND O:P **is** positive
THEN Action **is** Short

IF C:MA10 **is** slightlyNegative **AND** Close **is** around12000 **AND**
C:EMA10 **is** aroundZero
THEN Action **is** Long

IF dK:D **is** positive **AND** K **is** between50and70 **AND** dC:MA20 **is** smallPositive
AND O:P **is** positive **THEN** Action **is** Short

IF MFI **is** high **AND** OBV **is** positive **AND** dVolume **is** slightlyNegative
AND dK:D **is** positive **THEN** Action **is** Short

IF MFI **is** high **AND** OBV **is** positive **AND** dVolume **is** slightlyNegative
AND dK:D **is** positive **THEN** Action **is** Short

Fig. 8.3. The best model of the Dow Jones Industrial Average obtained from the short-term dataset in Run #1. The definitions of the linguistic values, which are an integral part of the model, are not shown for conciseness; instead, descriptive labels have been assigned by hand to improve clarity.

8.8 An Ensemble Technique for Improving Performance

Ensemble techniques are known to be very useful to boost the performance of several types of machine learning methods and their benefits have recently been proven experimentally (3) on Pittsburgh classifier systems, an approach to which the one proposed in this chapter is very closely related.

Ensemble learning, a family of techniques established for more than a decade in the Machine Learning community, provides performance boost and robustness to the learning process by integrating the collective predictions of a set of models in some principled fashion (2). In particular, *bagging* (9) (an abbreviation for *bootstrap aggregating*) is an ensemble technique which consists in training several models on different training sets (called bootstrap samples) sampled with replacement from the original training set, and combining such models by averaging the output or voting.

A simple but effective ensemble technique to greatly reduce the risk of the trading rules, similar to bagging, is a model-averaging approach whereby a number of models, discovered in independent runs of the algorithm, perhaps using different parameter settings but, unlike in bagging, the same training and test sets, are pooled and their output is combined by averaging. Table 8.4 reports the results obtained by such ensemble technique applied to the models resulting from the ten runs summarized in table 8.2 for each combination of instrument and dataset. These results show a dramatic reduction of risk, in the face of a modest reduction of the returns. Overall, the performance measures of the model pool are usually better than their averages over the members of the pool, with notable exceptions on those instruments, like the Dow Jones Industrial Average, where the performance of the individual model was already more than satisfactory on average. In some cases, the benefits of pooling do not manage to turn a negative result into a positive one; however, pooling makes the losses less severe.

8.9 Conclusions

This chapter has discussed an approach to a very specific class of day-trading problems, that we have called *single-position*, by means of a fuzzy-evolutionary modeling technique. A distributed evolutionary algorithm learns models, expressed in the form of sets of fuzzy IF-THEN rules, that relate the optimal trading signal for a given financial instrument on day $t+1$ to a set of 54 statistics and technical indicators for the same instrument measured on day t, which effectively filter and summarize the previous history of the relevant time series up to day t.

The capability of the approach to discover useful models, which can provide risk-adjusted returns in excess of a buy-and-hold strategy, has been demonstrated through a series of experiments, whose aim was also to assess the generalization capabilities of the models discovered. Finally, an ensemble technique has been suggested to further improve the performance of the trading rules generated by the modeling technique.

The strength of the approach described is its simplicity, which makes it accessible even to unsophisticated traders. Indeed:

- the data considered each day by a trading rule to make a decision about its action is restricted to those freely available on the Internet; visibility of the past time series is filtered through a small number of popular and quite standard technical indicators;
- the underlying trading strategy is among the simplest and most accessible even to the unsophisticated individual trader; its practical implementation does not even

Table 8.4. Results of the ensemble technique applied to the best models produced in ten independent runs of the island-based evolutionary algorithm, when applied to the corresponding validation set.

Performance Measure	Dataset		
	Long-Term	Medium-Term	Short-Term
Dow Jones Industrial Average			
Return*	0.1970	0.1272	0.3958
Sharpe Ratio	1.7094	0.8744	3.8192
Sortino Ratio	2.1619	1.0636	4.9320
Nikkei 225			
Return*	−0.0369	0.0175	0.0164
Sharpe Ratio	−0.4849	0.1238	0.1405
Sortino Ratio	−0.5878	0.1543	0.1767
ENI Stock			
Return*	0.0853	0.0057	−0.0622
Sharpe Ratio	0.9108	−0.7189	−1.2671
Sortino Ratio	1.2734	−1.2080	−1.5334
Deutsche Post World Net Stock			
Return*	−0.0776	−0.0188	0.0781
Sharpe Ratio	−2.2932	−2.4571	0.5558
Sortino Ratio	−2.3166	−2.6190	0.7174
Intel Co. Stock			
Return*	0.0946	0.2317	0.1041
Sharpe Ratio	0.5099	2.9882	0.5273
Sortino Ratio	0.6134	4.3481	0.6269

*) Annualized logarithmic return.

require particular kinds of information-technology infrastructures, as it could very well be enacted by placing a couple of orders with a broker on the phone before the market opens; there is no need to monitor the market and to react in a timely manner.

Nonetheless, the results clearly indicate that, despite its simplicity, such an approach may yield, if carried out carefully, significant returns, in the face of a risk that is, to be sure, probably higher than the one the average investor would be eager to take, but all in all proportionate with the returns expected.

To conclude, the issue of transaction costs is worth some discussion. Transaction costs are the sum of two parts: the commissions claimed by the broker or the dealer who execute the orders, and the price slippage that can occur in some circumstances, when the price moves before the order can be executed and the trader has to aggressively move the order price accordingly to get it executed. To begin with, no slippage can occur given the particular structure of the trading strategy (the position is opened with a market order placed before the opening auction, and closed either by the execution of a limit order entered when the position was opened or automatically by the broker at market close). Commissions are the only component that could have an impact on the profitability of the evolved trading rules. However, they can be made as low as desired simply by increasing the amount invested: under most commission structures applied by brokers, commissions are calculated as a percentage (e.g., 0.19%) of the order value with a maximum (e.g., $19); in other cases they are flat, independent of the order size; many brokers offer substantial discounts on commissions to frequent traders. Therefore, even if transaction costs have not been taken into account in reporting the experimental results, their impact would be negligible for an investor with sufficient capitalization.

References

[1] Allen F, Karjalainen R (1999) Using genetic algorithms to find technical trading rules. Journal of Financial Economics 51:245–271
[2] Various Authors (2006) Special issue on integrating multiple learned models. Machine Learning 36(1–2)
[3] Bacardit J, Krasnogor N (2006) Empirical evaluation of ensemble techniques for a pittsburgh learning classifier system. In Proceedings of the Ninth International Workshop on Learning Classifier Systems. Seattle, WA, USA, July 8–9 2006. ACM Press, New York
[4] Bäck T (1996) Evolutionary algorithms in theory and practice. Oxford University Press, Oxford
[5] Bäck T, Fogel D, Michalewicz Z (2000) Evolutionary Computation. IoP Publishing, Bristol
[6] Beretta M, Tettamanzi A (2003) Learning fuzzy classifiers with evolutionary algorithms. In Pasi G, Bonarini A, Masulli F, (ed) Soft Computing Applications. pp 1–10, Physica Verlag, Heidelberg
[7] Bodie Z, Kane A, Marcus A (2006) Investments (7th edn). McGraw-Hill, New York
[8] Brabazon A, O'Neill M (2006) Biologically Inspired Algorithms for Financial Modelling. Springer, Berlin
[9] Breiman L (1996) Bagging predictors. Machine Learning 24(2):123–140
[10] Cantú-Paz E (1997) A survey of parallel genetic algorithms. Technical Report IlliGAL 97003, University of Illinois at Urbana-Champaign
[11] DeJong K (2002) Evolutionary Computation: A unified approach. MIT Press, Cambridge, MA
[12] Dempster M, Jones C (2001) A real-time adaptive trading system using genetic programming. Quantitative Finance 1:397–413
[13] Dempster M, Jones C Romahi Y, Thompson G (2001) Computational learning techniques for intraday fx trading using popular technical indicators. IEEE Transactions on Neural Networks 12(4):744–754
[14] Harlow H (1991) Asset allocation in a downside-risk framework. Financial Analysts Journal pp. 30–40, September/October
[15] Hellendoorn H, Thomas C (1993) Defuzzification in fuzzy controllers. Intelligent and Fuzzy Systems 1:109–123

[16] Hellstrom T, Holmstrom K (1999) Parameter tuning in trading algorithms using asta. Computational Finance 1:343-357
[17] Mamdani E (1976) Advances in linguistic synthesis of fuzzy controllers. International Journal of Man Machine Studies 8:669–678
[18] Manderick B, Spiessens P (1989) Fine-grained parallel genetic algorithms. In Schaffer J (ed) Proceedings of the Third International Conference on Genetic Algorithms, San Mateo, CA, Morgan Kaufmann
[19] Marney J, Fyfe C, Tarbert H, Miller D (2001) Risk-adjusted returns to technical trading rules: A genetic programming approach. Computing in Economics and Finance 147, Society for Computational Economics, Yale University, USA, June 2001
[20] Mühlenbein H (1989) Parallel genetic algorithms, population genetics and combinatorial optimization. In Schaffer J (ed) Proceedings of the Third International Conference on Genetic Algorithms, pages 416–421, San Mateo, CA, Morgan Kaufmann
[21] Murphy J (1999) Technical Analysis of the Financial Markets. New York Institute of Finance, New York
[22] Neely C, Weller P, Dittmar R (1997) Is technical analysis in the foreign exchange market profitable? a genetic programming approach. Journal of Financial and Quantitative Analysis 32:405–26
[23] Potvin J-Y, Soriano P, Vallée M (2004) Generating trading rules on the stock markets with genetic programming. Computers and Operations Research 31(7):1033–1047
[24] Tettamanzi A, Poluzzi R, Rizzotto G (1996) An evolutionary algorithm for fuzzy controller synthesis and optimization based on SGS-Thomson's W.A.R.P. fuzzy processor. In Zadeh L, Sanchez E, Shibata T (ed) Genetic algorithms and fuzzy logic systems: Soft computing perspectives. World Scientific, Singapore
[25] Ross T (1995) Fuzzy Logic with Engineering Applications. McGraw-Hill, New York
[26] Sharpe W (1994) The Sharpe ratio. Journal of Portfolio Management 21(1):49–58
[27] Sortino F, van der Meer R (1991) Downside risk — capturing what's at stake in investment situations. Journal of Portfolio Management 17:27–31
[28] Subramanian H, Ramamoorthy S, Stone P, Kuipers B (2006) Designing safe, profitable automated stock trading agents using evolutionary algorithms. In Cattolico M (ed) Proceedings of Genetic and Evolutionary Computation Conference, GECCO 2006, Seattle, Washington, USA, July 8-12, 2006, pp. 1777–1784, ACM
[29] Tettamanzi A (1995) An evolutionary algorithm for fuzzy controller synthesis and optimization. IEEE International Conference on Systems, Man and Cybernetics, volume 5/5 pp. 4021–4026, IEEE Press
[30] Tettamanzi A, Carlesi M, Pannese L, Santalmasi M (2007) Business intelligence for strategic marketing: Predictive modelling of customer behaviour using fuzzy logic and evolutionary algorithms. In Giacobini M et al. (ed) Applications of evolutionary computing: Evoworkshops 2007, volume 4448 of

Lecture Notes in Computer Science, pp. 233–240, Valencia, Spain, April 11–13 2007, Springer

[31] Whitley D, Rana S, Heckendorn R (1999) The island model genetic algorithm: On separability, population size and convergence. Journal of Computing and Information Technology 7(1):33–47

[32] Wikipedia (2007) Technical analysis — wikipedia, the free encyclopedia. Online; accessed 5 June 2007

[33] Yao X, Xu Y (2006) Recent advances in evolutionary computation. Computer Science and Technology 21(1):1–18

[34] Yu T, Chen S-H, Kuo T-W (2006) Discovering financial technical trading rules using genetic programming with lambda abstraction. In: O'Reilly U-M, Yu T, Riolo R, Worzel B, (ed) Genetic Programming Theory and Practice II. pp 11–30, Springer, Berlin

[35] Zadeh L (1965) Fuzzy sets. Information and Control 8:338–353

[36] Zadeh L (1975) The concept of a linguistic variable and its application to approximate reasoning, i–ii. Information Science 8:199–249 & 301–357

[37] Zadeh L (1992) The calculus of fuzzy if-then rules. AI Expert 7(3):22–27

9

Strong Typing, Variable Reduction and Bloat Control for Solving the Bankruptcy Prediction Problem Using Genetic Programming

Eva Alfaro-Cid, Alberto Cuesta-Cañada, Ken Sharman, and Anna I. Esparcia-Alcázar

Instituto Tecnológico de Informática, Universidad Politécnica de Valencia, Valencia, Spain.
evalfaro@iti.upv.es

Summary. In this chapter we present the application of a genetic programming (GP) algorithm to the problem of bankruptcy prediction. To carry out the research we have used a database that includes extensive information (not only economic) from the companies. In order to handle the different data types we have used Strongly Typed GP and variable reduction. Also, bloat control has been implemented to obtain comprehensible classification models. For comparison purposes we have solved the same problem using a support vector machine (SVM). GP has achieved very satisfactory results, improving those obtained with the SVM.

9.1 Introduction

Bankruptcy prediction is a very important economic issue. It is of great significance for stakeholders as well as creditors, banks and investors, to be able to predict accurately the financial distress of a company. It is also beneficial for companies to be able to recognise on time the symptoms of economic failure in order to take measures to avoid failure. Given its relevance in real life, it has been a major topic in the economic literature. Many researchers have worked on this topic during the last decades; however, there is no generally accepted prediction model.

According to (6), a survey reviewing journal papers on the field in the period 1932-1994, the most popular methods for building quantitative models for bankruptcy prediction have been discriminant analysis (2) and logit analysis (20). Since the 1990s there has been an increasing interest in the application of methods originating from the field of artificial intelligence, mainly neural networks (14). Other methods from the artificial intelligence field, such as evolutionary computation have been scarcely used for the bankruptcy prediction problem. After an extensive (but not exhaustive) review of the literature, only a few papers could be found that applied evolutionary methods to the bankruptcy prediction problem.

Some authors have used genetic algorithms (GAs), either on its own (10), (23), (26) or in a hybrid method with a neural network (3) for the insolvency prediction problem. However, most of the approaches from the evolutionary computation field

E. Alfaro-Cid et al.: *Strong Typing, Variable Reduction and Bloat Control for Solving the Bankruptcy Prediction Problem Using Genetic Programming*, Studies in Computational Intelligence (SCI) **100**, 161–185 (2008)
www.springerlink.com

use genetic programming (GP)(12). The ability of GP to build functions make this approach more appropriate to the problem at hand than GA. In the literature we can find a couple of hybrid approaches that combine GP with another method, such as rough sets (18) and neural networks (24). Some authors have used GP on its own. In (27), the authors have used linear GP and have compared its performance to support vector machines and neural networks. In (13), the authors have used GP to predict bankruptcy on a database of Norwegian companies and in (22) GP has been used for the prediction of insolvency in non-life insurance companies, a particular case. Finally, grammatical evolution, a form of grammar-based genetic programming, has been used in (4) to solve several financial problems, corporate failure prediction among them.

One important advantage of the GP approach to bankruptcy prediction is that it yields the rules relating the measured data to the likelihood of becoming bankrupt. Thus a financial analyst can see what variables and functions thereof are important in predicting bankruptcy. Our approach differs from previous GP applications in several aspects:

- *Database*. Our database comprises data from Spanish companies from different industrial sectors. It is an extensive database that includes not only financial data from the companies but also general information that can be relevant when predicting failure. The database has some challenging features: it is highly unbalanced (only 5-6% of the companies go bankrupt) and some data is missing. Although this complicates the classification, it is an accurate reflection of the real world, where few companies go bankrupt in proportion and it is difficult to obtain all the relevant data from companies.
- *Strong typing*. Since the database comprises not only economic data we have used GP with strong typing to ensure that the prediction models evolved make sense from an economic point of view. Strong typing ensures that the functions used by GP take arguments of a given type and return values of the proper type as well.
- *Variable reduction*. Usually bankruptcy prediction models are based on 7 or 8 economic ratios. In our case, in order to handle the amount of data in the database and in order to ensure that the GP generated models are understandable, the prediction has been done in two steps. To start with, GP has been used to identify which data is relevant for solving the problem. This can be done since GP creates analytical models as a final result. Then the proper prediction models are evolved using those variables that had been identified as important in the first stage.
- *Bloat control*. In (15) code bloat is defined as: "the tendency of candidate program solutions to grow in size independent of any corresponding increase in quality". This flaw causes waste of computer resources, difficulties in the understanding of the final solutions and spoils the convergence by hampering the modification of trees in a meaningful way. We have undertaken a comparison study of various bloat control methods to find which one is best suited to the problem at hand. Our main goal while trying to control the bloat is for the GP to generate models that can be analysed and understood.

- *Data from a range of years*. In the literature of the field the prediction models are normally built using data from a single year to predict the insolvency in a certain time period. In this work we have also considered the possibility of evolving models that involve data from 3 consecutive years. We have adopted and compared two ways of handling the data: to present the data to the system as a vector (i.e. simultaneously) and to present the data from various years to the classifier as a time series (i.e. the evaluation process takes as many steps as time steps are in the data series).
- *Support Vector Machine*. We have also analysed the data using a support vector machine (SVM) classifier, and our results demonstrate that our proposed GP technique gives improved performance over the SVM.

9.2 Database

9.2.1 Data Description

The work presented in this chapter uses a database supplied by the Department of Finance and Accounting of the Universidad de Granada, Spain. The data were compiled by Axesor (www.axesor.es), a leading company in the business information industry. The database consists of a 2859×31 matrix comprising data from 484 Spanish companies from the year 1998 to the year 2003.[1] Each row of the matrix holds the data referent to a company during one year. The database includes not only financial data such as solvency, profit margin or income yield capacity, but also general information such as company size, age or number of partners. These independent variables are the inputs to the classifier. The desired output of the classifier is the variable that states whether the company was bankrupt in 2004 or not. We have used the data from years 2001, 2002 and 2003 to predict bankruptcy in the year 2004, that is 1, 2 and 3 years in advance. Table 9.1 shows the independent variables, their description and type.

9.2.2 Preparation of Data

As can be seen in table 9.1, the variables can take values from different numerical ranges: real, categorical, integer and boolean. Some of the non-financial data take categorical values; these are the size of the company, the type of company and the auditor's opinion. Usually, company size is a real variable but in this case the companies are grouped in three separated categories according to their size. Each categorical variable can take 3 different values. To work with them they have been transformed into 3 boolean variables each (instead of having one variable stating if the company has size 1, 2 or 3, this has been expressed using 3 boolean variables that answer the question: "Is the company's size equal to 1/2/3"). Therefore, after

[1] The number of rows in the data matrix should be 2904, i.e. 484×6, but some companies do not have available data for all the years

Table 9.1. Independent Variables

Financial Variables	Description	Type
Debt Structure	Long-Term Liabilities /Current Liabilities	Real
Debt Cost	Interest Cost/Total Liabilities	Real
Liabilities	Liabilities/Equity	Real
Cash Ratio	Cash Equivalent /Current Liabilities	Real
Working Capital	Working Capital/ Total Assets	Real
Debt Ratio	Total Assets/Total Liabilities	Real
Operating Income Margin	Operating Income/Net Sales	Real
Debt Paying Ability	Operating Cash Flow/Total Liabilities	Real
Return on Operating Assets	Operating Income/Average Operating Assets	Real
Return on Equity	Net Income/Average Total Equity	Real
Return on Assets	Net Income/Average Total Assets	Real
Asset Turnover	Net Sales/Average Total Assets	Real
Receivable Turnover	Net Sales/Average Receivables	Real
Stock Turnover	Cost of Sales/Average Inventory	Real
Current Ratio	Current Assets/Current Liabilities	Real
Acid Test	(Cash Equivalent + Marketable Securities + Net receivables) /Current Liabilities	Real
Continued Losses	If the company has suffered continued losses for 3 consecutive years	Boolean
Non-financial Variables	**Description**	**Type**
Size	Small/Medium/Large	Categorical
Type of company		Categorical
Auditor's opinion		Categorical
Audited	If the company has been audited	Boolean
Delay	If the company has submitted its annual accounts on time	Boolean
Linked in a group	If the company is part of a group holding	Boolean
Historic number of serious incidences	Such as strikes, accidents...	Integer
Historic number of judicial incidences	Since the company was created	Integer
Number of judicial incidences	Last year	Integer
Number of partners		Integer
Age of the company		Integer

this modification the available data set for each company includes 34 independent variables: 16 real, 5 integer and 13 boolean variables.

One of the problems with this database is that some of the data are missing. Specifically, around 16% of the companies in the database have one or more data values missing. To handle this we have used average values where missing data occurs. Although this may compromise the extrapolation of the evolved classifiers to new data, we have decided not to eliminate these companies from our experiments since our priority is to check the suitability and robustness of the proposed GP approach for the evolution of classifiers.

9.2.3 Training and Testing Sets

In order to apply GP to the prediction problem the data sets have been divided into two groups: the training and testing sets, which were selected randomly. Given that the data base is highly unbalanced (only 5-6% of the companies went bankrupt), this ratio needs to be reflected in the choice of the training set. The chosen training set consists of 367 companies (25 bankrupt vs. 342 healthy). The test set consists of the remaining 111 companies (8 bankrupt vs. 103 healthy).

9.3 Genetic Programming for Prediction

Genetic Programming (GP) is based on the idea that in nature structure undergoes adaptation. The structure created over a period of time is the outcome of natural selection and sexual reproduction. Thus, GP is a structural optimisation technique as opposed to a parametric optimisation technique. The individuals in GP are represented as hierarchical structures, typically tree structures, and the size and shape of the solutions are not defined a priori as in other methods from the field of evolutionary computation, but they evolve over time. The flow of a GP algorithm is as that of any evolutionary technique: a population is created at random, each individual in the population is evaluated using a fitness function, the individuals that performed better in the evaluation process have more possibilities of being selected for the new population than the rest and a new population is created once the individuals are subject to the genetic operators of crossover and mutation with a certain probability. The loop is run until a certain termination criterion is met.

In this section we briefly describe the GP framework that we have used for representing systems for bankruptcy prediction. Basically, the GP algorithm must find a structure (a function) which can, once supplied with the relevant data from the company, decide if this company is heading for bankruptcy or not. In short, it is a binary classification problem. One of the classes consists of the companies that will go bankrupt and the other consists of the healthy ones. For further information on classification using GP see (7), (11).

9.3.1 Classification

The classification works as follows. Let $X = \{x_0, \ldots, x_N\}$ be the vector comprising the data of the company undergoing classification. Let $f(X)$ be the function defined by an individual GP tree structure. The value y returned by $f(X)$ depends on the input vector X.

$$y = f(x_0, x_1, \ldots, x_N) \tag{9.1}$$

We can apply X as the input to the GP tree and calculate the output y. Once the numerical value of y has been calculated, it will give us the classification result according to

$$y > 0,\ X \in B \tag{9.2}$$

$$y \leq 0,\ X \in \overline{B} \tag{9.3}$$

where B represents the class to which bankrupt companies belong and $\overline{B}$ represents the class to which healthy companies belong. That is, if the evaluation of the GP tree results in a numerical value greater than 0 the company is classified as heading for bankruptcy, while if the value is less or equal to 0 the company is classified as healthy.

9.3.2 Fitness Evaluation

As mentioned previously, the database we are using is very unbalanced in the sense that only 5-6% of the companies included will go bankrupt. This is something to consider while designing the fitness function, otherwise the evolution may converge to structures that classify all companies as healthy (i.e. they do not classify at all) and still get a 95% hit rate.

We have addressed this problem by modifying the cost associated to misclassifying the positive and the negative class to compensate for the imbalanced ratio of the two classes (9). For example, if the imbalance ratio is 1:10 in favour of the negative class, the penalty of misclassifying a positive example should be 10 times greater. Basically, it rewards the correct classification of examples from the small class over the correct classification of examples from the over-sized class. It is a simple but efficient solution. Therefore, the fitness function for maximisation is

$$Fitness = \sum_{i=1}^{n} u_i \tag{9.4}$$

where

$$u_i = \begin{cases} 0 & : \text{ incorrect classification} \\ \frac{n_h}{n_b} & : \text{ bankrupt company classified correctly} \\ 1 & : \text{ healthy company classified correctly} \end{cases} \tag{9.5}$$

n_b is the number of bankrupt companies in the training set and n_h is the number of healthy companies in the training set.

9.3.3 Genetic Programming Algorithm

The GP implementation used is based on ECJ (http://cs.gmu.edu/~eclab/projects/ecj), a research evolutionary computation system developed at George Mason University's Evolutionary Computation Laboratory (ECLab). Table 9.2 shows the main parameters used during evolution.

Table 9.2. GP parameters

Initialisation method	Ramped half and half
Replacement operator	Generational with elitism (0.2%)
Selection operator	Tournament selection
Tournament group size	7
Cloning rate	0.1
Crossover operator	Bias tree crossover
Internal node selection rate	0.9
Crossover rate	0.8
Mutation rate	0.1
Tree maximum initial depth	7
Tree maximum depth	18
Population size	1000
Number of runs	30
Termination criterion	50 generations

9.3.4 Strong Typing

Strongly Typed GP (STGP) (19) is an enhanced version of GP that enforces data-type constraints, since standard GP is not designed to handle a mixture of data types. In STGP, each function node has a return-type, and each of its arguments also have assigned types. STGP permits crossover and mutation of trees only with the constraint that each node's return type matches the corresponding argument type in the node's parent. A STGP has been implemented in order to ensure that in the resulting classifying models the functions operate on appropriate data types so that the final model has a physical meaning. That is, the objective is to avoid results that operate on data which are not compatible, for instance, models which add up the liabilities and the age of a company. The terminal set used consists of 35 terminals: the independent variables from table 9.1 plus Koza's ephemeral random constant (12). Table 9.3 shows the function set used and the chosen typing.

9.4 Variable Reduction

In order to build models that can be applied to an extensive number of practical cases we need simple models which require the minimum amount of data.

Table 9.3. Function set

Functions	Number of arguments	Arguments type	Return
+, -, *, /	2	real	real
ln, exp	1	real	real
If $arg_1 \leq arg_2$ then arg_3 else arg_4	4	real	real
If arg_1 then arg_2 else arg_2	3	arg_1 is a boolean arg_2, arg_3 are real	real
If $arg_1 \leq int$ then arg_2 else arg_2 (*int* is randomly chosen)	3	arg_1 is an integer arg_2, arg_3 are real	real

Our initial predictions were made running the GP algorithm with all the available data for each year. Once these results were analysed we could observe that there were some variables used more frequently than others. We then ran a second set of experiments considering only those variables which appeared on, at least, 90% of the best-of-run trees obtained in the initial phase. This way the number of terminals was reduced drastically. In table 9.4 the variables found to be more relevant for solving the prediction problem are presented.

Table 9.4. Reduced Variables

Year	Variable	Type
2001	Debt Structure	Real
	Cash Ratio	Real
	Working Capital	Real
	Debt Ratio	Real
	Return on Operating Assets	Real
	Return on Equity	Real
	Continued Losses	Boolean
2002	Working Capital	Real
	Return on Equity	Real
	Return on Assets	Real
	Asset Turnover	Real
	Stock Turnover	Real
	Acid Test	Real
2003	Debt Cost	Real
	Return on Equity	Real
	Return on Assets	Real
	Asset Turnover	Real
	Current Ratio	Real

The variables found to be more relevant for the final model are different each year. This was expected since the indicators of economic distress vary with time. The only variable that appears in the reduced set of variables for every year is "return on equity”. Note that for years 2002 and 2003 it does not make sense to talk about STGP since all the variables in the reduced set are of the same type.

The reduction in the number of variables should not imply a loss of quality in the solutions. Therefore, in order to check that the reduction of variables does not compromise the quality of results we have performed a Kruskal-Wallis test (a non-parametric statistical test) for the global error for each of the 3 years studied. The outcome of the Kruskal-Wallis tests allows us to insure that there are not statistically significant difference between the classification errors (in-sample and out-sample) obtained using all the variables or the group of reduced variables in the evolution. Figs. 9.1, 9.2 and 9.3 plot these results. In the figures we can see a box and whisker plot for each column of data. The boxes have lines at the lower quartile, median, and upper quartile values. The whiskers are lines extending from each end of the boxes to show the extent of the rest of the data. Outliers are data with values beyond the ends of the whiskers. They are represented with a plus symbol. The notches in a box plot represent a robust estimate of the uncertainty about the medians for box-to-box comparison. Boxes whose notches do not overlap indicate that the medians of the two groups differ at the 5% significance level.

As important as getting good overall classification errors is to get results that minimise both the type I error and the type II error. In our case the type I error is the percentage of healthy companies that are classified as bankrupt and the type II error is the percentage of bankrupt companies that are classified as healthy. Although ideally both errors should be minimised for the bankruptcy prediction problem the error type II is of greater importance. Not being able to identify that a company is at risk causes problems to creditors and slows down the taking of measures that may solve the problem. Therefore the same statistical tests were carried out for the type I and type II errors. They led to the same conclusion: there was no statistically significant difference between both sets of results.

9.5 Bloat Control

Our aim when applying a bloat control method is for GP to generate comprehensible classification models. As an example of bloat fig. 9.4 shows the evolution of the average size of the population along the generations in a random run. In this work the size of a tree is measured as its number of nodes. After 50 generations the average size of the individuals doubles the initial size.

The most popular method to date for controlling bloat is the imposition of size limits proposed by Koza in (12): depth limitation for the generation of individuals in the initial population, depth limitation in the generation of subtrees for subtree mutation and restriction of the crossover and mutation operators so that children larger than the permissible size are not included in the population.

Recently, an extensive study has been published (17) where several approaches to bloat control are examined and conclusions are drawn on their success in reducing population size while maintaining the quality of the best-of-run result.

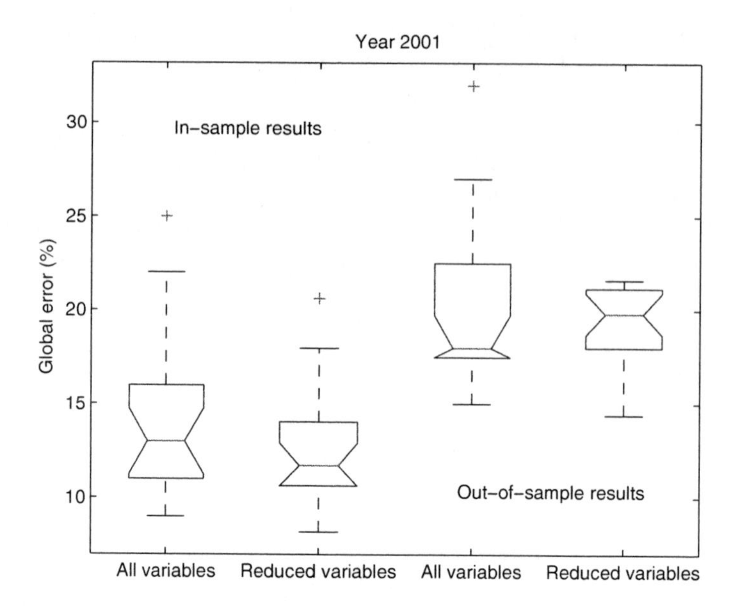

Fig. 9.1. Average best error obtained using all variables vs. average best error obtained using the reduced set of variables for year 2001 (prediction 3 years ahead)

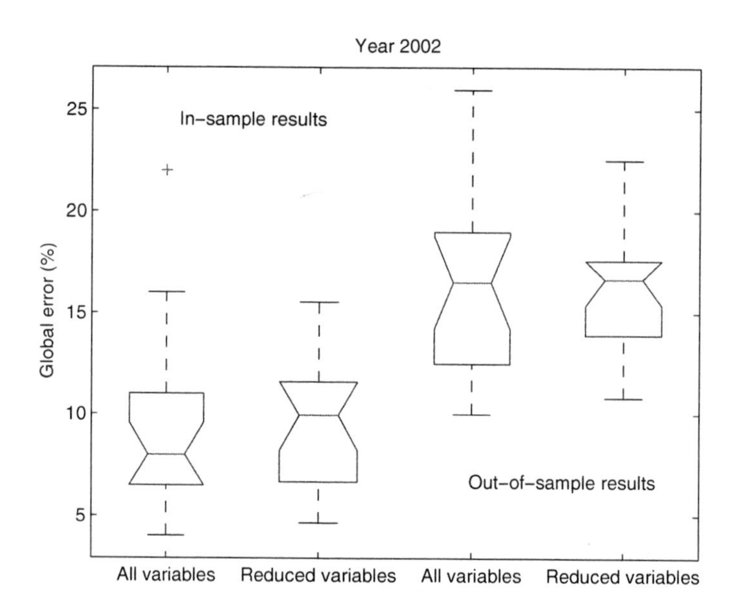

Fig. 9.2. Average best error obtained using all variables vs. average best error obtained using the reduced set of variables for year 2002 (prediction 2 years ahead)

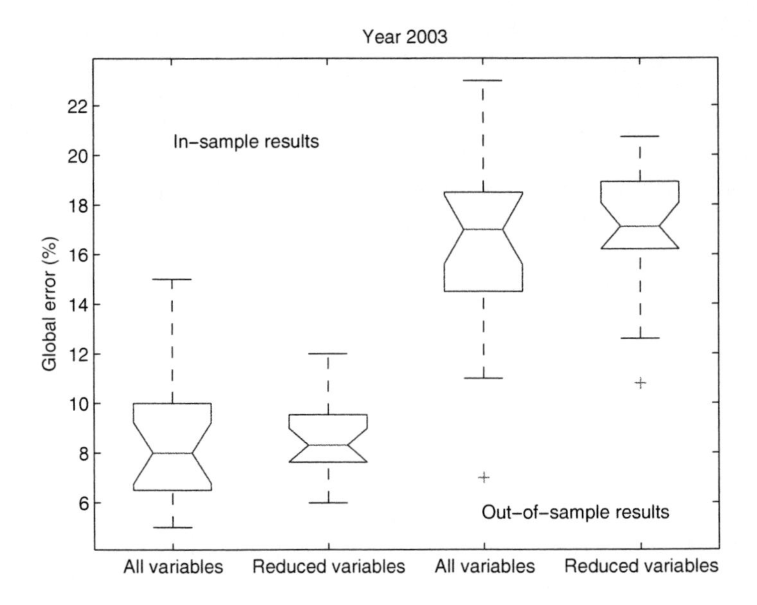

Fig. 9.3. Average best error obtained using all variables vs. average best error obtained using the reduced set of variables for year 2003 (prediction 1 year ahead)

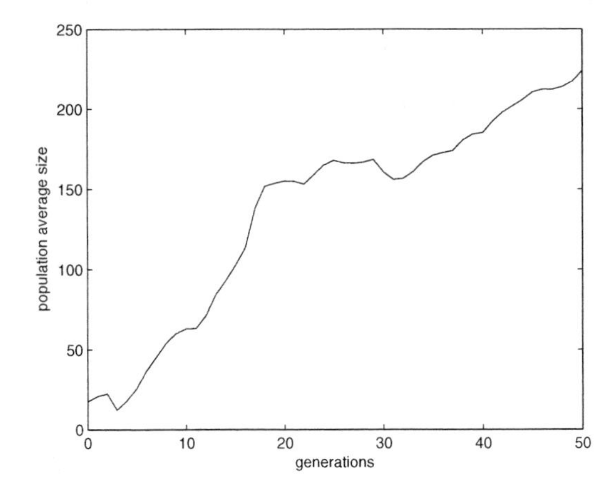

Fig. 9.4. Evolution of the average size of the population along the generations

We have carried out a comparison of bloat control methods to find out which one best suits our problem. We compare 4 bloat control methods versus Koza's method of restricting the tree depth. All the bloat control methods have been augmented with Koza's style tree depth restrictions, since it has been proved that any method augmented this way is superior to the bloat control method alone (16).

Three of the bloat control methods are chosen among those which provided better results in (17): *double tournament*, *tarpeian* and *lexicographic parsimony pressure*. The fourth method is called *prune and plant* (8). We have used this last method successfully in previous works on bankruptcy prediction (1).

9.5.1 Double Tournament

Double tournament is a method proposed in (17) that applies two layers of tournament in sequence, one for fitness and the other for size. The individuals in the tournament group are not chosen at random with replacement from the population, but they are the winners of a previous tournament. If the final tournament selects based on fitness then the previous tournaments select on size (or vice versa). Therefore the algorithm has three parameters: a fitness tournament size (F), a parsimony tournament size (P) and a switch (*do-fitness-first*) which indicates whether the qualifiers select on fitness and the final selects on size or the other way round. We have fixed F to 7 and P to 1.4 (i.e. two individuals participate in the tournament, and with probability $P/2$ the smaller individual wins, else the larger individual wins), a selection that proved to work well on a variety of problems in (17). Regarding the parameter *do-fitness-first* we have included both possibilities (true or false) in our comparison study.

9.5.2 Tarpeian Method

The Tarpeian method was introduced in (21). It limits the size of the population by making uncompetitive a fraction W of individuals with above-average size. Before the evaluation process those individuals are assigned a very bad fitness which reduces dramatically their chance of being selected. Since this happens before the evaluation it reduces the number of evaluations necessary. We have fixed W to 0.3 because this setting performed well across all problem domains in (17).

9.5.3 Lexicographic Parsimony Pressure

This technique (17) treats fitness as the first objective of the optimisation and size as a secondary objective. In plain lexicographic parsimony pressure an individual is considered superior to another if it is better in fitness; if they have the same fitness, then the smallest individual is considered superior. We have used a variation of the method especially suited for environments where few individuals have the same fitness: *lexicographic parsimony pressure with ratio bucketing*. Here the individuals from the population are sorted into ranked buckets and those individuals in the same bucket are treated as if they had the same fitness. The buckets are proportioned so that low-fitness individuals are placed into larger buckets than high-fitness individuals. A parameter *r* fixes the size of the buckets. The bottom *1/r* fraction of individuals of the population are placed in the bottom bucket. Of the remaining individuals, the bottom *1/r* fraction are placed into the next bucket and so on, until all individuals have been placed in a bucket. Again our choice of *r* equal to 2 was motivated by the good results across all problem domains that this setting obtained in (17).

9.5.4 Prune and Plant

This bloat control approach is described in (8) and is inspired in the strategy of the same name used in agriculture. It is used mainly for fruit trees and it consist of pruning some branches of trees and planting them in order to grow new trees. The idea is that the worst tree (in terms of fitness) in a population will be substituted by branches pruned from one of the best trees and planted in its place. This way the offspring trees will be of smaller size than the ancestors, effectively reducing bloat.

We have implemented this technique as a bloat-control crossover operator. The parameter to set is the probability of this kind of crossover. Since there are no previous studies on the amount of prune and plant to use we have compared 3 set of parameters (see table 9.5).

Table 9.5. Prune and plant settings

Prune & plant ratio	Crossover ratio	Cloning ratio	Mutation ratio
0.4	0.4	0.1	0.1
0.5	0.3	0.1	0.1
0.6	0.3	0.05	0.05

9.5.5 Bloat Control Results

Summarizing, table 9.6 shows all the methods included in the comparison study and the choice of settings for each of them. To draw our conclusions we have focused on the success of the bloat control methods in managing the average population size while retaining good best-fitness-of-run performance. Every method has been run 30 times. All the numerical results presented are averaged over the 30 runs and have been compared using a Kruskal-Wallis statistical test.

Regarding the size control (see fig. 9.5), all settings of prune and plant and double tournament decrease significantly the average population size as compared to Koza-style's depth limiting alone. The magnitude of the size reduction achieved by both settings of double tournament is remarkable. On the other hand, although lexicographic parsimony pressure with ratio bucketing and tarpeian both converge to

Table 9.6. Bloat control methods

Method	Settings
Koza's depth limitation	Tree max initial depth = 7; Tree max depth = 18
Double tournament	F = 7; P = 1.4; do-fitness-first = true
	F = 7; P = 1.4; do-fitness-first = false
Tarpeian	W = 0.3
Ratio bucketing	r = 2
Prune and plant	Prune and plant ratio = 0.4
	Prune and plant ratio = 0.5
	Prune and plant ratio = 0.6

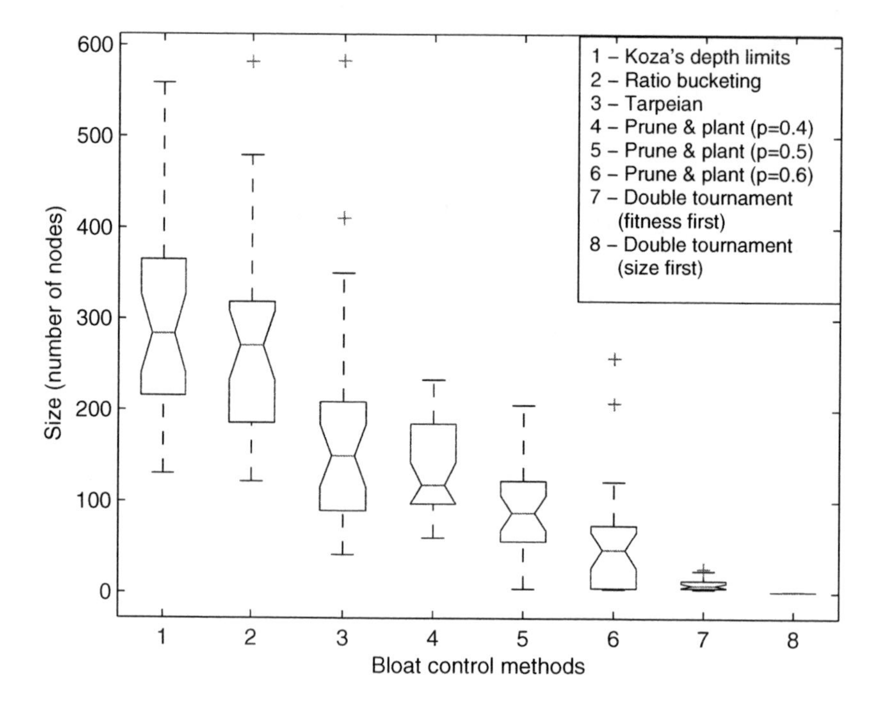

Fig. 9.5. Average size of final population averaged across 30 runs for each bloat control method

populations whose average size is smaller than Koza-style's, the statistical test shows there is no statistically significant difference.
Fig. 9.6 presents the average best-of-run fitness calculated over the training and testing sets and normalised in [0,1] to which each method has converged. Koza's-style depth limiting, lexicographic with ratio bucketing, tarpeian and prune and plant with probability 0.4, all obtained fitness results that are not statistically significantly different. All the other settings of prune and plant and both settings of double tournament get lower fitness values.

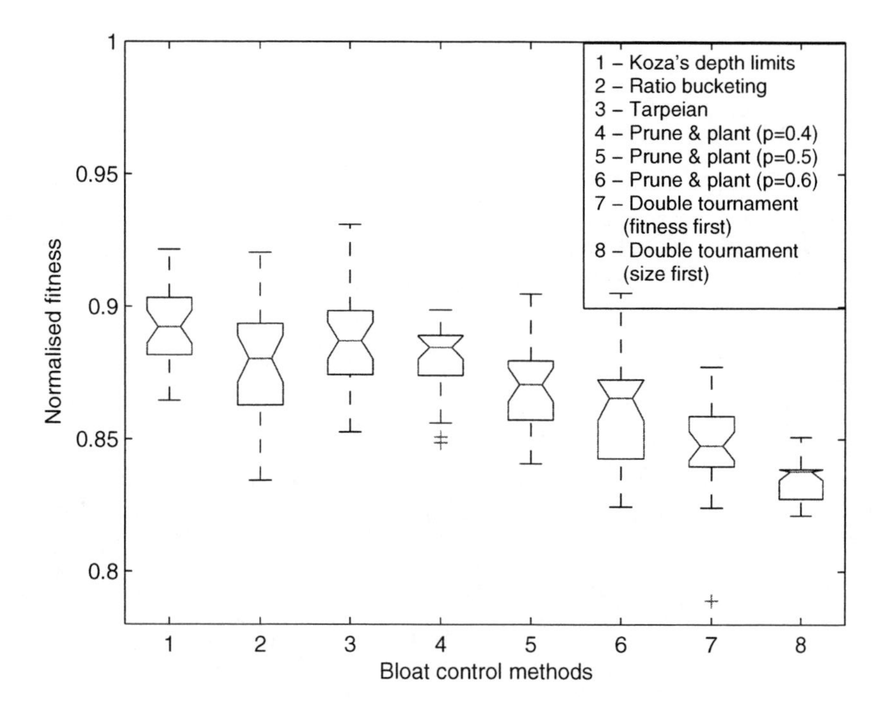

Fig. 9.6. Normalised best-of-run fitness averaged across 30 runs for each bloat control method

Overall the only bloat control method that reduces the final size of the population while maintaining the quality of the result is prune and plant with a probability of 0.4. Therefore that is the method we have used in the remainder of our experiments.

9.6 Final Results

9.6.1 Prediction Using Data from 1 Year

In this section we present the best results obtained using data from a single year for a prediction 3, 2 and 1 year ahead. Tables 9.7, 9.8 and 9.9 show these results.

The results of the prediction 3 and 1 years ahead (see Tables 9.7 and 9.9) are very similar, which is a little unusual, one would expect the accuracy to drop once you move three years away. The results of the prediction 2 years ahead are different (see table 9.8). The global error and the type I error are smaller but at the expense of a higher percentage of the type II error. As it has already been said in this particular application the type II error is more relevant so these results are a bit disappointing in that sense.

Table 9.7. Prediction results 3 years ahead

	Training	Testing	Overall
Global error (%)	15.26	19.82	16.32
Error type I (%)	16.37	20.39	17.3
Error type II (%)	0	12.5	3.03

Table 9.8. Prediction results 2 years ahead

	Training	Testing	Overall
Global error (%)	8.17	15.32	9.83
Error type I (%)	8.48	14.56	9.89
Error type II (%)	4	25	9.09

Table 9.9. Prediction results 1 year ahead

	Training	Testing	Overall
Global error (%)	13.62	18.92	14.85
Error type I (%)	14.62	19.42	13.42
Error type II (%)	0	12.5	3.03

This section includes an alternative set of results obtained using a support vector machine (SVM) (25). SVMs are learning machines that can perform binary classification and real valued function approximation tasks. SVMs non-linearly map their input space into a high dimensional feature space. Then, the SVM finds a linear separating hyperplane with the maximal margin in this higher dimensional space. In order to generate these results we have used LIBSVM (5), an integrated software for support vector classification. We have chosen this particular software because it supports weighted SVM for unbalanced data. Tables 9.10, 9.11 and 9.12 show the results obtained.

Table 9.10. Prediction results 3 years ahead using SVM

	Training	Testing	Overall
Global error (%)	27.52	27.03	27.41
Error type I (%)	29.24	29.12	29.21
Error type II (%)	4	0	3.03

Table 9.11. Prediction results 2 years ahead using SVM

	Training	Testing	Overall
Global error (%)	28.61	34.23	29.92
Error type I (%)	30.41	36.89	31.91
Error type II (%)	4	0	3.03

Table 9.12. Prediction results 1 year ahead using SVM

	Training	Testing	Overall
Global error (%)	26.98	27.03	26.99
Error type I (%)	26.32	27.18	26.52
Error type II (%)	36	25	33.33

The performance of the SVM is definitely inferior. Only the percentages of type II error for the prediction 2 and 3 years ahead are equivalent to those obtained with GP. The prediction results one year ahead leave much to be desired.

9.6.2 Prediction combining data from 3 years

Two further models have been constructed combining the data from all three years to obtain a global model that takes into account the evolution of the variables during this 3-year period. For these experiments we have also used a reduced set of variables (see table 9.13), those that were used more frequently in an initial set of runs.

Table 9.13. Reduced Variables

Variable	Type
Working Capital	Real
Debt Paying Ability	Real
Return on Equity	Real
Stock Turnover	Real
Continued Losses	Boolean

The results of the prediction 3 and 1 years ahead (see Tables 9.7 and 9.9) are very similar, which is a little unusual, one would expect the accuracy to drop once you move three years away. The results of the prediction 2 years ahead are different (see table 9.8). The global error and the type I error are smaller but at the expense of a higher percentage of the type II error. As it has already been said in this particular application the type II error is more relevant so these results are a bit disappointing in that sense. The first of these two models uses a parallel approach, i.e. all data from 3 years are fed into the GP model at the same time. This way the terminal set consists of 16 terminals (5 data items $\times$ 3 years + ephemeral random constant).

In the second model we have considered a sequential approach. Instead of feeding the model with all the data from the 3 years simultaneously, the evaluation takes place in three steps in which the model is fed with the data corresponding to that year. We have also added a further terminal that is the result of the evaluation of the model in previous years (a kind of feedback) since it seems very relevant for the analysis to know how the company was doing the previous years. In the first evaluation step (using data from year 2001) this figure is not available, but in the following evaluation steps it is.

In table 9.14 we can see the best results obtained when the data from the 3 years are combined together. The results are very promising. The sequential approach gave slightly better prediction rates than when introducing all the data in parallel, but the difference is not significant.

Table 9.14. Prediction results combining data from 3 years

		Training	Testing	Overall
Parallel approach	Global error (%)	14.44	19.82	15.69
	Error type I (%)	15.1	21.36	16.55
	Error type II (%)	4	0	3.03
Sequential approach	Global error (%)	12.16	18.02	13.52
	Error type I (%)	12.87	19.42	14.39
	Error type II (%)	4	0	3.03

Table 9.15 shows the results obtained when using a SVM for solving the problem. As in the previous section the error rates are larger than those obtained with GP.

Table 9.15. Prediction results combining data from 3 years using SVM

	Training	Testing	Overall
Global error (%)	33.51	35.14	33.89
Error type I (%)	35.67	37.89	36.18
Error type II (%)	4	0	3.03

9.6.3 Discussion

We begin this section with a comparison of the results obtained using different data for the prediction. From Tables 9.7, 9.8, 9.9 and 9.14 we can see that the differences are small. Fig. 9.7 plots the average best-of-run fitness calculated over the training and testing sets and normalised in [0,1] for each prediction year. It confirms that there is no statistically significant difference that could allow us to claim that any result is better than any other.

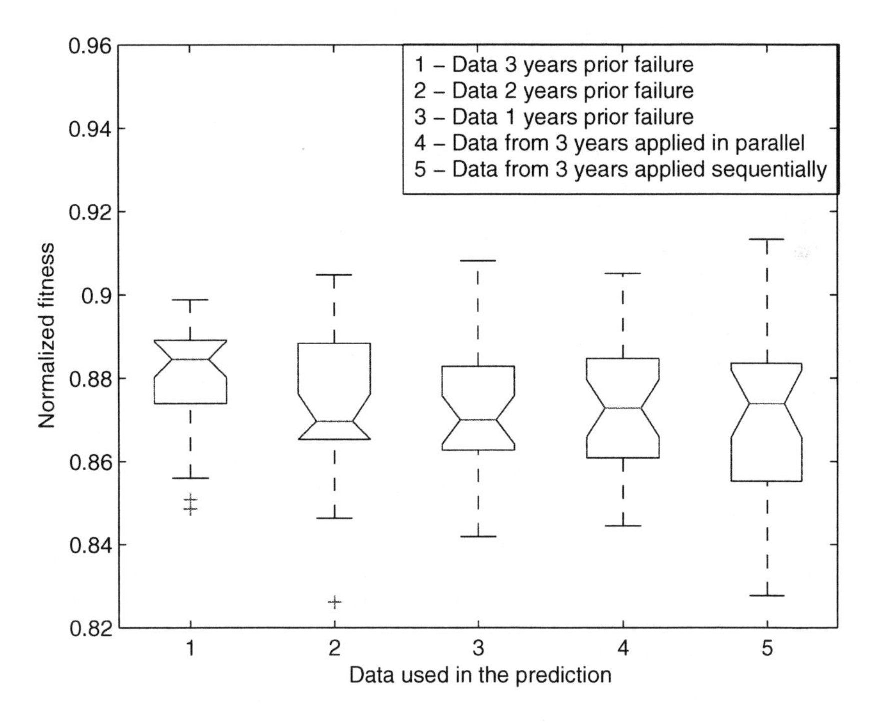

Fig. 9.7. Normalised best-of-run fitness averaged across 30 runs for each prediction year

Next, we present two examples of the type of classifier that GP evolves. The following equations state the classification model that has given the best results when predicting bankruptcy one year ahead (see table 9.9). The function consists of three nested conditional clauses:

$$y = \text{If} \quad \frac{x_{27}}{x_{23} - x_{24}} \leq x_{23}\exp(x_{22}) \quad \text{then} \quad if_0 - x_{22} - x_{27} - \exp(x_{14}) - \exp(x_{24}) \quad \text{else} \quad -1 \tag{9.6}$$

$$if_0 = \text{If} \quad x_{24} \leq x_{27} \quad \text{then} \quad \left(\text{If} \quad x_{24} \leq x_{22} + 43.45 \quad \text{then} \quad \frac{x_{22}}{\ln(x_{23}) + x_{24}} - 7.43 \quad \text{else} \quad \ln(x_{27}) \right) \quad \text{else} \quad x_{23} \tag{9.7}$$

where x_{14} is the debt cost, x_{22} is the return on equity, x_{23} is the return on assets, x_{24} is the asset turnover and x_{27} is the current ratio. Essentially, what this function does is to divide the set of companies in seven disjoint sets depending on which of the following clauses the company meets:

$$A \equiv \frac{x_{27}}{x_{23} - x_{24}} \leq x_{23}\exp(x_{22})$$
$$B \equiv x_{24} \leq x_{27}$$
$$C \equiv x_{23} - x_{22} - x_{27} - \exp(x_{14}) - \exp(x_{24}) \leq 0$$
$$D \equiv x_{24} \leq x_{22} + 43.45$$
$$E \equiv \frac{x_{22}}{\ln(x_{23}) + x_{24}} - 7.43 - x_{22} - x_{27} - \exp(x_{14}) - \exp(x_{24}) \leq 0$$
$$F \equiv \ln(x_{27}) - x_{22} - x_{27} - \exp(x_{14}) - \exp(x_{24}) \leq 0$$

Table 9.16 presents the classification sets and their classification as well as the percentage of companies in each set and the percentage of wrong classifications in each set. Belonging to one set is determined, as already mentioned, by meeting or not the inequalities. In the table the symbols '*F*', '*T*' and '-' stand for *false*, *true* and *not applicable*. The highest percentage of error is produced in sets 2, 4 and 6. This is due to the fact that the emphasis in the fitness function to avoid misclassifying a bankrupt company leads to an over classification of healthy companies as bankrupt.
Finally, the following equations state the classification model that has given the best results when predicting bankruptcy using the data from the three years in parallel (see table 9.14). Again, the function consists of three nested conditional clauses:

Table 9.16. Classification table according to the best prediction model evolved using data one year prior to bankruptcy

Set No.	A	B	C	D	E	F	Classification	Percentage of cases	Percentage of error
1	F	-	-	-	-	-	healthy	7.11	2.94
2	T	F	F	-	-	-	fails	10.88	76.92
3	T	F	T	-	-	-	healthy	40.79	0.00
4	T	T	-	T	F	-	fails	4.18	45.00
5	T	T	-	T	T	-	healthy	28.45	0.74
6	T	T	-	F	-	F	fails	7.53	55.56
7	T	T	-	F	-	T	healthy	1.05	0.00

$$\begin{aligned} y = \text{If} \quad & x_{12} \\ \text{then} \quad & if_0 \\ \text{else} \quad & \ln(\ln(x_{16_{01}})) - \exp(\exp(x_{16_{02}})) \end{aligned} \tag{9.8}$$

$$\begin{aligned} if_0 = \text{If} \quad & (2x_{26_{03}} > x_{22_{03}}) \wedge (-63.07 \leq x_{22_{03}}) \\ \text{then} \quad & (x_{22_{03}} + if_1)\left(x_{16_{02}} - \frac{x_{26_{02}} - x_{22_{01}}}{\ln(x_{22_{03}}) - \exp(\exp(x_{16_{02}}))}\right) \\ \text{else} \quad & x_{26_{03}} \end{aligned} \tag{9.9}$$

$$\begin{aligned} if_1 = \text{If} \quad & x_{20_{03}} - 30.18 \leq x_{22_{03}} \\ \text{then} \quad & \ln(x_{22_{03}})\left(x_{16_{02}} - \frac{49.91x_{16_{01}} - x_{26_{02}}}{3}\right) \\ \text{else} \quad & x_{26_{03}} \end{aligned} \tag{9.10}$$

where x_{12} indicates continued losses, $x_{16_{01}}$ is the working capital in year 2001, $x_{16_{02}}$ is the working capital in year 2002, $x_{20_{03}}$ is the debt paying ability in year 2003, $x_{22_{01}}$ is the return on equity in year 2001, $x_{22_{03}}$ is the return on equity in year 2003, $x_{26_{02}}$ is the stock turnover in year 2002 and $x_{26_{03}}$ is the stock turnover in year 2003. This time the prediction model classifies the companies in eight disjoint sets depending on whether they meet the following inequalities or not:

$$A \equiv x_{12} = 0$$
$$B \equiv \ln(\ln(x_{16_{01}})) - \exp(\exp(x_{16_{02}})) \leq 0$$
$$C \equiv (2x_{26_{03}} > x_{22_{03}}) \wedge (-63.07 \leq x_{22_{03}})$$
$$D \equiv x_{26_{03}} \leq 0$$
$$E \equiv x_{20_{03}} - 30.18 \leq x_{22_{03}}$$
$$F \equiv \left(x_{22_{03}} + \ln(x_{22_{03}})\left(x_{16_{02}} - \frac{49.91x_{16_{01}} - x_{26_{02}}}{3}\right)\right) a \leq 0$$
$$G \equiv (x_{22_{03}} + x_{26_{03}})a \leq 0$$
$$\text{where} \qquad a = x_{16_{02}} - \frac{x_{26_{02}} - x_{22_{01}}}{\ln(x_{22_{03}}) - \exp(\exp(x_{16_{02}}))}$$

Table 9.17 presents the classification sets for the model created combining data from the 3 years in parallel. Again the highest percentages of error are yielded by those sets whose companies are classified as bankrupt for the reason given before: the emphasis in the fitness function is to avoid misclassifying a bankrupt company, which leads to an overclassification of healthy companies as bankrupt.

Table 9.17. Classification table according to the best prediction model evolved using data from the three years in parallel

Set No.	A	B	C	D	E	F	G	Classification	Percentage of cases	Percentage of error
1	T	T	-	-	-	-	-	healthy	64.44	0.32
2	T	F	-	-	-	-	-	fails	0.21	0.00
3	F	-	F	T	-	-	-	healthy	3.97	0.00
4	F	-	F	F	-	-	-	fails	14.02	68.66
5	F	-	T	-	T	T	-	healthy	6.49	0.00
6	F	-	T	-	T	F	-	fails	3.77	77.78
7	F	-	T	-	F	-	T	healthy	29.29	0.00
8	F	-	T	-	F	-	F	fails	4.18	70.00

9.7 Conclusions

The application of GP to the bankruptcy prediction problem has yielded satisfactory results in spite of the imbalance and missing data in the database we have used for the analysis. The fact that GP generates results in the shape of an analytical function allows to analyse the relevance of the variables being used for building the prediction models. The reduction in the number of these variables makes possible the evolution of more intelligible models. From the analysis of the frequency of usage of the variables in the final models we can conclude that the non-financial variables do not play an important role in the building of the failure prediction models. On the other hand "return on equity" is the only variable that appears in all our reduced sets of variables, confirming its importance for the prediction.

Using Strongly Typed GP ensures that the resulting functions make sense from an economic point of view. Limiting the type of data that the functions can take and return also simplifies the task to be performed by the GP by limiting the size of the search space.

There are no statistically significant differences among the prediction results obtained using different sets of data from one year (1, 2 and 3 years ahead) or combining the data from the three years. Regarding the way of feeding the data in the system when using the data from the three years (i.e. sequentially or in parallel) the differences in the results have been minor but we are persuaded that this is due to the fact that we are considering very short time series (only three time steps). We are convinced that this approach will be very valuable in cases where the time span under consideration is longer or the data is available at a higher frequency (monthly, weekly).

The comparison study of bloat control methods has resulted in two methods that obtained fitness results as good as those obtained with the benchmark but did not reduce size efficiently (i.e. lexicographic parsimony pressure with ratio bucketing and tarpeian), four methods that reduced the average size of the population at the expense of inferior fitness values (i.e. both settings of double tournament and two settings of prune and plant) and one method that reduced the average size of the population while keeping the quality of the results: prune and plant with a probability of 0.4.

Finally, when compared with the numerical results obtained with SVM the GP results are clearly better. In addition GP provides us with a nonlinear function in a tree shape, which is easier to analyse and draw conclusions from than the SVM black box structure.

Acknowledgments

Thanks to Isabel Román Martínez, José M. de la Torre Martínez and Elena Gómez Miranda, from the Department of Finance and Accounting of the Universidad de Granada, for the financial database and to J. J. Merelo, from the Department of Computer Architecture and Technology of the University of Granada, for his useful suggestions. This work was supported by the Plan Nacional de I+D of the Spanish Ministerio de Educación y Ciencia (NADEWeb project - TIC2003-09481-C04).

References

[1] Alfaro-Cid E, Sharman K, Esparcia-Alcázar A (2007) A genetic programming approach for bankruptcy prediction using a highly unbalanced database. In: et al

MG (ed) Proceedings of the First European Workshop on Evolutionary Computation in Finance and Economics (EvoFIN'07), Springer-Verlag, Valencia, Spain, Lecture Notes in Computer Science, vol 4448, pp 169–178

[2] Altman EI (1968) Financial ratios, discriminant analysis and the prediction of corporate bankruptcy. Journal of Finance 23(4):589–609

[3] Brabazon A, Keenan PB (2004) A hybrid genetic model for the prediction of corporate failure. Computational Management Science 1:293–310

[4] Brabazon A, O'Neill M (2006) Biologically inspired algorithms for finantial modelling. Springer-Verlag, Berlin, Germany

[5] Chang CC, Lin CJ (2001) LIBSVM: a library for support vector machines. Software available at http://www.csie.ntu.edu.tw/cjlin/libsvm

[6] Dimitras AI, Zanakis SH, Zopounidis C (1996) A survey of business failures with an emphasis on predictions, methods and industrial applications. European Journal of Operational Research 90:487–513

[7] Eggermont J, Eiben AE, van Hemert JI (1999) A comparison of genetic programming variants for data classification. In: Hand DJ, Kok JN, Berthold MR (eds) Proceedings of the Third International Symposium on Advances in Intelligent Data Analysis (IDA'99), Springer-Verlag, Amsterdam, The Netherlands, Lecture Notes in Computer Science, vol 1642, pp 281–290

[8] Fernández de Vega F, Rubio del Solar M, Fernández Martínez A (2005) Implementación de algoritmos evolutivos para un entorno de distribución epidémica. In: Arenas MG, Herrera F, Lozano M, Merelo JJ, Romero G, Sánchez AM (eds) Actas del IV Congreso Español de Metaheurísticas, Algoritmos Evolutivos y Bioinspirados (MAEB'05), Granada, Spain, pp 57–62

[9] Japkowicz N, Stephen S (2002) The class imbalance problem: a systematic study. Intelligent Data Analysis 6(5):429–449

[10] Kim MJ, Han I (2003) The discovery of experts' decision rules from qualitative bankruptcy data using genetic algorithms. Experts Systems with Applications 25:637–646

[11] Kishore JK, Patnaik LM, Mani V, Agrawal VK (2001) Genetic programming based pattern classification with feature space partitioning. Information Sciences 131:65–86

[12] Koza JR (1992) Genetic Programming: On the programming of computers by means of natural selection. MIT Press, Cambridge, MA

[13] Lensberg T, Eilifsen A, McKee TE (2006) Bankruptcy theory development and classification via genetic programming. European Journal of Operational Research 169:677–697

[14] Leshno M, Spector Y (1996) Neural network prediction analysis: The bankruptcy case. Neurocomputing 10:125–147

[15] Luke S (2000) Issues in scaling genetic programming: Breeding strategies, tree generation, and code bloat. PhD thesis, University of Maryland, Maryland, USA

[16] Luke S, Panait L (2002) Lexicographic parsimony pressure. In: et al WBL (ed) Proceedings of the Genetic and Evolutionary Computation Conference (GECCO'02), New York, USA, pp 829–836

[17] Luke S, Panait L (2006) A comparison of bloat control methods for genetic programming. Evolutionary Computation 14(3):309–344
[18] McKee TE, Lensberg T (2002) Genetic programming and rough sets: A hybrid approach to bankruptcy classification. European Journal of Operational Research 138:436–451
[19] Montana DJ (1995) Strongly typed genetic programming. Evolutionary Computation 3(2):199–230
[20] Ohlson J (1980) Financial ratios and the probabilistic prediction of bankruptcy. Journal of Accounting Research 18(1):109–131
[21] Poli R (2003) A simple but theoretically-motivated method to control bloat in genetic programming. In: Ryan C, Soule T, Keijzer M, Tsang EPK, Poli R, Costa E (eds) Proceedings of the Sixth European Coference on Genetic Programming (EuroGP'03, Springer-Verlag, Essex, UK, Lecture Notes in Computer Science, vol 2610, pp 204–217
[22] Salcedo-Sanz S, Fernández-Villacañas JL, Segovia-Vargas MJ, Bousoño-Calzón C (2005) Genetic programming for the prediction of insolvency in non-life insurance companies. Computers and Operations Research 32:749–765
[23] Shin KS, Lee YL (2002) A genetic algorithm application in bankruptcy prediction modeling. Experts Systems with Applications 23:321–328
[24] Tsakonas A, Dounias G, Doumpos M, Zopounidis C (2006) Bankruptcy prediction with neural logic networks by means of grammar-guided genetic programming. Experts Systems with Applications 30:449–461
[25] Vapnik V (1998) Statistical Learning Theory. Wiley-Interscience, New York
[26] Varetto F (1998) Genetic algorithm applications in the field of insolvency risk. Journal of banking and Finance 22:1421–1439
[27] Vieira AS, Ribeiro B, Mukkamala S, Neves JC, Sung AH (2004) On the performance of learning machines for bankruptcy detection. In: Hand DJ, Kok JN, Berthold MR (eds) Proceedings of the IEEE Conference on Computational Cybernetics, Vienna, Austria, pp 323–327

10

Using Kalman-filtered Radial Basis Function Networks for Index Arbitrage in the Financial Markets

David Edelman

School of Business, University College Dublin, Ireland. davide@ucd.ie

Summary. A Kalman-filtered feature-space approach is taken to forecast one-day changes in the financial market indices using lagged returns from previous days as inputs. The resulting model is used to define a time-varying (adaptive) technical trading rule, one which, in cases examined, achieves an out-of-sample Sharpe ('reward-to-variability') Ratio far superior to the 'buy-and-hold' strategy and its popular 'crossing moving-average' counterparts. The approach is contrasted with recurrent neural network models and with other previous attempts to combine Kalman-filtering concepts with (more traditional) multi-layer perceptron models and is demonstrated on two stock indices, the Irish market index (ISEQ) and the FTSE 100 index. The new method proposed is found to be simple to implement, and represents what is argued to be the most natural melding to date of the general explanatory capability of neural networks with the proven adaptive properties of the Kalman filter. Preliminary results presented here, might be expected to perform well for index arbitrage and related types of problems.

10.1 Introduction

While the vast literature relating to attempts to gain excess returns from technical (i.e. price history) information is too diverse to summarise effectively here, that part of it having to do with machine learning is more limited. As mentioned in (3) and elsewhere, some of the key papers in the area by Lebaron, Lakonishok, and Lo (see (2) and (1) in particular) have suggested the possibility of weak form inefficiencies existing in some markets, consistent with the notion of bounded rationality (8), but the results obtained have stopped short of definitively asserting the affirmative result. In the manner presented here the notion of *efficiency* of an asset (or rule) will be interpreted as giving rise to no greater return, on average, than a standard 'buy and hold' position with a commensurate level of market risk, as measured by Beta. Specifically, if β or 'Beta', is defined as the coefficient of the regression for R_{asset}, the net return (above the interest rate) of an asset (or rule) as a linear function of market return R_{mkt}, then generally

$$E(R_{asset}) = \beta_{asset} E(R_{mkt}) \tag{10.1}$$

D. Edelman: *Using Kalman-filtered Radial Basis Function Networks for Index Arbitrage in the Financial Markets*, Studies in Computational Intelligence (SCI) **100**, 187–195 (2008)
www.springerlink.com

via the Capital Asset Pricing Model (CAPM) due to Sharpe (6) and others. Trivially, this implies that any asset uncorrelated with market return (i.e., with zero Beta) should have zero average net return. If, however, a zero-Beta asset is demonstrated to have positive net return, on average, then this theory (and by implication the assumption of efficient markets) is violated. Further, it is of interest to quantify the risk-return profile of the asset, as measured by the Sharpe Ratio (7), which is the average net return divided by the standard deviation of return, which may be thus compared with the same measure applied to other available investments.

If, as many believe, one supposes that no fixed technical trading rule could achieve excess returns over all time (otherwise it would be discovered and exploited to oblivion), the problem of tracking a potential time-varying rule arises. In this framework, the first method which would naturally present itself would be the Kalman filter. Unfortunately, for the most part, this would tend to limit the class of models to linear functions, which one would doubt as lacking the 'subtlety' which one might expect to be a property of an effective trading model.

Perhaps for this very reason, there has recently been a number of attempts to combine Kalman-filtering concepts with nonlinear models such as neural networks (4), arguably with limited success, due primarily to the fact that the concept of a unique, 'true', State or parameter vector does not really have any meaning for most types of neural network models. However, one approach which appears to have been overlooked (without the aforementioned drawback) is that of applying Kalman filters to the linear output layer of a network with fixed nonlinear feature space, prototypically, one based on radial basis functions at fixed, pre-determined centers. It is the latter approach (for which there appears to be no published precedent to date) which will be adopted here, with what appear to be very promising preliminary results from a very simple model.

10.2 Methods

In what follows, we outline what is a remarkably simple but effective method for fitting a Kalman-filtered radial basis function network, to the problem of forecasting daily changes in an equity or an index merely from a number of lagged daily returns of the same series. We begin by providing a brief review of radial basis function networks and Kalman filters, and then describe how they might be effectively combined. Following this, discussion of some more specifics of the problem at hand precede an exposition and analysis of the empirical results.

10.2.1 Radial Basis Function Networks

For a given dataset of input-output observations $\mathbf{x}_i$ and y_i $(i = 1, 2, \ldots)$ a radial basis function network is formed by first finding the Euclidean distances d_{ij} between each input observation vector x_i and each of a number of 'centers' c_j in the same space, and then computing $\exp(-\frac{1}{2}d_{ij}^2/b^2)$ for some 'bandwidth' b, to form a sort of 'sensor array', where for each j, proximity of an input vector to c_j would be indicated by a value close to unity, while distance from c_j would be indicated by a very small positive number. Thus a nonlinear *feature space* transformation P_{ij} of the vectors x_i is formed. Thus, the output y's may be regressed against a much richer set of inputs (typically of much larger dimensionality that the original dimension of $\mathbf{x}$), which may be better able to explain y in a linear sense. Thus, a radial basis network may be characterised by a two-step model, which results in a linear fit of variables which are (in general) a nonlinear transformation of the input. Conceptually, the 'feature space' referred to amounts to a sort of 'sensor array' in multiple dimensions, where each sensor corresponds to a certain location in hyperspace and where the sensor 'fires', either strongly or weakly, when a new input data point is near it.

Before determining the location of the 'sensors' a sense of scale of the hyperspace is needed. In what follows, our input space will consist of lagged daily returns (the differences in logarithm of level over the past five days), where it will be assumed that a crude approximate value for the daily standard deviation of return is available, so that all variables my be assumed to be scaled to approximately unit standard deviation.

Typically the simplest choice of centers c_j ('sensor locations') is just x_j, the (recycled) set of input observations themselves. Unfortunately, this leads to a linearly expanding approximation dimension as sample size increases, and a set of centers which in a sequential sampling setting, is unknowable for early observation, a property which makes this design inappropriate for financial time series.

As mentioned previously, the key to applying radial basis function networks in a time series setting is to keep the centers (and bandwidth) of a so-called feature space constant over time. In our example, the lagged returns (in percentage terms) forming the inputs therefore will have approximate mean zero and standard deviation approximately equal to 1, a set of centers formed by the $\{-1,0,1\}^m$-grid, where m is the number of lags being used on the input, might be likely to each be near the body of input points. To this end, let x_i and c_j again denote the ith input and j^{th} center point. Then (again) the feature space matrix $P(x;c)$ may be defined, where

$$P_{ij} = \exp(-\frac{1}{2}||x_i - c_j||^2/b^2) \tag{10.2}$$

$||\cdot||$ denoting the Euclidean distance (in m-space), for some suitably-chosen 'bandwidth parameter' b. Thus, the basic model in the non-dynamic case is

$$y_i = P(x_i;c)\cdot w + \varepsilon_i \tag{10.3}$$

where w is a fixed but unknown parameter vector, which may be determined easily via standard least-squares, generally with a penalty which is proportional to the squared Euclidean distance of w from the origin.

10.2.2 Kalman Filters

In many modeling applications, model parameter values are taken to be constant over the entire sample of interest. However, frequently, as with financial data, it is much more reasonable to assume that any parametric models should allow parameter values ('states') to vary over time. The simplest, arguably the most common, types of models used in a financial context are linear, such as the Capital Asset Pricing Model and Arbitrage Pricing Theory. Fortunately, time-varying linear models are well-developed in the literature, chief among these being a special case of the Kalman filter (5) with no control input. Specifically, an unobserved State process is assumed to vary according to a relationship

$$x_k = Ax_{k-1} + w_i \tag{10.4}$$

where w_i is assumed to be Normally distributed with mean of zero and a covariance matrix Q, where the only observation is via the process

$$y_k = Hx_k + v_k \tag{10.5}$$

where v_k is assumed to be Normally distributed with mean zero and covariance matrix R. The solution to such a Kalman system is well-known and will not be formally restated here, though notionally the solution may be described as daily 'update' application of the following four steps (to incorporate new information):

1. Compute current day's forecast percentage change
2. Add innovation component to state estimate covariance matrix
3. Use above results plus new observation for 'error-correction'
4. Update state estimate covariance matrix, given above

At any stage, the state estimate update process may be seen as analogous to a Bayes update step (there are indeed two parallel interpretations, though Bayes assumptions are not required), where the current state (i.e,, regression coefficient) estimate and its associated covariance matrix summarise all previous experience, which is combined with new observation to produce a new summary of previous experience. While the original formulation of Kalman specified normally distributed errors, as with most linear least squares methods, the validity of the procedures is much more general.

10.2.3 A Kalman-filtered Radial Basis Function Network

Having introduced radial basis function (RBF) networks and Kalman filters, it is worth imagining how they might be combined. Let the observed output at time i be denoted by y_i and $P(X_i;c)$ denote an RBF-feature-space transformation of input vector x_i about centers c. Then

$$y_i = P(x_i;c) \cdot w^{(i)} + \varepsilon_i \tag{10.6}$$

where the weights $w^{(i)}$ are assumed to be evolving over time in an unobservable fashion, with

$$w^{(i)} = w^{(i-1)} + \eta_i E(\eta_i) = 0, \;\; Cov(\eta_i) = \delta^2 I_M \tag{10.7}$$

M in this case representing the 3^m-strong dimension of the chosen basis and δ a tuning 'innovation' parameter (I_M the M by M Identity matrix). Of course, in practice, a system cannot be iterated from nothing, but must start somewhere. In light of this, in order to start a system we fit an initial period such as the first 250 days (corresponding to one year of trading) as a group using ordinary least squares, and then begin the Kalman iteration, forecasting the one-day return, crucially *out of sample* over the remaining days in the dataset.

10.2.4 Kelly Trading Rule

Next, given day-by-day forecast relative changes (which are 'out-of-sample', using only previous days' data) we use a simple trading rule based on investment in proportion to the predicted return $\hat{y}_i$ (this rule is based on the solution to a series approximation of Expected Logarithmic Utility, or so-called 'Kelly' criterion)

$$b_i = K\hat{y}_i \tag{10.8}$$

where K times the daily variance of return is a suitably chosen fraction of current level of wealth. The Kelly criterion entailing repeated myopic optimisation of logarithmic expectation has been proven (under quite general conditions) to maximise long-run capital growth, in an *almost sure* sense, and has been shown to be extremely effective in many practical applications. For any risky portfolio with net return R_P with finite moments, an investment of a units results in an expected net logarithmic return of

$$E\log(1+aR_P) \doteq a(\mu_P) - \frac{1}{2}a^2(\sigma_P^2+\mu_P^2) \tag{10.9}$$

as the time-interval of investment tends to zero, where μ_P an σ_P denote the mean and standard deviation of R_P. For a taken to be μ_P/σ_P^2, the expected net growth rate is optimised and the result well-approximated by

$$\frac{1}{2}(\frac{\mu_P}{\sigma_P})^2 \tag{10.10}$$

recognisable as one-half the square of the Sharpe Ratio. If, as in many trading applications, the variation of volatility is taken to be much smaller than that of expected return, an investment strategy based on proportional expected return emerges.

In order to evaluate the results of applying this rule to the Kalman filter forecasts, a plot of cumulative returns which would have been achieved via the trading rule will be produced, and a Sharpe ('Reward-to-Variability') calculated, along with a 'Beta' with respect to the index. As it happens, the empirical Beta for this rule will typically be close to zero (as will be discussed below), and the average net position (and hence cost of carry) zero. Hence, any significant return obtained, if it could be demonstrated to be net of transaction costs would constitute a violation of the weak form of the efficient markets hypothesis. While preliminary calculations suggest that a statistically significant inefficiency exists here 'in-house' (for brokers themselves, who are not charged transaction costs) this is not the emphasis of this chapter, and hence will be deferred to a future article.

10.3 Empirical Results

Below are two studies to test the out-of-sample effectiveness of the aforementioned methods on two data series, the (Irish) ISEQ and FTSE equity indices.

10.3.1 Daily Returns for the (Irish) ISEQ Index

For our first dataset, the (logarithmic) returns for 1306 trading days (from September 2001 to September 2005) of the Irish ISEQ Index index were computed and lags produced, resulting in an input series consisting of 5 lags by 1300 observations, with a corresponding output variable of 1300 daily returns. In this case, all of the variables, input and output, have similar character, being returns series with approximate mean zero and standard deviation 1.1%.

As has been mentioned previously, the method applied here is that of radial basis function networks, where a single hidden layer is used, as well as the Gaussian kernel. Most often with models of this type, the centers of the basis functions are taken to be the input datapoints themselves, thus ensuring they occur in the vicinity of the actual data. However, as the present problem requires the basis functions to remain constant and future input datapoints are not known in advance, this approach cannot be taken. Instead, a 5-dimensional grid (as centers for radial basis functions) is thought to constitute a sensible alternative, as each point in the input dataset can be expected to be relatively close to some point in the grid, while the grid points themselves can be expected to be fairly evenly spread throughout the data. For radial basis function networks which use input data points as centers, the bandwidth chosen always effectively decreases with sample size. In this case, however, the number of centers remains constant (at $3^5 = 243$) and the bandwidth also, here set at .01 (though sensitivity analysis performed indicates that the results are not too sensitive to this choice, so long as the basis functions are not too highly correlated nor highly concentrated).

Formally, the specification of an initial parameter value and covariance matrix at the start of a Kalman filter for regression parameters is equivalent to the specification of a ridge penalty configuration in regression, where the 'initial values' in the Kalman filter case are merely the values towards which coefficients are shrunk in ridge regression. In this case, then, it was decided to apply ridge regression to an initial sample, taken to be 250 days, the results of which were used to initialise a Kalman filter beginning at day 251. From this point onwards, the coefficients of the network merely follow from the standard Kalman update equations, where the only parameter which requires specification is the process innovation covariance matrix, which is taken to be diagonal with common standard deviation 0.1%. In each case, prior to performing the update, the prediction given all previous information is recorded, for purposes of ('out-of-sample') comparison with the actual realised values. The result is a prediction of daily return y_i, where it may be shown that investment (positive or negative) in proportion to the forecast $\hat{y}_i$ is optimal.

The cumulative returns (on the logarithmic scale) for the resulting filtered trading rule are summarised in Fig. 1, with the raw ISEQ itself over the same period included for comparison.

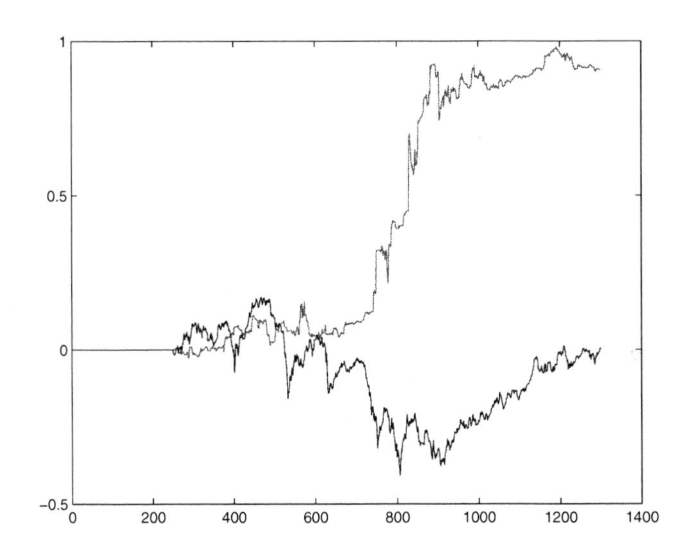

Fig. 10.1. Cumulative Returns for ISEQ (below) and RBF-Kalman fund (above) for period 2001-2005

The risk-adjusted Sharpe ('Reward to Variability') ratio is computed to be approximately 100%, nearly twice the value attained by mere long positions on most major world indices during a typical 'bull run' period.

10.3.2 Daily Returns on the FTSE

In order to test the methods, the same model is applied to the FTSE 100 series, beginning from 4 February, 1984 and continuing to 8 March, 2007, with cumulative returns for the raw FTSE 100 and resulting trading fund (log-scale) shown in Fig. 2. Following some initial volatility, the fund based on the forecasts soon stabilises and demonstrates a surprisingly steady growth, achieving a Sharpe Ratio in this case of approximately 200% over the final 20 year period. This would, for instance, suggest the safety of increasing, or 'leveraging' the overall level of investment in the trading fund, in order to achieve returns which are as high or higher than those of the raw FTSE series, but with much better Risk profile (graphically, this would result in a multiplicative increase in the lower series).

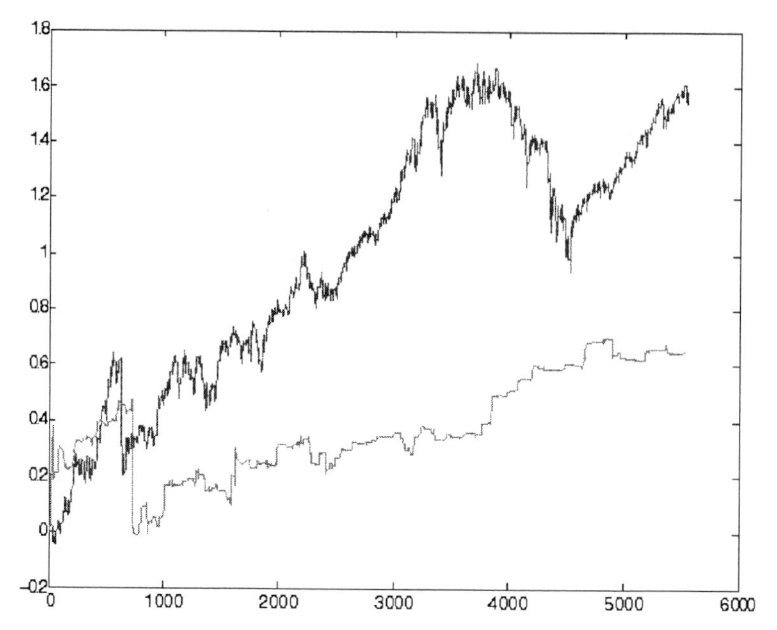

Fig. 10.2. Cumulative Returns for the FTSE 100 (above) and RBF-Kalman fund (below) for the period 1984-2007

It should be emphasised that the trading results presented above were (following an initial period of 1 year) all out-of-sample, by the nature of the Kalman Filtering framework, though transaction costs (as mentioned previously) have not been accounted for here.

10.4 Discussion and Conclusions

The primary contribution here is the suggestion of a simple paradigm which combines the power of neural network modeling (specifically, RBF networks with fixed centers) with the effectiveness of Kalman filtering for tracking time-varying unobservable systems.

The main user-specified parameters required for application of the method are the grid spacing, the RBF bandwidth, and the 'signal-to-noise' ratio of the system. These should be optimised during a 'pre-online' phase, here taken to be a time span of one year. While here this step has been carried out via trial-and-error on a small number of combinations only, it is believed that some form of evolutionary algorithm might be expected to greatly improve on this. Also worth noting is the special case of a static system, or 'signal-to-noise' ratio zero, which while not being presented here was found to be wholly inadequate for investment purposes.

Also worth noting is that as favourable as the performance presented in the previous section appears, for the case of the ISEQ, visual inspection of the graph of returns suggests some degree of serial correlation in performance, which might suggest scope for further modeling improvement for that dataset. Though on the other hand, if the short period high growth evident from the graph is omitted from the sample, the value of the Sharpe Ratio actually increases, suggesting a type of performance characterised by generally stable growth punctuated by a few sharp rises, in a manner presumably not unattractive to investors.

One extension to the current model was considered but dismissed after preliminary investigation, which involves using the input data points themselves as RBF centers, instead of the fixed ones used above. This would be more in keeping with the majority of RBF neural network and SVMR applications, and would have the advantage of ensuring that the RBF centers lie within the body of the input data. The disadvantage of this as an approach is that the Kalman filtering framework requires that the bases of the input data remain constant over time. The way to impose this is by completely rerunning the entire system from the beginning for each new basis at each new timestep, an experiment which was indeed performed, with disappointing results. It is felt that this may have been due to the (implicitly assumed) non-stationarity of the series, which would of course apply to the input data as well as the output, and might hence introduce a systematic bias to the system. For this reason, the above studies have been limited to the grid-center approach.

Finally, for completeness, it deserves mentioning that this model is a particular case of a recurrent neural network, but where the nature of the recurrence (i.e., the use of previous prediction in estimation and forecasting) has a clear formal 'filtering' interpretation, as opposed to a 'black box' one. On the whole, it is hoped that the preliminary findings reported here may suggest that this simple approach merits further investigation.

References

[1] Lo A, Mamaysky H, Wang J (2000) Foundations of technical analysis: Computational algorithms, statistical inference, and empirical implementation. Journal of Finance 55(4):1705–1765
[2] Brock W, Lakonishok J, Lebaron B (1992) Simple technical trading rules and the stochastic properties of stock returns. Journal of Finance 47(5):1731–1764
[3] Edelman D, Davy P (2004) Adaptive technical analysis in the financial markets using machine learning: a statistical view. In: Applied Intelligent Systems, Studies in Fuzziness and Applied Computing, Springer, pp 1–16
[4] Haykin S (2001) Kalman Filters and Neural Networks. New York: Wiley
[5] Kalman R (1960) A new approach to linear filtering and prediction problems. Transactions of the ASME D(82):35–45
[6] Sharpe WF (1964) Capital asset prices: A theory of market equilibrium under conditions of risk. Journal of Finance 19(3):425–442
[7] Sharpe WF (1966) Mutual fund performance. Journal of Business 39:119–138
[8] Simon HA (1990) A mechanism for social selection and successful altruism. Science 250(4988):1665–1668

11

On Predictability and Profitability: Would GP Induced Trading Rules be Sensitive to the Observed Entropy of Time Series?

Nicolas Navet[1,2] and Shu-Heng Chen[1]

[1] INRIA Lorraine, Campus-Scientifique, BP239, F-54506 Vandoeuvre, France, nnavet@loria.fr
[2] AI-ECON Research Center, Department of Economics, National Chengchi University, Taipei, Taiwan 11623, chchen@nccu.edu.tw

Summary. The entropy rate of a dynamic process measures the uncertainty that remains in the next information produced by the process given complete knowledge of the past. It is thus a natural measure of the difficulty faced in predicting the evolution of the process. The first question investigated here is whether stock price time series exhibit temporal dependencies that can be measured through entropy estimates. Then we study the extent to which the return of GP-induced financial trading rules is correlated with the entropy rates of the price time series. Experiments are conducted on end of day (EOD) data of the stocks making up the NYSE US 100 index during the period 2000-2006, with genetic programming being used to induce the trading rules.

11.1 Introduction

One fundamental issue which remains unclear for both financial econometricians and financial engineers is the relationship between the predictability and profitability of financial times series. The literature, so far, has reached no conclusion with regard to the proposition that a time series is profitable if it is predictable, and vice versa. While this proposition may sound obvious, it is not. What makes it subtle is that the two groups of researchers have employed different approaches to tackle the financial time series, and have caused predictability and profitability to be two separate entities. Financial econometricians are more concerned with predictability. For this purpose, formal statistical or information-theoretic approaches are applied to measure the predictability of financial time series. On the other hand, financial engineers or financial practitioners are more concerned with profitability. For that purpose, various heuristic trading algorithms have been used in an attempt to make proper market-timing decisions. There seems to have been a series of efforts made recently to bridge the gap between the two, but it is far from enough; more often than not what we see is that these two groups in the literature have developed without referring to or conforming to each other.

N. Navet and S.-H. Chen: *On Predictability and Profitability: Would GP Induced Trading Rules be Sensitive to the Observed Entropy of Time Series?*, Studies in Computational Intelligence (SCI) **100**, 197–210 (2008)
www.springerlink.com

This chapter, therefore, purports to shed some light on this issue by further cross-referencing the empirical evidence. In other words, we shall connect the empirical results on predictability more closely with the empirical results on profitability. The approach taken by us is, first, to determine the predictability of some sampled financial time series, and, second, to gauge the profitability of time series with different predictability. To do so, we need to choose one principal measure of predictability as well as a trading algorithm. For the former, we choose an information-based measure, i.e., an entropy-based measure, and for the latter we use genetic programming to induce trading algorithms.

Using entropy to measure the degree of randomness and the predictability of a series has a long history, that goes back almost to the very beginning of the development of communication and information theory. Its significance has been introduced to economists since the 1960s. In section 11.2, we shall give a brief review of the entropy measure and the associated estimator used in this paper. This set-up enables us to determine the degree of predictability of any time series coming later. In section 11.2.2, the reasonable behavior (performance) of this proposed measure (estimated entropy) is further illustrated with pseudo random series and financial time series. However, to show that there is no unique measure of predictability, in section 11.3 we further compare the results of our entropy-based measure with those from a well-known nonlinear dependence test, namely, the Brock, Dechert and Scheinkman (BDS) test (section 11.3.1), and the linear dependence test based on the familiar auto-correlation function (section 11.3.2). The purpose is to show that there are some discrepancies existing among different measures of predictability, which may become another obstacle to successfully establishing the connection between predictability and profitability.

Using genetic programming to evolve trading rules has gradually become a part of the practice of financial investment (22). In this chapter, we continue this trend and use genetic programming to exploit the potential profitable opportunities. We start section 11.4 with a simple review of genetic programming. The idea of using genetic programming to test the profitability performance is first established in (4, 5), where the random trading rule, known as lottery trading, is first formulated as a benchmark. This chapter applies the same idea to gauge the profitability of different financial time series. The experimental designs and results are given in sections 11.4.1 and 11.4.2, respectively, followed by concluding remarks in section 11.5.

11.2 Entropy Estimation

Entropy estimation is a field of investigation that has been very active over the last 10 years, one of the reasons being the crucial practical importance of information-theoretic techniques in the advances of neuroscience and, in particular, in the understanding of how the brain works. Methods for estimating the entropy rate can be roughly classified in two main classes (11):

- "Plug-in" (or maximum-likelihood) estimators that basically consist of evaluating the empirical distribution of all words of fixed length in the data, for instance by constructing an n-th order Markov chain, and calculating the entropy of its distribution. Unfortunately, the sample size that is needed increases exponentially in the length of the words and, in practice, plug-in methods are not well suited to capture medium or long range dependencies. In the context of financial time-series, we cannot rule out that there are medium or long range dependencies, after all this is the assumption underlying many trading strategies, and thus we choose to not measure entropy with an estimator belonging to that family.

- Estimators based on data compression algorithms, either estimators based on Lempel-Ziv (ZV, see (9) and (17)) or the Context-Tree Weighting algorithm (see (26) and (15)). Both approaches have been shown (10, 11) to have fast convergence rates (i.e., they are accurate even with a limited amount of observations) and to be able to capture medium and long-range dependencies.

11.2.1 $\hat{h}_{SM}$ entropy rate estimator

In this study, we use an estimator belonging to the Lempel-Ziv class that has been proposed in (17) (estimator a) from Theorem 1 in (17) - as in (16), it will be named $\hat{h}_{SM}$ in the following). Let n be the size of time series s and s_i the symbol at location i in s, the $\hat{h}_{SM}$ estimator is defined as:

$$\hat{h}_{SM} = \left(\frac{1}{n} \sum_{i=1}^{n} \Lambda_i \right)^{-1} \log_2 n \tag{11.1}$$

where Λ_i is the length of the shortest substring starting at position s_i that does not appear as a contiguous substring of the previous i symbols $s_0, ..., s_{i-1}$.

This estimator, which is well known and often used in the literature (see, for instance, (16)), has been shown in (17) to have better statistical properties and performances than earlier Lempel-Ziv estimators. To get further confidence in the efficiency of $\hat{h}_{SM}$, we measured the entropy rate of a sample made of independent draws of a uniform random variable P that takes its value in the set $\{1,2,...,8\}$. The theoretical entropy is equal to $H(P) = -\sum_{i=1}^{8}(1/8)\log_2(1/8) = 3$. The entropy estimate depends on the size of the sample, the quality of the random number generator and the efficiency of the entropy estimator. Using $\hat{h}_{SM}$ with a sample of size 10000, the entropy estimate is equal to 2.96 with the random generator from the boost C++ library (19) , which demonstrates the quality of the estimator since 3 is the best that can be obtained with a "perfect" random generator.

11.2.2 Entropy of NYSE US 100 stocks

Here we estimate the entropy of the daily price time series of the stocks that make up the NYSE US 100 index (the composition of the index can be found at url `http://www.nyse.com/marketinfo/indexes/nyid_components.shtml`). The data is processed so that the data points are the log ratios between consecutive daily closing prices: $r_t = \ln(p_t/p_{t-1})$ and points are then further discretized into 8 distinct states. The boundaries between states are chosen so that each state is assigned the same number of data points ("homogeneous" partitioning). This design choice has the advantage that the model is parameter free and thus no heuristic decision that may change the conclusion reached is required. Furthermore, this experimental setup proved to be very efficient at revealing the randomness of the original data, which is the main quality criterion for partition schemes (23).

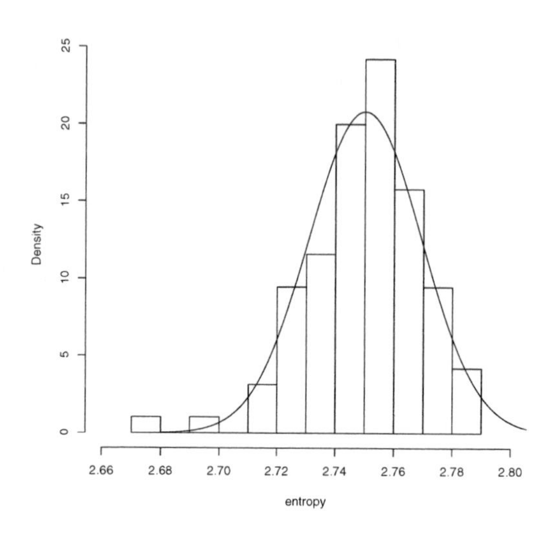

Fig. 11.1. Distribution of the entropy rate of the stocks that make up the NYSE US 100 index (log. ratios of consecutive daily closing prices). A normal distribution with the same mean and standard deviation is plotted for comparison. The reference period is 2000 – 2006.

The distribution of the entropy rate for the stocks of NYSE US 100 index between from 01/01/2000 to 31/12/2006 is shown in fig. 11.1. The minimum value is 2.68, the median 2.75, the mean 2.75 and the maximum value is 2.79. The time series have a high entropy since the theoretical upper bound is 3 and uniformly randomly generated samples achieve 2.90 with the same number of data points.[1] This is not very surprising per se since high entropy rates have been observed even for smaller time scales (see for instance (18)). The 5 stocks from NYSE US 100 index with the highest entropy, identified by their symbol, are *OXY* (2.789), *VLO* (2.787), *MRO* (2.785), *BAX* (2.78), *WAG* (2.776) and the five stocks with the lowest entropy are *TWX* (2.677), *EMC* (2.694), *C* (2.712), *JPM* (2.716), *GE* (2.723). These 10 stocks will be considered in the experiments in the next sections.

Although the entropy is high, there is evidence that the original time series are not random. Indeed, we compare the entropy of the original time series with the entropy

[1] A value of 2.90 is obtained using the boost C++ random generator, but with the standard *rand*() function from the *C* library, the entropy rate achieved is as low as 2.77, which is less that the entropy value of some stocks.

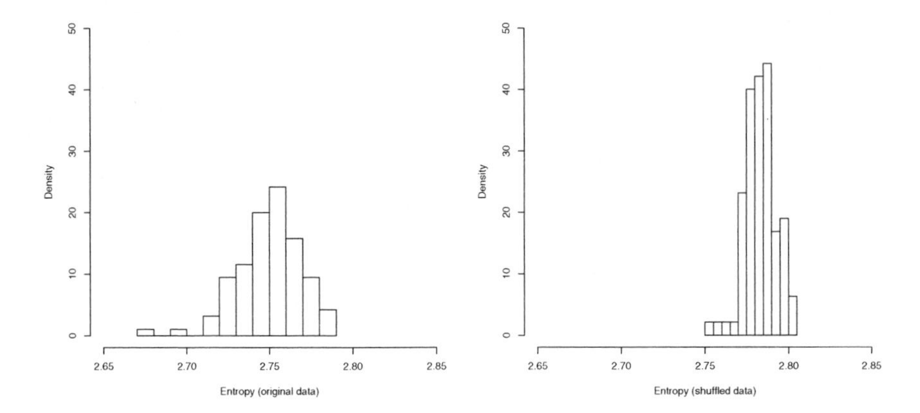

Fig. 11.2. Distribution of entropy rates of the original time series (left-hand graphics - rescaled with regard to Figure 11.1) and shuffled time series (right-hand graphics). The x-axis ranges from 2.65 to 2.85 on both graphics.

of randomly shuffled variants of the original data (surrogate testing). Precisely, 100 shuffled time series for each original time series (after the discretization step) are generated and their average entropy is measured. The complexity of the surrogate time series is greater (2.8 versus 2.75) with a lower standard deviation ($9 \cdot 10^{-3}$ versus $1.9 \cdot 10^{-2}$) and distributed differently as can be seen in fig. 11.2. This provides evidence that, at least for some stocks, there are (weak) temporal dependencies in the original time series.

11.3 Linear and Nonlinear Dependencies

Since some limited temporal dependences have been highlighted we now wish to estimate more precisely their extent through the Brock, Dechert and Scheinkman test (BDS), and to try to find out whether they are linear or nonlinear by performing an autocorrelation analysis.

11.3.1 BDS test statistics

We employ the BDS statistics (see (2)) to test the null hypothesis that the daily log price changes are independent and identically distributed (i.i.d.). The BDS is a widely used and powerful test serving to identify departure from independence and identical distribution caused by non-stationarity, as well as linear and nonlinear dependencies. The test can be applied to raw data or residuals of an estimated model to test for omitted dynamics (see, for instance, (3)), and thus help to decide the model's relevance.

The reader is referred to (2) for the description of the test as well as an assessment of its power, and to (8, 13) for some applications to financial data.

The samples under study are sufficiently large for the test to be accurate (*i.e.*, with a sample size greater than 500, see (13)). The two parameters of importance are the "embedding dimension", parameter m in what follows, and parameter δ, the maximal distance (expressed in terms of sample standard deviations) between points that are considered to be "close". Both parameters are used in the test for the computation of the correlation integral, which is a measure of serial dependence originating from physics (12). The parameter values used in the following analysis are classical in the BDS literature and conform to the recommendations given by the authors of the test.

Table 11.1. BDS test statistics of daily log price changes for the highest entropy stocks. All values are significant at the 1% level.

m	δ	*OXY*	*VLO*	*MRO*	*BAX*	*WAG*
2	0.5	5.68	4.23	6.60	7.38	6.78
3	0.5	6.40	5.38	9.52	11.17	7.79
5	0.5	9.86	7.18	13.62	17.73	9.94
2	1	5.66	4.17	6.69	8.13	7.45
3	1	6.61	5.35	9.40	11.11	8.89
5	1	9.04	6.88	13.08	15.31	11.17
2	1.5	5.34	4.01	6.17	8.33	7.63
3	1.5	6.53	5.34	8.75	10.46	9.52
5	1.5	8.66	6.86	12.08	13.30	11.55
2	2	4.81	3.48	5.32	7.29	6.82
3	2	6.08	4.80	7.97	8.57	9.00
5	2	8.24	6.39	10.86	10.32	10.64

Table 11.1 presents the BDS values of the daily log price changes between 01/01/2000 and 12/31/2006 for the highest entropy stocks, while table 11.2 shows the BDS values for the lowest entropy stocks. The significance levels of the statistics are 1.645 (10%), 1.96 (5%) and 2.576 (1%). The first observation is that, regardless the stock and the BDS parameters, the null hypothesis that daily log price changes are i.i.d. should be rejected at the 1% significance level. This suggests that price time series are not stochastic i.i.d. processes.

What is striking however is that the values of the statistics are much larger for the lowest entropy stocks than for the highest. For a given set of parameters (e.g., $m = 2$ and $\delta = 0.5$), the smallest BDS value among the set of lowest entropy stocks is larger than the highest value among the set of the highest entropy stocks. This shows that the departure from the i.i.d. property is much more important when the entropy is weaker, which conforms to what was expected, and provides evidence that the price dynamics are different between the stocks composing the two sets.

Table 11.2. BDS test statistics of daily log price changes for the lowest entropy stocks. All values are significant at the 1% level.

m	δ	*TWX*	*EMC*	*C*	*JPM*	*GE*
2	0.5	19.04	14.57	14.60	12.94	11.29
3	0.5	26.84	22.15	20.91	18.12	17.12
5	0.5	51.26	39.00	38.57	35.70	29.48
2	1	18.06	14.21	13.9	11.82	11.67
3	1	22.67	19.54	18.76	16.46	16.34
5	1	34.18	29.17	28.12	26.80	24.21
2	1.5	14.67	10.97	12.57	9.5	10.72
3	1.5	17.30	14.59	16.16	13.25	14.24
5	1.5	23.41	20.15	21.54	19.26	18.67
2	2	12.07	8.51	11.40	7.47	9.27
3	2	13.87	11.61	14.25	10.95	13.04
5	2	17.57	15.53	17.72	15.17	16.20

11.3.2 Autocorrelation analysis

The BDS test enables us to reject the null hypothesis that price changes are i.i.d. but it does not provide us with the precise cause of the rejection. In particular, we would like to know whether the rejection is caused by linear or nonlinear dependencies in the time series because this would have implications in terms of forecasting and trading strategies. Here, to obtain some insight into this question, we analyze the autocorrelation of the daily log price changes.

Low Entropy Stocks		High Entropy Stocks	
C	8	*BAX*	2
EMC	5	*MRO*	1
GE	1	*OXY*	1
JPM	5	*VLO*	1
TWX	4	*WAG*	4

Fig. 11.3. Number of autocorrelation coefficients up to a lag of 100 that are significant at the 1% level.

What can be observed is that the autocorrelation of the log-returns is much larger for the lowest entropy stocks than for the highest entropy stocks. As shown in table 11.3, up to a lag of 100, there are on average 4.6 autocorrelations that are significant at the 1% level for the lowest-entropy stocks versus 1.8 for the highest-entropy stocks. However, even for low-entropy stocks, the autocorrelation is very limited. Let us consider the CITIGROUP stock (symbol *C*) whose autocorrelation coefficients up to a lag 100 are shown in fig. 11.4. This stock has 8 autocorrelation coefficients that are significant at the 1% level, which is the highest count among all stocks under study.

However, no single coefficient is larger than 0.08, which is too weak to be of any value for the purpose of forecasting.

Given the limited serial correlation that can be found in the time series, we cannot conclude that the departure from i.i.d. is solely caused by linear dependencies, and, most likely, it should be explained by a combination of linear and nonlinear dependencies. The question addressed in the next section is whether GP is able to take advantage of these temporal dependencies, be they linear or nonlinear, and come up with profitable trading strategies.

11.4 Experiments with Genetic Programming

Genetic programming (GP) applies the idea of biological evolution to a society of computer programs. Specifically, in financial trading, each computer program represents a trading system - a decision rule - which when applied to the market provides trading recommendations. The society of computer programs evolves over the course of the successive generations until a termination criterion is fulfilled, usually a maximum number of generations or some property of the best individuals (e.g., stagnation for a certain number of generations, or a minimum performance threshold is reached). Classical genetic operators, namely, mutation, crossover and reproduction, are applied at each generation to a subset of individuals and the selection among the programs is biased towards the individuals that constitute the best solutions to the problem at hand. The reader may for instance refer to (7, 20) for GP applied to trading in foreign exchange markets, (1, 7, 21) in stock markets, (25) in futures markets and (6, 14) for GP used for pricing options.

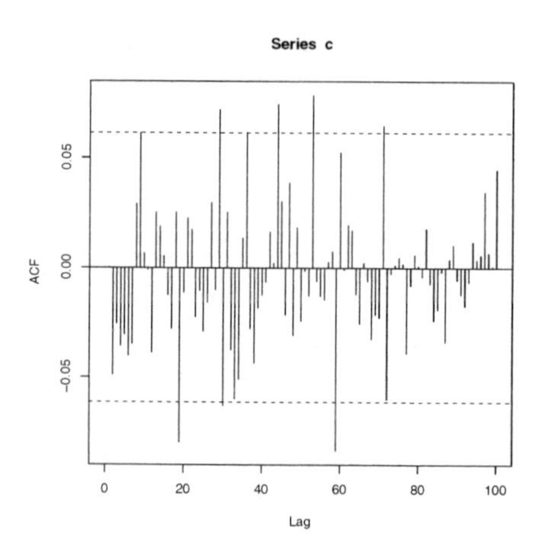

Fig. 11.4. Autocorrelation coefficients of the CITIGROUP stock (symbol C) daily log price changes up to a lag of 100. The points outside the region comprised of the area between the horizontal dotted lines are significant at the 1% level.

The aim of the experiments is to evaluate whether there is a link between the entropy of the time series and the profitability of the GP-induced trading rules. To assess its efficiency, GP is tested against a strategy that would consist of making the investment decision randomly (*"Lottery Trading"*). We follow the methodology proposed in (4, 5) and, in particular, we constrain the randomness so that the expected number of transactions for lottery trading is the same as for GP in order to allow a fair comparison. Hypothesis testing is performed with the Student's t-test at a 95% confidence level.

11.4.1 Experimental setup

Experiments are conducted for the period 2000 – 2006, which is divided into three sections: the training (2000 – 2002), validation (2003 – 2004) and out-of-sample test periods (2005 – 2006). The trading rules are created by genetic programming on the training set, and a subset of top-performing rules are further selected on the unseen data (validation set). The best rule on the validation set is then evaluated for the out-of-sample test period. As classically done in the literature in terms of data-preprocessing, data is normalized with a 100-day moving average. The individuals of GP are trading rules that decide when to enter a long position (no short selling allowed). Exits are decided by a maximum stop loss (−5%), a profit target stop (10%) and a 90-day stop (exit from a position that has been held for the last 90 days). The performance metric is the net profit, with a starting equity of $100,000 and the size of each position equal to 100% of the current equity. The functions, terminals and parameters of the GP runs are described in table 11.5.

Population size	1000
Number of generations	50
Maximum tree depth	7
Function set	+, -, *, /, norm, average, max, min, lag, and, or, not, >, <, if-then-else, true, false
Terminal set	daily closing prices, real and integer ephemeral constants
Value range for real constants	[-1,1]
Value range for integer constants	[0,1000]
Offsprings created by:	
crossover	50%
standard mutation	20%
swap mutation	15%
reproduction	10%
ephemeral constant mutation	5%
Initialization	ramp-half-and-half
Evolution scheme	generation replacement strategy
Elitism	10 best individuals are kept for the next generation
Selection scheme	tournament selection of size 3
Fitness function	accumulated return
Transaction costs	0.1%
Validation	
number of best trees saved	10 individuals per run are saved for validation

Fig. 11.5. GP control parameters

The price time series used in the experiments are shown in figs. 11.6 and 11.7. The first section of each graphic is the training period, the second section is the validation period and the last is the test period.

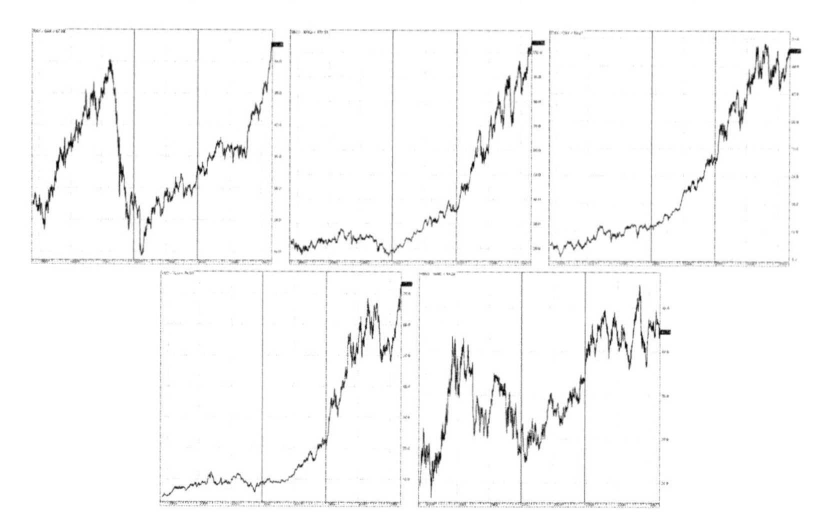

Fig. 11.6. Price time series of stocks having the highest entropies : *BAX*, *MRO*, *OXY*, *VLO*, *WAG*.

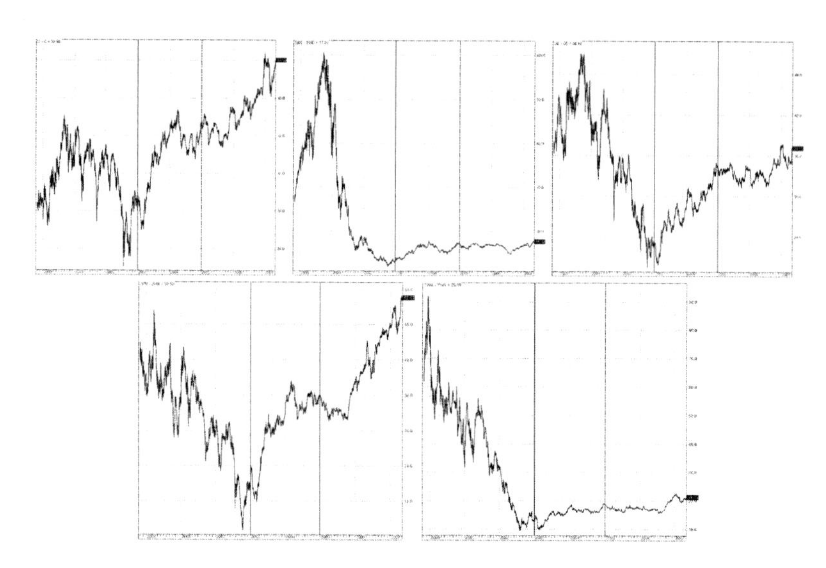

Fig. 11.7. Price time series of stocks having the lowest entropies : *C*, *EMC*, *GE*, *JPM*, *TWX*.

11.4.2 Experimental results

The outcomes of applying GP to the highest entropy stocks are shown in table 11.3. What can be first observed is that GP always leads to profitable strategies but this is not very informative since the price time series of these stocks, except for *WAG*, exhibit a strong upward trend during the test period (see fig. 11.6) and short selling is not possible in these experiments. What is more significant is that GP outperforms Lottery trading (LT) 2 times out of the 5 experiments, while LT is never better than GP.

Table 11.3. Net return of GP and Lottery trading (LT) on the highest entropy stocks (rounded to the nearest 500$). The first two columns are the average profit with GP (20 runs) and Lottery Trading (1000 runs). The third (resp. fourth) column indicates whether one should reject the hypothesis that GP (resp. LT) does not outperform LT (resp. GP) at the 95% confidence level.

	GP net profits	LT net profits	GP>LT?	LT>GP?
OXY	15.5*K*$	14*K*$	No	No
VLO	7*K*$	11.5*K*$	No	No
MRO	15*K*$	18.5*K*$	No	No
BAX	24*K*$	13*K*$	**Yes**	No
WAG	6*K*$	−0.5*K*$	**Yes**	No

Table 11.4 shows the results of GP on the lowest entropy stocks. It turns out that GP is never better than LT while LT outperforms GP 2 times out of 5. This suggests to us that the evolution process is not efficient here, and might even be detrimental.

Table 11.4. Net return of GP and Lottery trading (LT) on the lowest entropy stocks (same settings as table 11.3). Experiments conducted with the possibility of selling the stocks short do not show significant improvements.

	GP net profits	LT net profits	GP>LT?	LT>GP?
TWX	−9*K*$	−1.5*K*$	No	**Yes**
EMC	−16.5*K*$	−11*K*$	No	**Yes**
C	15*K*$	18.5*K*$	No	No
JPM	6*K*$	10*K*$	No	No
GE	−0.5*K*$	0.5*K*$	No	No

From the results shown in Tables 11.3 and 11.4, it should be concluded that, with our experimental setup, selecting the stocks with the lowest entropy does not lead to a better profitability for the GP induced trading rules. We actually observe the opposite which can be explained, as highlighted in (7), because GP is usually not efficient when the training interval exhibits a time series pattern which is significantly different from the out-of-sample period (e.g., "bull" versus "bear", "sideways" versus "bull", etc.). This is exactly what happens here for the lowest entropy stocks as can be seen in fig. 11.7.

On the contrary, in the two cases (*BAX* and *WAG*, see fig. 11.6) where the training period is very similar to the test period, GP clearly outperforms Lottery Trading. This suggests to us that improvements can be made by rethinking the data division scheme and coming up with criteria to select stocks that would integrate a measure of the dissimilarity between current and past market conditions. A contribution that might prove useful in that regard is given in (24) where the authors show that the Haar wavelet transform is suitable for providing estimates of similarity between time series.

11.5 Conclusion and Future Work

It has been shown that the EOD price time series of the NYSE U.S. 100 stocks do not all have equal entropies and, based on surrogate testing, that there are some weak temporal dependencies in the time series. The BDS statistics and the autocorrelation suggest that these temporal dependencies are both linear and nonlinear. The next step has been to test the hypothesis that selecting the stocks with the lowest entropy - the ones with the most predictable price time series - would lead to less risky investments. In the experiments, however, we did not observe that this hypothesis holds.

Recent studies (e.g., (11, 16)) have shown that Context Tree Weighting (CTW) entropy estimators often lead to faster convergence rates than Lempel-Ziv-based estimators. Since the samples of daily data are small in size, the use of CTW may lead to some improvements, although what is really crucial here is not the precise entropy estimate but the relative ordering between distinct time series.

Here, the empirical evidence suggests that predictability is neither a necessary not a sufficient condition for profitability. The predictability test only tells us about the existence of temporal patterns, but it does not give further information on how easy or difficult it is to discover the pattern. Therefore, predictability may not necessarily lead to profitability. On the other hand, we observed on two series with high entropy that it was possible to come up with efficient trading rules. As the large literature on the subject suggests, predictability has a multi-dimensional description, and only one measure of predictability may not be enough to capture all of its attributes. We think that further study regarding the relationship between predictability and profitability should not rest only upon a single measure.

In this study we limit ourselves to the stocks making up the NYSE US 100 because they are of primary interest to investors. The stocks are very liquid and have huge capitalizations (47% of the entire market capitalization of US companies). It is possible that the price time series of these stocks share many common structural characteristics, and so would not be not good candidates for a selection technique based on entropy. Future experiments should include stocks of lower entropy that do not belong to the NYSE US 100, and other time scales should be considered. In particular, higher frequency data would enable us to study the variations in entropy over time.

Acknowledgements

We would like to thank the anonymous reviewers for their helpful suggestions. For the second author, the research support in the form of NSC grant No. NSC. 95-2415-H-004-002-MY3 is gratefully acknowledged.

References

[1] Allen F and Karjalainen R (1999) Using genetic algorithms to find technical trading rules. Journal of Financial Economics 51:245–271
[2] Brock W, Dechert W, LeBaron B and Scheinkman J (1995) A test for independence based on the correlation dimension. Working papers 9520, Wisconsin Madison - Social Systems
[3] Caporale G, Ntantamis C, Pantelidis T and Pittis N (2004) The BDS test as a test for the adequacy of a GARCH(1,1) specification. A Monte Carlo study. Institute for Advanced Studies, 3(2):282–309
[4] Chen S H Navet N (2006) Pretests for genetic-programming evolved trading programs: zero-intelligence strategies and lottery trading. In I King, J Wang, L Chan and D L Wang (eds) Neural Information Processing, 13th International Conference, ICONIP 2006, Proceedings, Part III, volume 4234 of Lecture Notes in Computer Science, pages 450–460, Hong Kong, China. Springer.
[5] Chen S H and Navet N (2007) Evolutionary Computation in Economics and Finance - Volume 2, chapter Failure of Genetic-Programming Induced Trading Strategies: Distinguishing between Efficient Markets and Inefficient Algorithms. Springer
[6] Chen S H, Yeh C H and Lee W C (1998) Option pricing with genetic programming. In J R Koza, W Banzhaf, K Chellapilla, K Deb, M Dorigo, D B Fogel, M H Garzon, D E Goldberg, H Iba and R Riolo (eds) Genetic Programming 1998: Proceedings of the Third Annual Conference, pp. 32–37, University of Wisconsin, Madison, Wisconsin, USA. Morgan Kaufmann
[7] Chen S H, Kuo T W and Hoi K M (2007) Genetic programming and financial trading: How much about "what we know". In C Zopounidis, M Doumpos and P M Pardalos (eds), Handbook of Financial Engineering. Springer (forthcoming)
[8] Chu P (2001) Using BDS statistics to detect nonlinearity in time series. In 53rd session of the International Statistical Institute (ISI)
[9] Farach M, Noordewier M, Savari S, Shepp L, Wyner A and Ziv J (1995) On the entropy of DNA: algorithms and measurements based on memory and rapid convergence. In SODA'95: Proceedings of the Sixth Annual ACM-SIAM Symposium on Discrete Algorithms, pp. 48–57, Philadelphia, PA, USA. Society for Industrial and Applied Mathematics
[10] Gao Y, Kontoyiannis Y and Bienenstock E (2003) Lempel-Ziv and CTW entropy estimators for spike trains. Slides presented at the NIPS03 Workshop on Estimation of entropy and information of undersampled probability distributions - Theory, algorithms, and applications to the neural code

[11] Gao Y, Kontoyiannis I and Bienenstock E (2006) From the entropy to the statistical structure of spike trains. In 2006 IEEE International Symposium on Information Theory, pp. 645–649
[12] Grassberger P and Procaccia I (1983) Measuring the strangeness of strange attractors. Physica D 9:189–208
[13] Hsieh D (1993) Implications of nonlinear dynamics for financial risk management. Journal of Financial and Quantitative Analysis 28(1):41–64
[14] Keber C (1999) Option pricing with the genetic programming approach. Journal of Computational Intelligence 7(6):26–36
[15] Kennel M B and Mees A I (2002) Context-tree modeling of observed symbolic dynamics. Physical Review E 66(5), 056209
[16] Kennel M B, Shlens J B, Abarbanel H D I and Chichilnisky E J (2005) Estimating entropy rates with Bayesian confidence intervals. Neural Computation 17(7):1531–1576
[17] Kontoyiannis I, Algoet P H, Suhov Y M and Wyner A J (1998) Nonparametric entropy estimation for stationary processes and random fields, with applications to English text. IEEE Transactions on Information Theory 44(3):1319–1327
[18] Lee J W, Park J B, Jo H H, Yang J S and Moon H T (2006) Complexity and entropy density analysis of the Korean stock market. In Proceedings of the 5th International Conference on Computational Intelligence in Economics and Finance (CIEF2006)
[19] Maurer J (2007) Boost random number library. Available at url `http://www.boost.org/libs/random/index.html`
[20] Neely C, Weller P and Dittmar R (1997) Is technical analysis in the foreign exchange market profitable? A genetic programming approach. Journal of Financial and Quantitative Analysis 32(4):405–427
[21] Potvin J Y, Soriano P and Vallée M (2004) Generating trading rules on the stock markets with genetic programming. Comput. Oper. Res. 31(7):1033–1047
[22] Smith S N (1998) Trading applications of genetic programming. Financial Engineering News 2(6)
[23] Steuer R, Molgedey L, Ebeling W and Jimnez-Montao M (2001) Entropy and optimal partition for data analysis. The European Physical Journal B - Condensed Matter and Complex Systems 19(2):265–269
[24] Struzik Z and Siebes A (1999) The Haar wavelet transform in the time series similarity paradigm. In PKDD '99: Proceedings of the Third European Conference on Principles of Data Mining and Knowledge Discovery, pp. 12–22, London, UK. Springer-Verlag
[25] Wang J (2000) Trading and hedging in S&P 500 spot and futures markets using genetic programming. Journal of Futures Markets 20(10):911–942
[26] Willems F M J, Shtarkov Y M Tjalkens T J (1995) The context-tree weighting method: basic properties. IEEE Transactions on Information Theory 41(3):653–664

12

Hybrid Neural Systems in Exchange Rate Prediction

Andrzej Bielecki[1], Pawel Hajto[2], Robert Schaefer[3]

[1] Institute of Computer Science, Jagiellonian University, Kraków, Poland.
bielecki@softlab.ii.uj.edu.pl
[2] Motorola Polska, Kraków, Poland.
hajto@salbis.net
[3] AGH University of Science and Technology, Kraków, Poland.
schaefer@agh.edu.pl

Summary. In this chapter, a new hierarchical hybrid wavelet - artificial neural network strategy for exchange rate prediction is introduced. The wavelet analysis (the Mallat's pyramid algorithm) is utilised for separating signal components of various frequencies and then separate neural perceptrons perform prediction for each separate signal component. The strategy was tested for predicting the US dollar/Polish zloty average exchange rate. The achieved accuracy of prediction of value alterations direction is equal to 90%.

12.1 Introduction

Generally, two hypothesis concerning predictability of markets exist. The first one states that it is impossible to predict market behaviour whereas according to the second hypothesis, markets can be described by their own statistical dynamics modelled by walk-type processes with a memory (see (28)). Generally, a market which is weak dependent on political decisions, ecological catastrophes, which has a great number of participants, a great inertia of its processes, is usually, partially predictable. There are several tools, based on statistical methods, dynamical systems theory, approximation theory and artificial intelligence, used for forecasting time-series behaviour. The review of these methods, areas of applications, and obtained results, can be found in (40). In this chapter, time series prediction of the currency market is considered. The specifics of the currency market is discussed in detail in section 12.3.

The aim of this chapter is to present results obtaining by using a hybrid artificial intelligence (AI) system for forecasting of the Polish zloty - US dollar exchange rate. A hybrid artificial neural network (ANN) - wavelet analysis system is used.

In the next section the specifics of hybrid AI neural systems is discussed. Section 12.3 is devoted to describing the currency market and previous literature on currency market prediction. Later sections describe the methodology used and the results obtained - see also (10). Finally, some conclusions are presented.

A. Bielecki et al.: *Hybrid Neural Systems in Exchange Rate Prediction*, Studies in Computational Intelligence (SCI) **100**, 211–230 (2008)
www.springerlink.com

12.2 Neural Networks and Hybrid Systems

An artificial neural network is a cybernetic system whose structure and activity is modelled after animals' and humans' nervous systems, in particular brains. A neuron is a basic signal processing unit. The first neural model, very simplified in comparison with a biological neural cell, was described in 1943 (see (25)). According to this model, a neuron is a module having a few weighted inputs and one output - see fig.12.1. Input signals, say $x_1,...,x_M$, and weights $w_1,...,w_M$ constitute vectors $\mathbf{x} = [x_1,...,x_M] \in \mathbb{R}^M$ and $\mathbf{w} = [w_1,...,w_M] \in \mathbb{R}^M$ respectively. Then a scalar product $s = \mathbf{x} \circ \mathbf{w}$ is calculated. An output signal $y = f(s)$, where $f : \mathbb{R} \to \mathbb{R}$ is called an activation function.

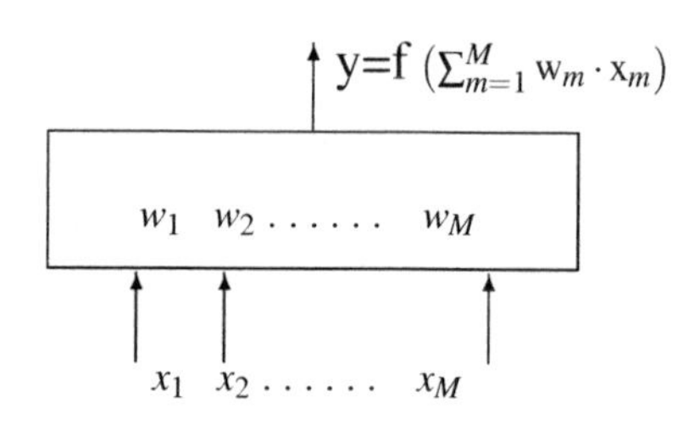

Fig. 12.1. Neuron model

An ANN is a structure consisting of neurons connected to each other in such a way that output signals of some neurons are input signal of others. Furthermore, some neurons constitute an input to the whole system, processing signals from the external environment. The structure of inter-neural connections and weight values determine the properties of the ANN. In this chapter multi-layer ANNs are used. In this kind of network, neurons constitute layers in which neurons are not connected but a neuron belonging to the kth layer is connected with each neuron of the $k+1^{\text{th}}$ layer - see fig. 12.3. The last layer is the output of the whole system.

There are several methods of artificial neural networks learning i.e. setting their weights. Most of them are iterative processes. In order to explain a perceptron learning process in detail, assume that a finite sequence $\left((\mathbf{x}^{(1)},\mathbf{z}^{(1)}),...,(\mathbf{x}^{(N)},\mathbf{z}^{(N)})\right)$, called the learning sequence, is given, where $\mathbf{z}^{(n)}$ is a desired response of the perceptron if the vector $\mathbf{x}^{(n)}$ is put to its input and N is a number of input vectors used in the learning process. Let a real function E be a criterion how correctly all weights of the perceptron are set. It should have nonnegative values and exactly one global minimum having a value equal to zero at the point $\mathbf{w}_0$ such that $\mathbf{y}^{(n)}(\mathbf{w}_0) = \mathbf{z}^{(n)}$ for each $n \in \{1,...,N\}$. Furthermore, greater differences between responses $\mathbf{y}^{(n)}$ of the perceptron and the proper responses $\mathbf{z}^{(n)}$, greater value of the function E. Assuming that the perceptron has J weights and that a learning sequence is given, the function

$E : \mathbb{R}^J \to \mathbb{R}$. Most often the square criterial function is used, which is defined by the formula

$$E(\mathbf{w}) = \frac{1}{2} \sum_{n=1}^{N} \left[\mathbf{y}^{(n)}(\mathbf{w}) - \mathbf{z}^{(n)} \right]^2, \tag{12.1}$$

where $\mathbf{y}^{(n)}(\mathbf{w})$ is the output signal of the perceptron if the vector $\mathbf{x}^{(n)}$ is put to its input. Assuming that the activation functions of each neuron is a mapping of the class $\mathscr{C}^r(\mathbb{R},\mathbb{R})$, where $r \geq 1$ - most types of activation functions used in practice, for instance bipolar and unipolar sigmoid functions and most radial functions, satisfy this assumption - the criterial function E is also of the class $\mathscr{C}^r(\mathbb{R},\mathbb{R})$. The simplest differential one-step method - so called the gradient descent method - leads to the iterative variation of synapses given by the following difference scheme

$$\mathbf{w}(p+1) = \mathbf{w}(p) - h \cdot grad\, E(\mathbf{w}(p)), \tag{12.2}$$

where $\mathbf{w} = [w_1, ..., w_J]$ is a vector of all weights of a perceptron (see fig.12.2) whereas p numerates steps of the learning process. The formula 12.2 describes a process of finding a local minimum of the function E using the Euler method which is a Runge-Kutta method of order 1. The Runge-Kutta methods of order 2 are also sometimes considered as learning processes (see (13)). The formula 12.2 can be rewritten in the following form

$$\Delta\mathbf{w}(p) = -h \cdot grad\, E(\mathbf{w}(p)), \tag{12.3}$$

where $\Delta\mathbf{w}(p) := \mathbf{w}(p+1) - \mathbf{w}(p)$. In order to accelerate the learning process an additional term can be added

$$\Delta\mathbf{w}(p) = -h \cdot grad\, E(\mathbf{w}(p)) + \eta \cdot \Delta\mathbf{w}(p-1). \tag{12.4}$$

The formula 12.4 is called a gradient descent method with momentum. It is a two-step method of order 1. This method was applied for training neural networks in the system described in this chapter.

Every AI system has its own specific advantages and limitations. Considering neural networks, multi-layer ones have universal approximation capabilities (see (6), (11), (12), (14), (15), (16), (19)). Furthermore, neural networks can model automatically complex, nonlinear, relations between its input data and output signals basing only on sample training set of input data. On the other hand, a single ANN often cannot achieve satisfactory accuracy, especially when a very high-level of accuracy is demanded.

Because of AI system limitations, there is a demand for complex systems combining various approaches. Generally, AI systems can be combined with other AI systems or with mathematical tools. Hybrid systems consisting of a neural network aided by genetic algorithms, fuzzy or rule systems, have been become a standard approach. Dividing a task into subtasks in such a way that a single module solves a single subtask in order to organise a neural system as a modular one, is the second approach to a neural system improving.

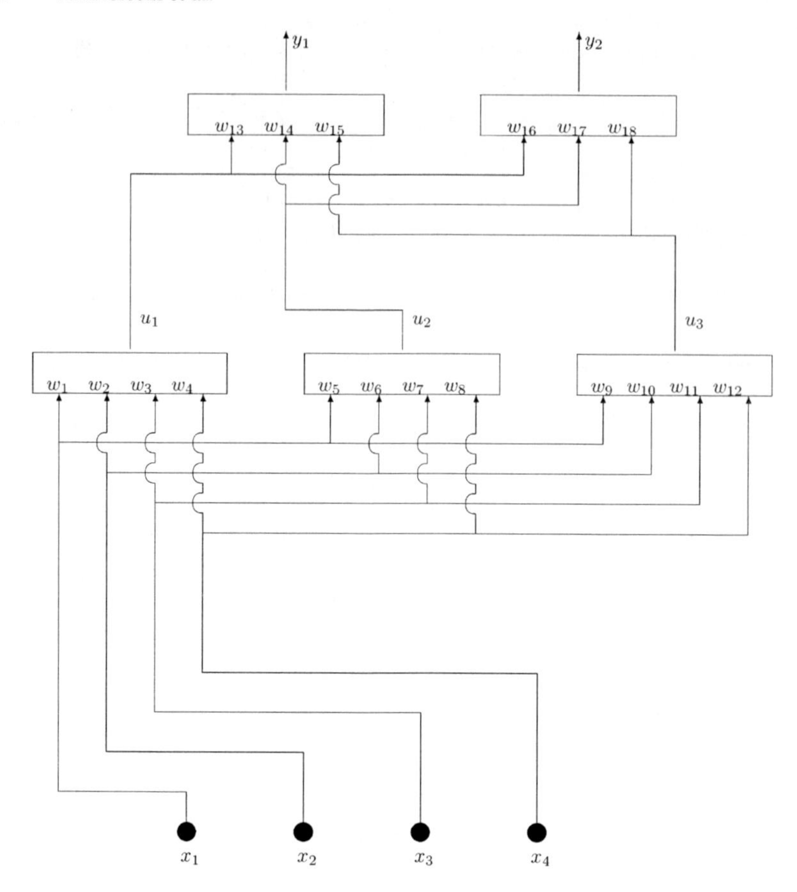

the perceptron input vector $\vec{x} = [x_1, x_2, x_3, x_4]$

the perceptron output vector $\vec{y} = [y_1, y_2]$

the hidden layer output vector $\vec{u} = [u_1, u_2, u_3]$

the perceptron weight vector $\vec{w} = [w_1, w_2, ..., w_{18}]$

the criterial function

$$E(\vec{w}) = E(\vec{y}(\vec{w})) = E(\ y_1(w_{13}, w_{14}, w_{15}, \vec{u}(w_1, ...w_{12})),\ y_2(w_{16}, w_{17}, w_{18}, \vec{u}(w_1, ...w_{12}))\ )$$

Fig. 12.2. An example of a perceptron (a two-layer one)

The partial solutions are combined in order to obtain the complete solution of the task. In such a case, deep analysis of the task is necessary. The third possibility is combining a neural approach with statistical analysis or approximation models, including wavelet analysis. In this case a mathematical tool is used to study properties of training sample in order to prepare the optimal one or to decompose it into subsets.

The mathematical analysis of the problem can, sometimes, help to optimise a neural network architecture (see (33)).

In this chapter a hybrid wavelet-neural system for exchange rate forecasting is presented - see section 12.3. It should be mentioned that ANNs has been widely used for capital markets modelling, for example (35, 36, 38, 41).

12.3 Exchange Rate Prediction

Characteristics of currency markets are well described in (4) and (5). A few key points from these papers are described here. Foreign exchange markets are among the deepest financial markets in the world. Currency liberalisation and technological innovations have led to enormous levels of currency trading. The behaviour of currency markets affects trade flows, international investment, key determinants of economic performance and also political outcomes.

The efficient markets hypothesis states that the forward exchange rate is an unbiased predictor of the future spot exchange rate. Empirical studies, however, conclude that the forward rate is often a biased predictor of future exchange rate. Economists offers a variety of explanations for this bias existence. One of the most prevalent arguments attributes the bias to the existence of a risk premium. Many of studies point to politics as a source of exchange rate risk. In parliamentary systems, for instance, four distinct event-periods when political news affects currency markets, can be identified: election campaigns, post-elections negotiations, cabinet dissolutions, and the immediate month after government formation. This means, for example, that the forward exchange rate is biased more often during these periods. However many other factors, including political ones, can influence currency markets. For instance, news about macroeconomic variables is rapidly incorporated into exchange rates. Summing up, exchange rates are affected by many highly correlated economic, political and even psychological factors. These factors interact in a very complex fashion. Exchange rate series exhibit high volatility, complexity and noise that result from an elusive market mechanism generating daily observation. Therefore studies concerning exchange rate markets behaviour, including prediction of exchange rates values, are in constant demand.

A short review of a sample of the large literature concerning currency market predictability is provided below. Both mathematical models and applications of artificial intelligence systems, as well as their effectiveness are briefly outlined. It should be stressed, however, that it is not possible to fully describe all the literature on this topic, and the survey presented below concerns only selected examples.

In (26), (28) and (31), the predictability of the Yen currency market is discussed. The authors analysed the Japanese yen - US dollar exchange rate with a focus of its up and down movement using probabilistic and stochastic models. A time series of $D(t) := Y(t+1) - Y(t)$, where $Y(t)$ denotes a value of yen at the time t, is composed and it is shown that there exists a significant negative correlation between $D(t+u)$ and $D(t)$ if u is very small - ranging from about 10 seconds (see (28)) to 1 minute (see (26)). A time series X, reflecting only information on whether the price of yen increases ($X(t) = +1$) or decreases ($X(t) = -1$), is constructed as well. A one-dimensional random-walk model $Z(t+1) = Z(t) + X(t)$ is studied. It is shown that a high frequency yen-dollar exchange dynamics is not completely random and it is possible to design a profitable strategy using higher order statistical analysis of the model (see (28)). Furthermore, some interesting phenomena related to the stochastic resonance appears (see (31)).

Fractal analysis was applied to analyse Malaysian currency exchange rate behaviour in (27). The time series was examined using the fractal model based on Brownian motions. The authors applied not only the mono-fractal model with the Hurst exponents determined using R/S analysis, detrended analysis and the method of second moment, but a new approach based on multi-fractional Brownian motions with time-varying Hurst exponents was proposed as well. Long-range fluctuations Malaysian ringgit price were investigated for a number of leading currencies: US dollar, Japanese yen and Singapore dollar. The applied methodology allowed to observe the scaling behaviours in the time series.

The paper (2) is an example of the application of both mathematical and AI tools for nonlinear time-series analysis of the Greek exchange-rate market. Dynamical systems theory, including the theory of chaos, was used to study four major currencies - US dollar, German mark, French franc and British pound - against Greek drachma. Generally, long memory components were found in mark and franc series whereas pound and US dollar behaviours were random. The results were confirmed using a multi-layer ANN. In their conclusions, the authors stressed that the analysis showed that a chaotic explanation could not be ruled out for currencies in general. Each currency should be analysed separately, considering any special conditions issuing in the market.

Simultaneous nearest-neighbour predictors were used for the analysis of nine currencies participating in the exchange rate mechanism of the European Monetary System at the end of the twentieth century: the Belgian franc, Danish crown, Portuguese escudo, French franc, Dutch guilder, Irish pound, Italian lira, Spanish peseta and British pound - see (7). Evaluating the forecasting performance using Theil's U statistic, the nonlinear simultaneous nearest-neighbour performed marginally better than both a random walk model and linear auto-regressive integrated moving average (ARIMA) predictors. The Diebold Mariano test suggested that in most cases, the nearest-neighbour predictor outperforms the random walk at the 1% significance level. Moreover, the Pesaran Timmermann test (see (30)) showed that the probability of correctly predicting the sign of change was higher for the nearest-neighbour predictions than the ARIMA case. Thus, the presented analysis confirmed the presence of predictable components in exchange rates.

The approach described in (33) is a combination of three methods: wavelet analysis, genetic algorithms and neural networks. Wavelet analysis processes information at different scales and therefore is useful for feature detection in complex and chaotic time series. Detecting significant patterns from historical data is crucial for good performance in time-series forecasting. The proposed methodology consists of four phases. First, the studied time-series is decomposed into different time and frequency components using the discrete wavelet transform. Then the refined highpass, lowpass and bandpass filters are extracted from the decomposed time-series that are based on the feedback from the third phase - genetic learning of the neural network. Its input is stated as a multi-scale extraction layer giving data obtained in the second phase to the ANN's first layer. The desired multi-scaled input structure in a neural network is near-optimally extracted by a genetic algorithm. In the fourth phase weights of the ANN are fixed by the hill climbing algorithm. The introduced method was applied to Korean won / US dollar exchange rate forecasting.

A wide review of ANNs applications for exchange rate forecasting is presented in (17). First of all, data preprocessing is crucial in exchange rate prediction. Many kinds of ANNs are used for prediction including feedforward, radial basis function, recurrent, competitive and modular networks. ANNs can also be combined with other approaches including fuzzy inference, genetic algorithms and approximation methods. Comparing neural network results for exchange rate prediction with those of other forecasting methods is inconclusive. It is clear, however, that ANNs can be effective tools for time-series prediction.

12.4 Discrete Wavelet Transformation

Foundations of wavelet analysis can be found in (37), where it is compared with the Fourier transform method. The monograph (39) provides an advanced presentation of the topic.

12.4.1 Theoretical foundations

Let us recall briefly foundations of wavelet theory and its applications to signal processing. The wavelet theory evolved in the mid-1980s (see (3, 21, 24, 34)), though some constructions and theoretical results were discovered much earlier (see (8, 9, 32, 34)). It can be regarded as an extension of Fourier analysis, specially in the scope of signal processing. Wavelets are functions, whose localisations in time and frequency can be fully controlled. This leads to improved and new signal processing applications. Wavelet transforms are used in physics, geophysics, astronomy, biology, chemistry, image processing (NMR, tomography), sound processing, data compression and economics.

Definition 1 *A function $\Psi \in L_2(\mathbb{R})$ is a* wavelet, *if the functions $\Psi_{j,k}$, such that*

$$\Psi_{j,k}(t) := 2^{\frac{j}{2}}\Psi(2^j t - k), \quad j,k \in \mathbb{Z}$$

almost everywhere in $\mathbb{R}$, create an orthonormal basis in $L_2(\mathbb{R})$, where $L_2(\mathbb{R})$ denotes the set of functions $f : \mathbb{R} \to \mathbb{C}$, such that:

$$\int_{-\infty}^{\infty} |f(t)|^2 dt < \infty$$

with the inner product defined by:

$$f \circ g = \int_{-\infty}^{\infty} f(t)\overline{g(t)}dt.$$

An example is the Haar wavelet, defined as follows:

$$\Psi := \begin{cases} 1 & \text{for } t \in [0, \frac{1}{2}), \\ -1 & \text{for } t \in [\frac{1}{2}, 1], \\ 0 & \text{otherwise.} \end{cases} \tag{12.5}$$

Definition 1 *A* multi-resolution analysis (MRA) *is a nested sequence*

$$\ldots \subset V_{-1} \subset V_0 \subset V_1 \subset \ldots$$

of subspaces of $L_2(\mathbb{R})$ satisfying

1. $\bigcup_{n\in\mathbb{Z}} V_n$ *is dense in* $L_2(\mathbb{R})$,
2. $\bigcap_{n\in\mathbb{Z}} V_n = \{0\}$,
3. $f(t) \in V_n$ *if and only if* $f(2^{-n}t) \in V_0$,
4. *there exists a function $\Phi(t)$, called* a scaling function, *such that* $\{\Phi(t-k)\}_{k\in\mathbb{Z}}$ *is an orthonormal basis for* V_0.

Property 1 *Because $\Phi \in V_0 \subset V_1$, condition 3 of the MRA definition implies, that $\Phi(x/2) \in V_0$. This leads to*

$$\Phi(x/2) = \sum_{n\in\mathbb{Z}} a_n \Phi(x-n).$$

We define m_Φ:

$$m_\Phi(\xi) = \frac{1}{2}\sum_{n\in\mathbb{Z}} a_n e^{-in\xi}.$$

There exists a relationship between wavelets and a multi-resolution analysis ((39), section 3.4):

Lemma 1 *Let us suppose, that we have a MRA. Let, furthermore, W_0 be given by the condition $V_0 \oplus W_0 = V_1$. A function $\Psi \in W_0$ is a wavelet if and only if*

$$\hat{\Phi}(\xi/2) = e^{i\xi/2} v(\xi)\overline{m_\Phi}(\xi/2+\pi)\hat{\Psi}(\xi/2),$$

where $\hat{\Phi}$ and $\hat{\Psi}$ are Fourier transforms of Φ and Ψ respectively, $v(\xi)$ is a 2π-periodic function such that $|v(\xi)| = 1$. Additionally, for Ψ and every $s \in \mathbb{Z}$ span $\{\psi_{j,k}\}_{k\in\mathbb{Z}, j<s} = V_s$.

If $v = 1$, the wavelet Ψ is defined by:

$$\Psi(x) = \sum_{n\in\mathbb{Z}} \overline{a_n}(-1)^n \Phi(2x+n+1),$$

where $a_n = \int_{-\infty}^{\infty} \Phi(x/2)\overline{\Phi}(x-n)dx$.

Definition 1 *Having a MRA, we define an* orthogonal *subspace* $V_j^{\perp} \subset L_2(\mathbb{R})$ *to subspace* $V_j \subset L_2(\mathbb{R})$ *with the following condition:*

$$V_j \oplus V_j^{\perp} = V_{j+1}.$$

The MRA definition implies (see (39), section 3.40) that

$$L_2(\mathbb{R}) = \oplus_{j \in \mathbb{Z}} V_j^{\perp}.$$

The theory of a multi-resolution analysis states that if a MRA is given, we can find a function Ψ, which generates an orthonormal wavelet basis for V_s for all $s \in \mathbb{Z}$, in other words, $\text{span}\{\psi_{j,k}\}_{k \in \mathbb{Z}, j<s} = V_s$. In practical applications we are interested in examining the orthogonal projections $P_n(f)$ of a function $f \in L_2(\mathbb{R})$ onto wavelet spaces V_n. This process is realized by using *wavelet filters* (see (1), p. 70 and (18) 7.1-7.8).

12.4.2 Signal processing by wavelets

Let us assume that signals are given as a bi-infinite time series. Such signal $s = [\ldots, s_{-1}, s_0, s_1, \ldots]$ defines a function $f \in V_n$ by

$$f = \sum_{k \in \mathbb{Z}} s_k \cdot \psi_{k,n}. \tag{12.6}$$

Now the wavelet filters process this signal by using two operators, H (the low-pass filter) and G (the high-pass filter), where

$$H(s)_k = \sum_{j \in \mathbb{Z}} h_{j-2k} \cdot s_j$$

and

$$G(s)_k = \sum_{j \in \mathbb{Z}} g_{j-2k} \cdot s_j.$$

The sequences $\{h_k\}$, $\{g_k\}$ arise from MRA and inner product properties (see (1), p. 70) and are unique for every wavelet family. Having a signal s, and the associated function $f \in V_n$ (as in 12.6), $H(s)$ are coefficients of the orthogonal projection $P_{n-1}(f)$ onto V_{n-1} and $G(s)$ coefficients of $P_{n-1}(f)$ onto $V_{n-1}^{\perp}$. A good practical interpretation of this is that $H(s)$ and $G(s)$ contain the low and the high frequencies respectively.

Once we know how to decompose a signal s, it is equally important to have a tool to recompose it. Each of the operators H and G has a so-called dual operator, denoted H^* and G^* respectively, defined by

$$H^*(s^*)_k = \sum_{j \in \mathbb{Z}} h_{k-2j} \cdot s_j^*$$

and

$$G^*(d^*)_k = \sum_{j \in \mathbb{Z}} g_{k-2j} \cdot d_j^*.$$

The filters and their dual operators act as follows

$$s = H^*(H(s)) + G^*(G(s)).$$

In real world we cannot deal with sequences of the infinite length. The wavelet families that are used (Daubechies, CDF, etc.) have a finite number of non-zero $\{h_k\}$, $\{g_k\}$ filter coefficients. And the solutions for the assumption of the infinite length of the signal s are periodization, mirroring, Gram-Schmidt boundary filters and zero-padding (see (18), section 10).
The algorithm for processing a signal using wavelet filters is called *a Mallat's pyramid algorithm*. Let us consider a finite signal $s = [s_0, s_1, \ldots, s_{2^n-1}]$, and wavelet filters H, G with $\{h_k\}$, $\{g_k\}$ coefficients from a chosen wavelet family. Frequencies in s range from 0 to f_N, where f_N is the Nyquist frequency, the highest frequency one can observe in a signal sampled with sampling frequency f_S, $f_N = \frac{f_S}{2}$.

We compute $s^1 = H(s)$ and $d^1 = G(s)$. The length of s^1, d^1 is 2^{n-1} (see (1), p.72). The frequencies contained in s^1 range from 0 to $\frac{f_N}{2}$ (the low part) while in d^1 from $\frac{f_N}{2}$ to f_N (the high part). Then we apply the same procedure to s^1, obtaining s^2 and d^2, each of length 2^{n-2}. The available frequencies are: 0 to $\frac{f_N}{4}$ (s^2) and $\frac{f_N}{4}$ to $\frac{f_N}{2}$ (d^2). After n steps the algorithm stops and we get a vector

$$s^* = [s_0^n, d_0^n, d_0^{n-1}, d_1^{n-1}, \ldots, d_{2^{n-2}-1}^2, d_0^1, \ldots, d_{2^{n-1}-1}^1].$$

This is the discrete wavelet transform (DWT) of s. To this form of s one can apply some operations like zero-padding of high-frequency coefficients for noise reduction or to separate only the desirable frequencies in order to get data to train an ANN, which was important in the described application. Obviously an inverse process is also possible, using H^* and G^* operators and a reversed version of the Mallat's algorithm. It is called the inverse discrete wavelet transform (iDWT).

12.5 Hybrid Wavelet-neural System

In economic time series prediction a typical set of data is a signal $x = [x_1, \ldots, x_k]$, containing e.g. stock market index values or currency exchange rates. Each entry comes from another time point, which means, that x_1 is the exchange rate value at the beginning, x_2 the value on the next day and so on.

12.5.1 The basic approach

The problem of time-series forecasting can be defined in the following way. Having the values $x_1, \ldots, x_k$ of economic data at consecutive time points (e.g. stock index at day no. 1, day no. 2, day no. 3) it is desired to estimate its unknown value on the forthcoming day. In other words, we would like to have a function $g : \mathbb{R}^k \to \mathbb{R}$, such as:

$$x_{k+1} = g(x_1, \dots, x_k),$$

where x_{k+1} is the next step value of the time-series. Obviously, the future is generally unpredictable. And to find such a function g is not possible. What can be done, to try to observe as much regularity of our data as possible and to look for a function $f : \mathbb{R}^k \to \mathbb{R}$, such that:

$$\tilde{x}_{k+1} = f(x_1, \dots, x_k)$$

where the distance $|x_{k+1} - \tilde{x}_{k+1}|$ is small enough. A first idea could be splitting our data into patterns for a multi-layer perceptron (MLP). Assuming the data is a vector $x = [x_1, x_2, ..., x_m]$ the patterns look like this:

$$\begin{array}{l} ([x_1, x_2, \dots, x_k], x_{k+1}) \\ ([x_2, x_3, \dots, x_{k+1}], x_{k+2}) \\ ([x_3, x_4, \dots, x_{k+2}], x_{k+3}) \\ \dots \\ ([x_{m-k}, x_{m-k+1}, \dots, x_{m-1}], x_m), \end{array}$$

where $k < m$. Next, an architecture for an ANN must be chosen. Lula designed a network ((23), p. 158) for testing the market efficiency hypothesis based on Warsaw Stock Exchange index data. The author used an MLP with three layers, 6 neurons in the input layer, 6 neurons with a tangensoidal activation function and 1 neuron in the output layer with a linear activation function. The value of $k = 6$ is estimated with a BDS input data test, described in (22). After this MLP is trained, it realises the function f for $k = 6$

$$\begin{array}{l} x_7 \approx f(x_1, \dots, x_6) \\ x_8 \approx f(x_2, \dots, x_7) \\ x_9 \approx f(x_3, \dots, x_8) \\ \dots \\ x_m \approx f(x_{m-6}, x_{m-5}, \dots, x_{m-1}). \end{array}$$

That is just an input (known) data approximation. But now we can try to estimate the *unknown* values:

$$\begin{array}{l} \tilde{x}_{m+1} = f(x_{m-5}, x_{m-4}, \dots, x_m) \\ \tilde{x}_{m+2} = f(x_{m-4}, x_{m-3}, \dots, x_m, \tilde{x}_{m+1}) \\ \tilde{x}_{m+3} = f(x_{m-3}, x_{m-2}, x_{m-1}, x_m, \tilde{x}_{m+1}, \tilde{x}_{m+2}). \\ \dots \end{array}$$

However, this basic "one-network" idea has not been used in this work, because of the poor results Lula achieved with the Warsaw Stock Exchange index. Despite using sophisticated training algorithms the *DIR* coefficient (the percent of correctly guessed directions of market fluctuations) on the testing patterns was only 61% (see (23), p. 159). These results are of low practical usefulness. In contrast, the MLPs used in the application described in this chapter achieved a *DIR* on testing patterns of ca. 86%-90%. Crucially, the patterns used for forecasting contained wavelet-filtered oscillations, not raw economic data.

12.5.2 The wavelet approach

The wavelet approach is based on applying Mallat's pyramid algorithm to the given data, splitting the data into separated frequency bands, approximating each band by an ANN and predicting their values as described above. The input data is a vector $x = [x_0, \dots, x_{2^n-1}]$. The assumption of its length is important because of Mallat's algorithm. In practical applications zero-padding can be used to achieve this. We compute the DWT of x, getting a vector

$$x^* = [x_0^n, d_0^n, d_0^{n-1}, d_1^{n-1}, \dots, d_{2^{n-2}-1}^2, d_0^1, \dots, d_{2^{n-1}-1}^1].$$

In order to split x into different frequency ranges we need to set all entries in x^* responsible for unwanted frequencies to zero.

Range	Vector
$\frac{f_N}{2}$ to f_N	$x^{(n)*} = [0, \dots, 0, d_0^1, \dots, d_{2^{n-1}-1}^1]$
$\frac{f_N}{4}$ to $\frac{f_N}{2}$	$x^{(n-1)*} = [0, \dots, 0, d_0^2, \dots, d_{2^{n-2}-1}^2, 0, \dots, 0]$
...	...
$\frac{f_N}{2^n}$ to $\frac{f_N}{2^{n-1}}$	$x^{(1)*} = [0, d_0^n, 0, \dots, 0]$
0 to $\frac{f_N}{2^n}$	$x^{(0)*} = [x_0^n, 0, \dots, 0]$.

Now the inverse DWT of each $x^{(i)*}$ is computed:

$$y^{(i)} = IDWT(x^{(i)*}),$$

where $i = 0, \dots, n$. Note that $y^{(i)}$ contains a range of frequencies from x as shown above and its length is 2^n. To approximate and predict $y^{(i)}$ for $i = 1 \dots n$ MLPs are used with the same three layer architecture as shown in the basic approach. The patterns are given as vectors

$$\begin{gathered}
([y_1^{(i)}, y_2^{(i)}, \dots, y_6^{(i)}], y_7^{(i)}) \\
([y_2^{(i)}, y_3^{(i)}, \dots, y_7^{(i)}], y_8^{(i)}) \\
([y_3^{(i)}, y_4^{(i)}, \dots, y_8^{(i)}], y_9^{(i)}) \\
\dots \\
([y_{n-6}^{(i)}, y_{n-5}^{(i)}, \dots, y_{n-1}^{(i)}], y_n^{(i)}),
\end{gathered}$$

where $i = 1, \dots, n$. There is no need to build an ANN to approximate $y^{(0)}$ since all the entries in this vector are equal to the mean value of $x_0, \dots, x_{2^n-1}$. Let $N^{(i)}$ denote the ANN used to approximate $y^{(i)}$. Unknown values of $y^{(i)}$ can be predicted:

$$\begin{aligned}
\tilde{y}_{n+1}^{(i)} &= N^{(i)}(y_{n-5}^{(i)}, y_{n-4}^{(i)}, \dots, y_n^{(i)}) \\
\tilde{y}_{n+2}^{(i)} &= N^{(i)}(y_{n-4}^{(i)}, y_{n-3}^{(i)}, \dots, y_n^{(i)}, \tilde{y}_{n+1})^{(i)} \\
\tilde{y}_{n+3}^{(i)} &= N^{(i)}(y_{n-3}^{(i)}, y_{n-2}^{(i)}, y_{n-1}^{(i)}, y_n^{(i)}, \tilde{y}_{n+1}^{(i)}, \tilde{y}_{n+2}^{(i)}), \\
&\dots
\end{aligned}$$

where $i = 1, \ldots, n$. Thus

$$\tilde{x}_{n+j} = \sum_{i=1}^{n} \tilde{y}_{n+j}^{(i)} + M,$$

where $j > 0$ and $M = y_0^{(0)}$ is the average value of $x_0, \ldots, x_{2^n-1}$. This is a consequence of wavelet filter properties and the Orthogonal Decomposition Theorem ((1), p. 101). The structure of the forecasting system is shown in fig.12.3.

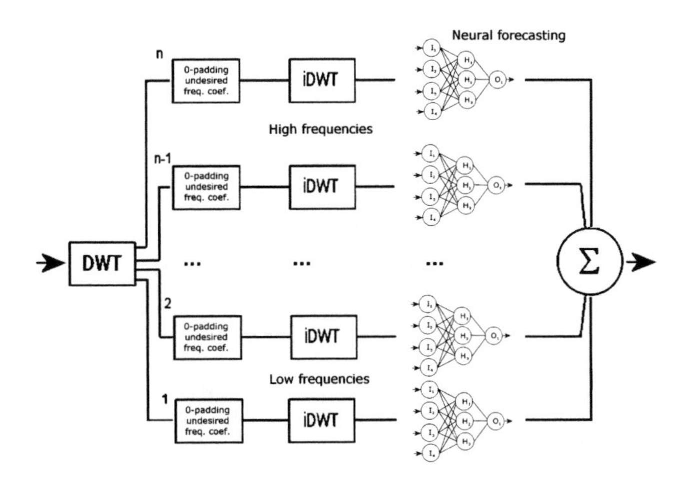

Fig. 12.3. A scheme of the forecasting wavelet-neural system

Each MLP is forecasting a next sample of a time series which comes from decomposing the original data with discrete wavelet transform, then removing undesired frequency coefficients and finally applying the reverse wavelet transform. This approach, based on Mallat's pyramid algorithm, leaves only a range of frequencies from the original data in the new data series. The MLP approximates a next sample using the nonlinear autoregression model.

12.5.3 A small improvement

There exists a simple method of improving the wavelet-neural prediction. It can be easily observed that there is no need to approximate low frequency ranges with ANNs if it is intended to forecast just a few values.

In the application example developed below, a data vector of length 1561 is used. It was intended to predict just the next 5 observations from this time series. The data was zero padded to achieve a length of 2^{11} and split into 11 frequency ranges.

Range	*Oscillations length*
$\frac{f_N}{2}$ to f_N	2-4 samples
$\frac{f_N}{4}$ to $\frac{f_N}{2}$	4-8 samples
...	...
$\frac{f_N}{2^{11}}$ to $\frac{f_N}{2^{10}}$	2048-4096 samples.

Let us denote with $s = [s_1, s_2, \ldots, s_{1561}, 0, \ldots, 0]$ the first data set of length 2^{11} and with $t = [s_1, s_2, \ldots, s_{1561}, s_{1562}, \ldots, s_{1566}, 0, \ldots, 0]$ the other, where $s_{1562}, \ldots, s_{1566}$ are the desired real, not forecasted values. As there is no way for this future 5 entries $s_{1562}, \ldots, s_{1566}$ to generate long oscillations (i.e. 1024–2048, $512-1024, \ldots, 64-128$ samples) they have very little or no effect on low and medium frequency wavelet coefficients.

So having the $s^* = DWT(s)$ only high frequency bands are separated and used as samples for ANNs (as above). The high frequency coefficients in s^* are zero padded and the *IDWT* is applied. The resulting signal $\tilde{s}$ is a rough approximation of s *and* of t. The unknown values $s_{1562}, \ldots, s_{1566}$ are approximated in the following way:

$$s_{1561+j} \approx \sum_{i=k}^{n} \tilde{y}^{(i)}_{1561+j} + \tilde{s}_{1561+j},$$

where in the described application $j = 1, \ldots, 5$, $n = 11$ (the number of frequency ranges). Ranges $k, k+1, \ldots, n$ are approximated by ANNs ($\tilde{y}^{(i)}_{1561+j}$) and $1, \ldots, k-1$ are contained in $\tilde{s}$. $k = 8$ gave the best results (lowest error) for forecasting the next 5 values.

The described improvement helped to remove errors generated by ANNs predicting low frequencies and to reduce time needed to train all networks.

12.5.4 An application

In this section results obtained by Hajto are described (see (10)). The described wavelet-neural method was applied to a USD/PLN average exchange rate. The archival data was downloaded from National Bank's of Poland web site (http://www.nbp.pl) and covered the period 1996.01.02 – 2002.03.08, that is 1561 values. To test the prediction method the following procedure was developed and repeated 5 times:

1. Let $k = 100$.
2. Let $s = [s_1, \ldots, s_{1561-k-5}, 0, \ldots, 0]$ be the vector containing the exchange rates, zero padded to fulfill the Mallat's algorithm assumptions (length: 2^{11}).
3. Five consecutive values: $\tilde{s}_{1561-k-4}, \ldots, \tilde{s}_{1561-k}$ are forecasted using the improved wavelet-neural method on s.
4. Predicted data is saved.
5. If $k > 1$ then $k := k-1$ and go to step 2.
6. End.

In step 3 four MLPs were used to approximate the four highest frequency ranges, since this number of MLP forecasted ranges generated the smallest prediction error.

The filter coefficients came from the Daubechies 4 wavelet family. Optimal ANN architectures were estimated using JavaNNS (a Java interface to SNNS kernel) and its Optimal Brain Surgeon algorithms. The networks had an input layer (6 input neurons), one hidden layer and an output layer (1 neuron). The hidden and output neurons used the logistic activation function. Table 12.1 contains details about architectures and frequency ranges.

Table 12.1. Frequency ranges and ANN architectures

Network	Range	Oscillations length	Hidden neurons
1	$\frac{f_N}{16}$ to $\frac{f_N}{8}$	16–32 samples	1
2	$\frac{f_N}{8}$ to $\frac{f_N}{4}$	8–16 samples	2
3	$\frac{f_N}{4}$ to $\frac{f_N}{2}$	4–8 samples	6
4	$\frac{f_N}{2}$ to f_N	2–4 samples	6

In order to explain the term *oscillation length* let us notice that the wavelet transform is in fact an orthogonal projection of a vector (input time series) onto vector spaces generated by wavelet functions. This process can be regarded as decomposing the input series into new series (new vectors) each containing a frequency range of variations.

The highest possible frequency that can be recognised in a discrete signal sampled with the frequency Fs is Fs/2 (Nyquist Theorem). Fs/2 in a periodic signal means a period of 2 samples length. Due to wavelet transform properties (see e.g. (18), page 109) the frequency ranges in which the signal is split are: Fs/2-Fs/4 (oscillations 2-4 samples long), Fs/4-Fs/8 (oscillations 4-8 samples long) and so on. Please note, that this decomposition is not "sharp", because wavelet functions are not sharp localised in frequency, nor in time. It depends on the properties of the chosen wavelet family.

The MLPs were trained with the back-propagation algorithm with momentum and the ANN patterns were split into learning (L) and testing (T) sets. The testing set contained 80 randomly selected patterns, the learning set 1380 – 1480 (depending on k). A typical learning result during the prediction test procedure (for a particular k) is shown in table 12.2.

Note that the error measures are computed using learning and testing patterns, but not *prediction* errors of the whole, aggregated wavelet-neural model. These are the error measure definitions:

1. Sum of Squares Error

$$SSE = \sum_{i=1}^{N} (y_i - \tilde{y}_i)^2.$$

2. Mean Squared Error

$$MSE = \frac{1}{N} \sum_{i=1}^{N} (y_i - \tilde{y}_i)^2.$$

Table 12.2. A typical learning result during the prediction test procedure (*L denotes learning set, T denotes testing set*)

Net	Set	SSE	MSE	RMSE	NRMSE	R2	DIR
1	L	0,0787	0,00005	0,0073	0,2828	0,920	85,36%
2	L	0,0717	0,00005	0,0069	0,3858	0,851	84,34%
3	L	0,0285	0,00002	0,0044	0,3391	0,884	89,85%
4	L	0,0428	0,00003	0,0054	0,4825	0,767	88,49%
1	T	0,0316	0,00040	0,0198	0,3299	0,891	90,00%
2	T	0,0051	0,00006	0,0080	0,5451	0,702	86,25%
3	T	0,0053	0,00007	0,0081	0,4577	0,790	86,25%
4	T	0,0016	0,00002	0,0044	0,3533	0,875	90,00%

3. Root of MSE
$$RMSE = \sqrt{MSE}.$$
4. Normalised RMSE
$$NRMSE = \frac{RMSE}{\sqrt{\sigma}}.$$
5. R2
$$R2 = 1 - \frac{MSE}{\sigma},$$

where $\sigma = \frac{1}{N}\sum_{i=1}^{N}(y - y_i)^2$, $y = \frac{1}{N}\sum_{i=1}^{N} y_i$. y_i, $\tilde{y}_i$ denote the expected and obtained MLP's output value on i-th pattern, respectively. *DIR* is the percentage of correctly predicted directions of value alteration.

After the prediction testing procedure was 5 times repeated, 2500 of predicted exchange rates were obtained. They were divided into 5 groups containing the 1st, 2nd, 3rd, 4th and 5th forecasted rate. In each of these groups all predicted values were compared to the real data to estimate the prediction error. The following error measures were used:

1. Root Average Square Error
$$RASE = \sqrt{\frac{1}{N}\sum_{i=1}^{N}(s_i - \tilde{s}_i)^2}.$$
2. Mean Absolute Percentage Error
$$MAPE = \frac{1}{N}\sum_{i=1}^{N}\left|\frac{s_i - \tilde{s}_i}{s_i}\right| * 100.$$
3. Theil's information coefficient
$$T^R = \frac{\sqrt{\sum_{i=1}^{R}(s_i - \tilde{s}_i)^2}}{\sqrt{\sum_{i=1}^{R}(s_i - s_{i-1})^2}},$$

where s_i is the real value, $\tilde{s}_i$ its prediction, N the number of predictions of a value. Note that $RASE$ and $MAPE$ are applied to all $N = 500$ values in each of 5 groups, while T^R to results of each forecasting. It means that having forecasted values $\tilde{s}_{k+1}, \ldots, \tilde{s}_{k+5}$ and real data $s_{k+1}, \ldots, s_{k+5}$ five Theil's coefficients are computed:

$$T^R = \frac{\sqrt{\sum_{i=1}^{R} (s_{k+i} - \tilde{s}_{k+i})^2}}{\sqrt{\sum_{i=1}^{R} (s_{k+i} - s_{k+i-1})^2}},$$

where $R = 1, \ldots, 5$. The purpose is to focus on the relationship between a prediction's length and quality. The forecasting method results are presented in table 12.3.

Table 12.3. Prediction test procedure's results (*the predicted value's number equals to R in* T^R, $\sigma(.)$ *denotes the standard deviation.*)

	Predicted value's number				
Error measure	1	2	3	4	5
$RASE$	0,028	0,041	0,048	0,056	0,061
$MAPE$	0,544%	0,789%	0,960%	1,107%	1,227%
$Avg(T^R)$	2,281	1,363	1,492	1,670	1,808
$T^R < 1$	51,4%	33,8%	29,4%	20,8%	16,6%
$Avg(T^R < 1)$	0,322	0,240	0,211	0,155	0,131
$\sigma(T^R < 1)$	0,278	0,298	0,293	0,279	0,276
DIR	55%	58,6%	64%	51,2%	49,2%

Values of $T^R < 1$ are exposed in Table 12.3 because of their importance. $T^R = 0$ means there was no prediction error, $T^R > 1$ means it was worse than the trivial "forecasting with the previous value".

12.6 Concluding Remarks

The test results presented above indicate that the developed prediction algorithm works pretty well while generating values for short time periods. The errors rise as the forecast period is extended, which is intuitive. Simultaneously the amount of $T^R < 1$ falls. The $MAPE$, $RASE$ errors and direction coefficients from the 1st, 2nd and 3rd forecasted exchange rate are very satisfactory. The fact that DIR rises achieving the maximum value at the 3rd rate is rather surprising. This value of 64% may make some practical applications possible. However, $DIR's$ next values: 51,2% and 49,2% indicate that there is no possibility to trust the forecasted 4th and 5th value of the exchange rate direction change forecast.

It is likely that an improvement of prediction could be achieved adding some additional economic data to the learning patterns (like stock market indices, interest rates, inflation rates etc.) on which the USD/PLN exchange rate may depend. To reiterate the most important results:

1. A high (64%) direction coefficient while forecasting the future 3rd exchange rate.
2. A low (0,544%, 0,789%) *MAPE* error while forecasting the future 1st and 2nd rates.
3. A satisfactory (51,4%) amount of good-quality (low T^R and its standard deviation) predictions of the 1st rate.
4. The designed MLPs achieved a high *DIR* coefficient (86,25% – 90.00%) on testing patterns.

Summing up, the applied methods gave satisfactory results comparable with the ones described in literature. For instance in (20), a neuro-fuzzy system was applied for the one-day ahead Australian dollar/US dollar exchange rate prediction. The RMSE belonged to the interval $(0.0023, 0.0083)$ in dependence of the day for which the prediction was done. In (29), the Indian rupia/US dollar weekly exchange rate was predicted using multi-layered ANNs. The RMSE of the applied neural systems varied from 0.0235 to 0.0522 in dependence of the ANN's architecture.

Both our results and those of other authors seem to indicate that artificial neural networks are effective in forecasting market behaviour particularly if they constitute a hybrid system consisting of both neural and non-neural modules. The neuronally aided wavelet approach used in this chapter turned to give quite good results whereas a complex, rule-neural system gave results comparable with the best ones described in the bibliography. Prospects for application of hybrid systems to market prediction seem therefore to be promising. It is out of discussion that, till now, neural networks have been a widely used tool for time-series prediction, including economic ones (see (40)). Probably, artificial intelligence, particularly in combination with mathematical approach, in particular statistical, approximation, fuzzy and logical tools will be widely developed both in theoretical and practical aspects.

References

[1] Aboufadel E, Schlicker S (1999) Discovering Wavelets, Wiley, Chichester
[2] Andreou AS, Pavlides G, Karytinos A, (2000) Nonlinear time-series analysis on the Greek exchange-rate market Int. Jour. Bifurcation and Chaos 10:1729-1758
[3] Battle G (1987) A block spin construction of ondelettes. Part I: Lemarie functions, Commun. Math. Phys. 110:601-615
[4] Bernhard W, Leblang D (2002) Democratic processes, political risk, and foreign exchange markets, American Journal of Political Science 46:316-333
[5] Cheung YW, Chinn MD (2001) Currency traders and exchange rate dynamics: a survey of the US market, Journal of International Money and Finance 20:439-471

[6] Cybenko G (1989) Approximation by Superposition of a Sigmoidal Function, Mathematics of Control, Signals and Systems 2:303-314.
[7] Fernández-Rodriguez F, Sosvilla-Rivero S, Andrada-Félix J (1999) Exchange-rate forecasts with simultaneous nearest-neighbour methods: evidence from the EMS, Int. Jour. of Forecasting 15:383-392
[8] Franklin P (1928) A set of continues orthogonal functions, Math. Ann. 100:522-29
[9] Haar A (1910) Zur Theorie der orthogonalen Funktionensysteme, Math. Ann. 69:331-371
[10] Hajto P (2002) A neural economic time series prediction with the use of the wavelet analysis, Schedae Informaticae 11:115-132
[11] Hecht-Nielsen R (1987) Kolmogorov's Mapping Neural Network Existence Theorem, Proceedings of the International Conference on Neural Networks, Part III, IEEE, New York
[12] Hecht-Nielsen R (1990) Neurocomputing. Addison-Wesley Publ., Reading
[13] Hertz J, Krogh A, Palmer RG (1991) Introduction to the Theory of Neural Computation. Addison-Wesley Publ., Massachusetts
[14] Hornik K, (1991) Approximation Capabilities of Multilayer Feedforward Networks, Neural Networks 4:251-258.
[15] Hornik K, (1993) Some New Results on Neural Network Approximation, Neural Networks 6:1069-1072
[16] Hornik K, Stinchcombe M, White H (1989) Mulitilayer Feedforward Networks Are Universal Approximators, Neural Networks 2:359-366
[17] Huang W, Lai KK, Nakamori Y, Wang S (2004) Forecasting foreign exchange rates with artificial Neural networks: a review, Int. Jour. of Information Technology and Decision Making 3:145-165
[18] Jensen A, Cour-Harbo A (2001) Ripples in Mathematics. The Discrete Wavelet Transform. Springer-Verlag, Berlin Heidelberg
[19] Kurkova V (1992) Kolmogorov's Theorem and Multilayer Neural Networks, Neural Networks 5:501-506
[20] Lee VCS, Wong HT (2007) A multivariate neuro-fuzzy system for foreign currency risk management decission making, Neurocomputing 70:942-951
[21] Lemarie PG (1998) Ondelettes à localisation exponentielle, J. Math. Pures Appl 67:227-236
[22] Lin K (1997) The ABC's of BDS, Journal of Computational Intelligence in Finance 5:23-26
[23] Lula P (1999) Feedforeward neural networks for economic phenomena modelling. Wydawnictwo Akademii Ekonomicznej w Krakowie, Kraków (in Polish)
[24] Mallat S (1989) Multiresolution approximation and wavelet orthonormal bases of $L_2(\mathbb{R})$, Trans. Am. Math. Soc. 315:69-88
[25] McCulloch WS, Pitts W (1943) A logical calculus for the ideas immanent in nervous activity, Bull. of Math. Biophysics 5:115-133

[26] Mizuno T, Kurihara S, Takayasu M, Takayasu H (2003) Analysis of high-resolution foreign exchange data of USD-JPY for 13 years, Physica A 324:296-302
[27] Muniandy SV, Lim SC, Murugan R (2001) Inhomogeneous scalling behaviors in Malaysian foreign currency exchange rates, Physica A 301:407-428
[28] Ohira T, Sazuka N, Marumo K, Shimizu T, Takayasu M, Takayasu H (2002) cit Predictability of currency market exchange, Physica A 308:368-374
[29] Panda C, Narasimhan V (2007) Forecasting exchange rate better with artificial neural network, Jour. Policy Modeling 29:227-236
[30] Pesaran MH, Timmermann A (1992) A simple non-parametric test of predictive performance, Jour. Business and Economic Statistics 10:461-465
[31] Sazuka N, Ohira T, Marumo K, Shimizu T, Takayasu M, Takayasu H (2003) A dynamical stucture of high frequency currency exchange market, Physica A 324:366-371
[32] Schauder MJ (1928) Einige Eigenschaften der Haarschen Orthogonalsysteme, Math. Zeit. 28:317-320
[33] Shin T, Han I (2000) Optimal signal multi-resolution by genetic algorithms to support artificial neural networks for exchange-rate forecasting, Expert Systems with Applications 18:257-269
[34] Strömberg JO (1983) A modified Franklin system and higher order spline systems on $\mathbb{R}^n$ as unconditional bases for Hardy spaces, Proc. Conference in Harmonic Analysis in Honor of A. Zygmund, vol. II, Wadsworth, Belmont 475-493
[35] Steiner M, Wittkemper HG (1995) Neural Networks as an Alternative Stock Market Model. In: Refenes APN (ed) Neural Networks in the Capital Markets. Wiley, Chichester
[36] Tsibouris G, Zeidenberg M (1995) Testing the Efficient Markets Hypothesis with Gradient Descent Algorithm. In: Refenes APN (ed) Neural Networks in the Capital Markets. Wiley, Chichester
[37] Walker JS (1997) Fourier analysis and wavelet analysis, Notices of the AMS 44:658-670
[38] White H (1988) Economic Prediction Using Neural Networks: The Case of IBM Daily Stock Returns, Proceedings of the IEEE International Conference of Neural Networks, San Diego
[39] Wojtaszczyk P (1987) A Mathematical Introduction to Wavelets. Cambridge University Press, Cambridge
[40] Zhang G, Patuwo BE, Hu MY (1998) Forecasting with artificial neural networks: the state of the art, Int. Jour. of Forecasting 14:35-62
[41] Zirilli JS (1966) Financial Prediction Using Neural Networks. International Thomson Computer Press, London

Part III

Agent-based Modelling

13

Evolutionary Learning of the Optimal Pricing Strategy in an Artificial Payment Card Market*

Biliana Alexandrova-Kabadjova[1], Edward Tsang[2], and Andreas Krause[3†]

[1] Centre for Computational Finance and Economic Agents (CCFEA), University of Essex, UK. balexa@essex.ac.uk
[2] Department of Computer Science, University of Essex, UK. edward@essex.ac.uk
[3] School of Management, University of Bath, UK. mnsak@bath.ac.uk

Summary. This chapter introduces an artificial payment card market in which we model the interactions between consumers, merchants and competing card issuers with the aim of determining the optimal pricing structure for card issuers. We allow card issuers to charge consumers and merchants fixed fees, provide net benefits from card usage and engage in marketing activities. The demand by consumers and merchants is only affected by the size of the fixed fees and the optimal pricing structure consists of a sizeable fixed fee to consumers, no fixed fee to merchants, negative net benefits to consumers and merchants as well as a high marketing effort.

13.1 Introduction

Payment cards - more commonly referred to as credit and debit cards - are of ever increasing importance for making payments. In 2002 (7) report that 1.8 billion cards were used to buy products and services worth more than US$ 2.7 trillion with high growth rates since then. Despite the importance of payment cards the competition between the different card issuers, most prominently Mastercard, Visa, American Express, Discovery, JCB and Diners Club, is not well understood. In this paper we provide a model of this competition by using an agent-based approach allowing us to introduce complex interactions between the various market participants which is not easily possible using other modeling approaches. In our model we are able to derive the main driving factors of the demand for payment cards and the profits made by card issuers, as well as the optimal pricing strategy.

What distinguishes the market for payment cards from most other markets is that it is a two-sided market, i.e. both partners in the transaction, consumers and merchants, using a payment card need a subscription to a specific payment card. Modeling such markets is challenging as the behavior of market participants is determined

* We acknowledge the financial support of the Consejo Nacional de Ciencia y Tecnologa (CONACYT).

† Corresponding author

B. Alexandrova-Kabadjova et al.: *Evolutionary Learning of the Optimal Pricing Strategy in an Artificial Payment Card Market*, Studies in Computational Intelligence (SCI) **100**, 233–251 (2008)
www.springerlink.com

by a set of complex interactions between consumers and merchants as well as within the group of consumers and the group of merchants. Consumers and merchants face network externalities as a larger number of merchants and consumers using a certain card makes a subscription to it more valuable and card issuers will also affect behavior by changing subscription fees and benefits associated with the cards. In order to capture these numerous interactions we have developed a novel approach to the payment card market using an agent-based model.

Agent-based models study dynamic systems of interacting agents, where agents can be any participants in a system. In economic systems such agents might be consumers, merchants or investors. A good introduction to agent-based modeling in general is given in (15). A main characteristic of agent-based models is that each agent can be given his own behavioral rule, they are generally interacting with a small fraction of available agents and for that reason exhibit significant heterogeneity. Agent-based models in economics and finance have become more popular in recent years, in particular as they have been able to provide insights into the complex dynamics of economic systems, financial markets in particular. These advances, as summarized in (11), have given rise to the insight that the interactions between agents are of central importance for the emergence of realistic properties. As traditional economic and financial models do not consider such interactions between (mostly homogenous) agents, they fail to derive such properties. Similarly such models have brought important insights into macroeconomics, the spatial development of economies as well as the structure of organizations, among many others, see (16) for an overview of the current literature. The interactions between agents and heterogeneity of their behavioral rules makes it in most cases impossible to obtain an analytical solution and therefore requires the use of computer experiments to analyze their properties.

Most models of the payment card market only give cursory considerations to these complex interactions and how they affect competition; the literature focuses on a peculiarity of the payment card market, the so called interchange fee (7, 9, 12–14, 17, 18). This fee arises as follows: card issuers do not directly issue payment cards to customers but rather allow banks to distribute them in their own name; card issuers only provide a service in form of administering the payments made using these cards. Similarly, merchants have a contract with a bank that allows them to accept payments made using a specific payment card. In the majority of cases the consumer will have been given his card from one bank with the merchant having a contract with another bank. In this case the bank of the merchant will have to pay the bank of the consumer a fee for making the payment, which is called the interchange fee. Not only does much of the academic literature focus on the interchange fee, it is also the focus of regulators (4–6, 8).

With the focus on the interchange fee the literature makes a number of very simplifying assumptions on the behavior of consumers and merchants. In contrast, we explicitly model the behavior of consumers and merchants and concentrate on the competition between payment cards to attract subscribers and transactions. We abstract from the interchange fee by implicitly assuming that payment cards are directly issued by card issuers, i.e. neglecting the role of banks in the market. This approach allows us to analyze all the fees paid by consumers and merchants using payments cards rather than only the interchange fee. This will enable us to gain an understanding of the competitive forces in the payment card market and how the competition between different payment cards affects consumers, merchants and the payment card issuers themselves. So far no other study has investigated this issue adequately.

The remainder of this chapter is organized as follows: the coming section introduces the artificial payment card market with its elements and interactions, section 13.3 then briefly introduces the learning algorithm used to optimize the card issuers' strategies and discusses the parameter constellation used in the computer experiments. The results of the computer experiments are presented in section 13.4, where we evaluate the demand and profits functions as well as the optimal pricing structure by card issuers. Finally section 13.5 concludes.

13.2 The Artificial Market

In this section we introduce our model of an artificial payment card market by describing in detail the market participants - consumers, merchants and card issuers - and how they arrive at their decisions through interactions with each other.

13.2.1 Model Elements

In this subsection we formally introduce the three key elements of the model - merchants, consumers and payment cards - with their attributes.

Merchants

Suppose we have a set of merchants $\mathscr{M}$ with $|\mathscr{M}| = N_{\mathscr{M}}$, who are offering a homogeneous good at a common price and face marginal cost of production lower than this price. With the elimination of price competition among merchants, we can concentrate on the competition among payment card providers and how the card choice affects merchants. The merchants are located at random intersections of a $N \times N$ lattice, where $N^2 \gg N_{\mathscr{M}}$, see fig. 13.1 and for each computer experiment we use different random locations of merchants. Let the top and bottom edges as well as the right and left edges of this lattice be connected into a torus.

Consumers

Consumers occupy all the remaining intersections of the above lattice. The set of consumers is denoted $\mathscr{C}$ with $|\mathscr{C}| = N_{\mathscr{C}}$, where $N_{\mathscr{C}} \gg N_{\mathscr{M}}$ and $N^2 = N_{\mathscr{C}} + N_{\mathscr{M}}$. Each consumer has a budget constraint that allows him in each time period to buy exactly one unit of the good offered by the merchants in a single interaction with one merchant. By making this transaction the utility of the consumer increases. In order to obtain the good any consumer $c \in \mathscr{C}$ has to travel to a merchant $m \in \mathscr{M}$. The distance imposes travel costs on consumers, which reduces the attractiveness

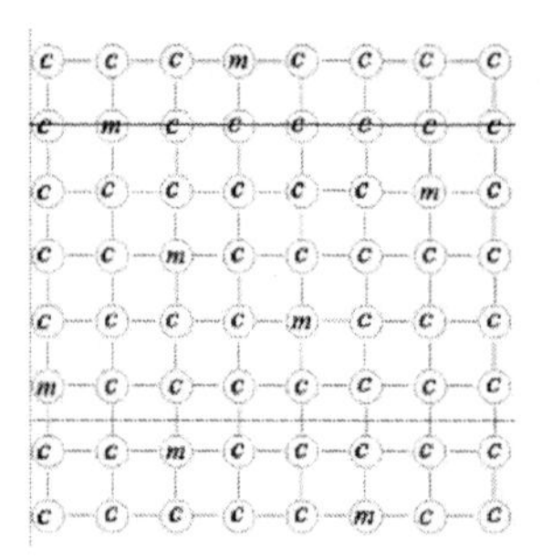

Fig. 13.1. Sample of a lattice with consumers (c) and merchants (m)

of visiting a merchant. We have explored the case where the connections among consumers and merchants are local and the distance traveled by a consumer c to a merchant m, is measured by the "Manhattan distance" $d_{c,m}$ between the intersections on the lattice. The distance between two neighboring nodes has been normalized to one. We further restrict the consumer to visit only the nearest m_c merchants and denote by $\mathscr{M}_c$ the set of merchants a consumer considers going to.

Payment Cards

We consider a set of payment methods $\mathscr{P}$ with $|\mathscr{P}| = N_{\mathscr{P}} + 1$ and $N_{\mathscr{P}} \ll N_{\mathscr{M}}$. The first payment method is the benchmark and can be interpreted as a cash payment, whereas all other payment forms are card payments. Cash is available to all consumers and accepted by all merchants. For a card payment to occur, the consumer as well as the merchant must have a subscription to the card in question. We assume that card payments, where possible, are preferred to cash payments by both, consumers and merchants. In each time period a fixed subscription fee of $F_p \geq 0$ is charged to the consumer, and $\Gamma_p \geq 0$ to the merchant. Cash payments do not attract any fees.

For each unit of goods sold using a payment card $p \in \mathscr{P}$, a merchant $m \in \mathscr{M}$ receives net benefits of $\beta_p \in \mathbb{R}$. Such benefits may include reduced costs from cash handling, and could differ across payment cards and are assumed to be identical for all merchants for any given card. Note that the benefits β_p could have a negative value. This means that the variable fee paid by the merchant to the card issuer is bigger than the benefits he received from the same payment card in which case they can be interpreted as a transaction fee. Cash payments do not provide any net benefits.

Consumers also receive net benefits from paying by card, $b_p \in \mathbb{R}$, but no net benefits from cash payments. Here, the benefits may arise from delayed payment, insurance cover or cash-back options. As with the benefits to merchants, the benefits to consumers can also be negative and again represent a transaction fee.

In addition, the issuer of the payment card has to decide how much he should spend on marketing effort $l_p \geq 0$, in order to increase the awareness by the consumers and the merchants for the payment card that he is providing.The strategy employed by a payment card issuer is defined as the set of variables controlled by them: $\mathbb{S} =$

$\{F_p, \Gamma_p, \beta_p, b_p, l_p\}$. It is this set of variables that we will be optimizing for payment cards in section 13.4.

13.2.2 Decision-making of market participants

Decisions by market participants are arrived at through interactions with each other. This section sets out how these interactions drive decisions by consumers and merchants. The decisions on the strategies chosen by card issuers are considered in section 13.3.

Decisions by consumers

Consumers face three important decisions: which merchant to choose, which payment card to use in the transaction with the merchant, and to which payment cards to subscribe to. This section addresses each of these decisions in turn.

The consumers' choice of a merchant

We assume that when deciding which merchant to visit, the consumer has not yet decided which of the cards he holds will be used. Suppose $\mathscr{P}_{c,m}$ is the set of cards consumer $c \in \mathscr{C}$ and merchant $m \in \mathscr{M}$ have in common and let $|\mathscr{P}_{c,m}| = N_{\mathscr{P}_{c,m}}$. The more payment cards the merchant and the consumer have in common, the more attractive a merchant becomes, as the consumer always carries all his cards with him. Additionally the smaller the distance $d_{c,m}$ between the consumer and the merchant, the more attractive this merchant will be to the consumer. From these deliberations we propose to use a preference function for the consumer to visit the merchant as follows

$$v_{c,m} = \frac{\frac{N_{\mathscr{P}_{c,m}}}{d_{c,m}}}{\sum_{m' \in \mathscr{M}_c} \frac{N_{\mathscr{P}_{c,m'}}}{d_{c,m'}}}. \tag{13.1}$$

Each consumer $c \in \mathscr{C}$ chooses a merchant $m \in \mathscr{M}$ with probability $v_{c,m}$ as defined in equation (13.1). The consumers will continuously update their beliefs on the number of common payments they share with a particular merchant, by observing the number of common payments of all shops they can visit - i.e. not only those actually visited - as subscriptions change over time in the way introduced below.

The consumers' choice of a payment card

The consumer decides which payment card he wants to use with the merchant he has selected. We assume a preferred card choice in which he chooses the card with the highest benefits b_p from the set $\mathscr{P}_{c,m}$; if there are multiple cards with the highest net benefits the card is chosen randomly from them. In cases where the merchant does not accept any of the consumers' cards, the transaction is settled using cash payment.[1]

[1] Please note that even for a negative b_p consumers prefer to use payment cards. Without changing the argument we also could associate a large transaction fee with cash payments to justify our previous assumption that card payments are preferred.

Consumer subscriptions

Initially consumers are allocated payment cards such that each consumer is given a random number of randomly assigned payment cards. Periodically consumers have to decide whether to cancel a subscription to a card they hold and whether to subscribe to new cards. The frequency with which consumers take these decisions is defined by a Poisson distribution with a mean of λ time periods between decisions. The reasoning behind this assumption is that consumers (and similarly merchants as outlined below) will not constantly consider their subscriptions but do so only at certain times, which we here assume to be random.

In order to make decisions on subscriptions, every consumer $c \in \mathscr{C}$ keeps track of whether the cards he owns, $\mathscr{P}_c$, are accepted by a merchant or not. If a card $p \in \mathscr{P}_c$ is accepted by the merchant $m \in \mathscr{M}_c$ he is visiting, the consumer increases the score of the card $\omega_{c,p}^-$ by one.[2] Assume that he cancels his subscription to a card with probability[3]

$$\pi_{c,p}^- = \frac{x_c^- k}{x_c^- k + e^{\frac{\omega_{c,p}^-}{\omega_c}}}, \tag{13.2}$$

where ω_c denotes the number of merchants visited and $x_c^- k$ accounts for the propensity of the consumer to cancel his subscription of the payment card. We define $k = 1 + F_p + N_{\mathscr{P}_c} + \frac{\varepsilon}{\kappa + b_p}$, ε and x_c^- are constants and κ is another constant with the restriction that $\kappa + b_p > 0$. A larger value for $x_c^- k$ implies that for a given number of merchants accepting the card, the consumer is more likely to cancel his subscription. As long as $x_c^- k < 1$ we can interpret the influence of this term as the inertia to cancel a subscription. The parameter constellation used below ensures that with optimized strategies we find $x_c^- k < 1$ and obtain the realistic case of inertia in consumers with respect to changing their *status quo*.

The decision to cancel a subscription is also affected by the fees and benefits associated with a payment card. A card becomes more attractive to subscribe and existing subscriptions are less likely to be canceled if the fixed fee charged is low and the net benefits from each transaction are high. Furthermore, the more cards a consumer holds, the less attractive it becomes to maintain a subscription as the consumer has many alternative payment cards to use with merchants.

Let $\mathscr{P}_c^- = \mathscr{P} \setminus \mathscr{P}_c$ denote the set of cards the consumer does not subscribe to, with $|\mathscr{P}_c^-| = N_{\mathscr{P}_c^-}$. If the merchant and the consumers have no payment card in common, i.e. $\mathscr{P}_{c,m} = \emptyset$, and the merchant accepts at least one payment card, i.e. $\mathscr{P}_m \neq \emptyset$, the consumer increases the score $\omega_{c,p}^+$ by one for all $p \in \mathscr{P}_m \cap \mathscr{P}_c^-$. With x_c^+ a constant, the probability of subscribing to a card not currently held by the consumer is then determined by

[2] Please note that here consumers only take into account the merchant he actually visits. This is in contrast to the decision which merchant he visits where he is aware of the number of common cards for potential merchants.

[3] The probabilities defined in equations (13.2) and (13.3) are also affected by the marketing effort of each payment card provider.

$$\pi^+_{c,p} = \frac{e^{\frac{\omega^+_{c,p}}{\omega_c}}}{x^+_c k + e^{\frac{\omega^+_{c,p}}{\omega_c}}}. \tag{13.3}$$

This probability uses the inertia of consumers to subscribe to new cards through the use of $x^+_c k$. A large value of this term implies that consumers are less likely to subscribe to new cards for a given number of merchants accepting the payment card.

Decisions by merchants

The decisions of merchants are limited to the choice of card subscriptions. Similar to consumers the frequency with which merchants review their subscriptions is governed by a Poisson distribution specific to each individual with a common mean of λ time periods, the same as for the subscription decisions of consumers. As with consumers the initial subscriptions of merchants are a random number of randomly selected payment cards.

Merchants keep track of all cards presented to them by consumers. Every time a card $p \in \mathscr{P}$ is presented to the merchant $m \in \mathscr{M}$ and he has a subscription to this card, i.e. $p \in \mathscr{P}_m$, he increases the score of $\theta^-_{m,p}$ by one. With $|\mathscr{P}_m| = N_{\mathscr{P}_m}$ the probability of canceling this subscription[4] is given by

$$\pi^-_{m,p} = \frac{x^-_m q}{x^-_m q + e^{\frac{\theta^-_{m,p}}{\theta_m}}}, \tag{13.4}$$

where θ_m denotes the number of cards presented and $x^-_m q$ represents the propensity to cancel the subscription similar to that of consumers with x^-_m being a constant and $q = 1 + \Gamma_p + N_{\mathscr{P}_m} + \frac{\varepsilon}{\kappa+\beta_p}$. κ takes the same value as for consumers and has to fulfill the additional restriction that $\kappa + \beta_p > 0$. The interpretation of the term $x^-_m q$ follows the same lines as for consumers and the parameter setting ensures inertia by merchants to cancel their subscriptions with the optimized payment card strategies.

Similarly, if the merchant does not have a subscription to the card, i.e $p \in \mathscr{P}^-_m$, the score of $\theta^+_{m,p}$ is increased by one and the probability of subscribing to a card is given by

$$\pi^+_{m,p} = \frac{e^{\frac{\theta^+_{m,p}}{\theta_m}}}{x^+_m q + e^{\frac{\theta^+_{m,p}}{\theta_m}}}, \tag{13.5}$$

where once again x^+_m is a constant.

Decisions by card issuers

Card issuers have to decide on all variables in their strategy space $\mathbb{S}$, i.e. decide on the fees and net benefits of consumers and merchants as well as the marketing expenses.

[4] The probabilities defined in equations (13.4) and (13.5) are also affected by the marketing effort of each payment card provider.

While optimizing these variables will be the main subject of the following sections, we want to establish the impact these variables have on the profits of card issuers as well as the impact of the marketing effort on the decisions of consumers and merchants. The total profit Φ_p of a card issuer is calculated applying the following equation

$$\Phi_p = \Phi_{\mathscr{C}_p} + \Phi_{\mathscr{M}_p} - \mathscr{L}_p, \tag{13.6}$$

where $\Phi_{\mathscr{C}_p}$ are the profits received from consumers and $\Phi_{\mathscr{M}_p}$ those from merchants. These profits are given by

$$\Phi_{\mathscr{C}_p} = \sum_{t=1}^{I} N_{t,\mathscr{C}_p} F_p - \sum_{t=1}^{I} N_{t,T_p} b_p, \tag{13.7}$$

$$\Phi_{\mathscr{M}_p} = \sum_{t=1}^{I} N_{t,\mathscr{M}_p} \Gamma_p - \sum_{t=1}^{I} N_{t,T_p} \beta_p, \tag{13.8}$$

where the additional index t denotes the time period, I the number of time periods considered by the card issuer and N_{T_p} the number of transactions using card p. The fees and net benefits set by the card issuers will affect the number of subscriptions and transactions using a card, which then determine the profits for the card issuers. Thus we have established a feedback link between the behavior of card issuers on the one hand and consumers and merchants on the other hand. The sum of all publicity cost is denoted $\mathscr{L}_p$ and is calculated as

$$\mathscr{L}_p = \sum_{t=1}^{I} l_p = I l_p, \tag{13.9}$$

where l_p denotes the publicity costs for each time period, which we assume to be constant. These publicity costs now affect the probabilities with which consumers and merchants maintain their subscriptions and subscribe to new cards. The probabilities, as defined in equations (13.2) - (13.5), are adjusted due to these publicity costs as follows

$$\xi = \tau\pi(1-\pi), \tag{13.10}$$

where π represents , $\pi^+_{c,p}$, $\pi^-_{c,p}$, $\pi^+_{m,p}$, or $\pi^-_{m,p}$, as appropriate and $\tau = \alpha\left(\varphi - e^{-l_p}\right)$. The constants α and φ satisfy the constraint $0 \leq \pi + \xi \leq 1$. The revised probabilities as used by consumers and merchants are then given by $\pi' = \pi + \xi$.
Fig. 13.2 summarizes the structure of our model by showing the dependencies of the model elements. Card issuers now seek to maximize their profits and market share as measured through the number of transactions conducted by optimally choosing their strategies. The way this optimization is accomplished by card issuers is discussed in the coming section.

13.3 Set-up of the Computer Experiments

The above model is implemented computationally and the optimization of the strategies chosen by card issuers conducted using machine learning techniques.

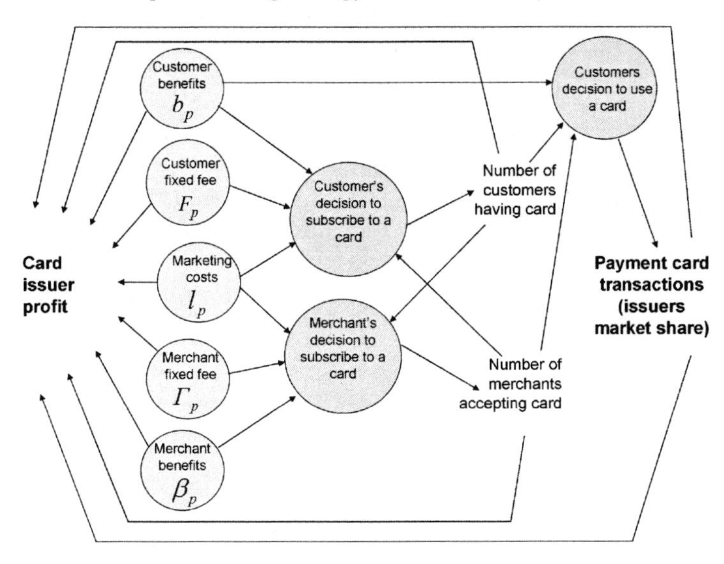

Fig. 13.2. Structure of the model

13.3.1 The optimization procedure of card issuers

For our optimization procedure we consider an Estimation of Distribution Algorithm (EDA) which analyzes the population of agents with the aim of identifying the best-performing parameter constellation. A distribution of the parameter constellations in the population is built with the next generation being drawn from a revised distribution where the likelihood of selecting a well-performing parameter constellation is increased and that of selecting a poorly performing parameter constellation reduced. Different versions of how to build and modify this distribution have been proposed, where our model will be based on the Population-based Incremental Learning algorithm (PBIL) developed in (3).

The PBIL algorithm in its original form assumes the parameters to be binary and attempts to find the optimal binary combination. By changing the probabilities for each parameter for the coming generation according to the relative performance of the agent, the algorithm slowly identifies the best performing parameter constellation. The randomness of parameter choices ensures that all possible combinations are eventually searched to obtain the global optimum. Various extensions of the PBIL have been introduced, most notably for our purpose the introduction of continuous rather than binary parameters.

In our model we use a combination of the PBIL with the Multiple Algorithms Parameter Adaptation algorithm (MAPA), introduced in (10) as the Generalized Population-based Incremental Learning algorithm (GPBIL). In a MAPA algorithm two different criteria to assess the performance are combined. This algorithm divides the domain of a variable x, $[a;b]$, into n sub-domains $a \leq a_1 < a_2 < \cdots <$

$a_{n-1} < a_n \leq b$. We can now define subintervals as $\left[a; \frac{a_1+a_2}{2}\right), \left[\frac{a_1+a_2}{2}; \frac{a_2+a_3}{2}\right), \ldots,$ $\left[\frac{a_{i-1}+a_i}{2}; \frac{a_i+a_{i+1}}{2}\right), \ldots, \left[\frac{a_{n-1}+a_n}{2}; b\right]$.

Each subinterval is equally likely to be selected, i.e. with probability $\frac{1}{n}$. The algorithm changes the location of the parameters a_i such that the subintervals with the best performance are selected with a higher likelihood. This learning is achieved through a positive and a negative feedback mechanism. Suppose we have a value of $x \in [a;b]$; we can then determine the new value of a_i with the help of a_j, the value closest to x. If the outcome associated with x is positive we then determine the updated $\widehat{a}_i$ as follows

$$\widehat{a}_i = a_i + \zeta \nu_x h_\delta(i,j)(x - a_i), \tag{13.11}$$

where ζ denotes the learning rate, the role of ν_x is explained below and

$$h_\delta(i,j) = \begin{cases} 1 \text{ if } |i-j| \leq \delta \\ 0 \text{ if } |i-j| > \delta \end{cases} \tag{13.12}$$

denotes the neighborhood in which learning occurs, where δ denotes cylinder size of the kernel. This ensures that values close to x get chosen more frequently. In the case of a negative outcome we want values on either side of x to be chosen less frequently and therefore use the following rule on updating the values of a_i

$$\widehat{a}_i = \begin{cases} a_i + \zeta \nu_x h_{\delta'}(i,j)(a_{i-\delta'} - a_i) \text{ if } a_i \leq x \\ a_i + \zeta \nu_x h_{\delta'}(i,j)(a_{i+\delta'} - a_i) \text{ if } a_i > x \end{cases}. \tag{13.13}$$

If $a_{i-\delta'}$ or $a_{i+\delta'}$ are not defined we set them as a and b, respectively. In our model a positive outcome is achieved if the market share of the payment card as determined by the number of transactions using the payment card is higher than the average market share, i.e. $\frac{1}{N_{\mathscr{P}}}$; otherwise it is regarded as a negative outcome.

Once it has been determined whether an outcome is positive or negative from its market share, the positive and negative outcomes are ordered ascending according to the profits achieved from the strategy. The position of a strategy x determines its weight in the updating of the values through ν_x. If we denote by ϕ the number of positive or negative outcomes, respectively, and $1 \leq \rho(x) \leq \phi$ the position, we define $\nu_x = \frac{\rho(x)}{\phi}$. The domain of the strategy variables as well as the parameters of the learning algorithm are shown in table 13.1. We employ the GPBIL algorithm to optimize the strategy vector $\mathbb{S}$.

13.3.2 Parameter investigation

The model is characterized by a large number of free parameters which need to be exogenously fixed in the experiments. Table 13.2 provides an overview of the values chosen for further analysis. An analysis of a wide range of parameter constellations has shown the results to be not very sensitive to these values and we can thus treat them as qualitatively representative examples for the remainder.

It might be noted that the inertia resulting from net benefits, ε, is relatively small compared to the fixed fee. We can justify this choice by pointing out that consumers

Table 13.1. Domains of the strategy variables

Description	Symbol	Value range
Consumer fixed fee	F_p	$[0;10]$
Merchant fixed fee	Γ_p	$[0;10]$
Net benefits of consumers	b_p	$[-1;1]$
Net benefits of merchants	β_p	$[-1;1]$
Publicity costs	l_p	$[0;20]$
Number of subintervals	n	5
Learning rate	ζ	0.1
Kernel size for positive outcomes	δ	2
Kernel size for negative outcomes	δ'	1

Table 13.2. Parameter settings

Description	Symbol	Value
Network size	N	35
Number of consumers	$N_{\mathscr{C}}$	1100
Number of merchants	$N_{\mathscr{M}}$	125
Number of payment cards	$N_{\mathscr{P}}$	9
Number of merchants considered by each consumer	$N_{\mathscr{M}_{\mathscr{C}}}$	5
Inertia/propensity with respect to net benefits	ε	1
Inertia/propensity with respect to net benefits	κ	1.1
Propensity of consumers to cancel their subscriptions	x_c^-	0.05
Inertia with respect to consumers making new subscriptions	x_c^+	2
Propensity of consumers to cancel their subscriptions	x_m^-	0.05
Inertia with respect to merchants making new subscriptions	x_m^+	9
Size of the probability adjustment due to marketing effort	α	0.1
Size of the probability adjustment due to marketing effort	φ	0.05
Expected time between subscription decisions	λ	20
Number of time steps	I	20000

and merchants will in many cases not be aware of the size of these benefits because they are not commonly recognized, e.g. small charges for overseas usage is hidden in a less favorable exchange rate. Empirical evidence suggests that such hidden charges and benefits are much less relevant than fees directly charged to customers. It is also for this reason that we limit the domain of the net benefits to $[-1;1]$ such that we avoid them becoming too visible to consumers and merchants relative to the fixed fee. In doing so we willingly accept a possible corner solution in the optimal pricing strategy.

13.4 Outcomes of the Computer Experiments

Using the model of the payment card market as developed in the previous sections, we can now continue to analyze the resulting properties of the market. Before evaluating the optimal strategies chosen by payment card issuers, we will assess the resulting demand function for the payment cards by consumers and merchants as well as the profit function of card issuers, which we then can use to interpret the optimal pricing strategies.

13.4.1 Demand for payment cards

We evaluate the demand for payment cards by assigning each card a random strategy as detailed in table 13.3. Using these fixed strategies we conduct a single computer experiment from which we estimate the demand function at the end of the experiment; it has to be noted that the results from this single experiment is representative and was confirmed for other random strategies.[5] Estimates of the demand for payment cards held by consumers $N_{\mathscr{C}_p}$, merchants $N_{\mathscr{M}_p}$ and the number of transactions N_{T_p} as well as the profits made by the card issuers, Φ_p, are given as follows:

$$\begin{aligned}
\ln N_{\mathscr{C}_p} &= 6.433 - 0.156 F_p, \\
\ln N_{\mathscr{M}_p} &= 4.339 - 0.088 F_p - 0.0222 \Gamma_p, \\
\ln N_{T_p} &= 10.837 - 0.208 F_p - 0.244 \Gamma_p, \\
\ln \Phi_p &= 16.769 + 0.054 F_p - 0.091 \beta_p.
\end{aligned}$$

We only show those strategy variables which were found to have a significant impact on the demand or profits. The equations presented above provide a nearly perfect fit of the data and the coefficients are highly significant. It is interesting to note that the demand is not affected by the net benefits consumers and merchants receive from each transaction; instead the demand is entirely driven by the fixed fees. We also observe a feature of two-sided markets as the demand by merchants depends on both the consumer and merchant fixed fee, where the consumer fixed fee is much more relevant than the merchant fixed fee. The reason for this outcome can be found in the importance of consumer demand and usage for the subscription of merchants. For the transaction demand we observe that both fees are of similar importance.

Interestingly, the profits made by card issuers only depend on the consumers fixed fee and the net benefits given to merchants; the increased revenue of a potential fixed fee to merchants is offset by a reduced usage resulting in its insignificance for the outcome. It has also to be noted that while these outcomes are statistically significant, their economic impact is relatively small, e.g. by increasing the fixed fee for consumers from zero to 10 (the maximum value), the profits would only increase by about 3% and an increase of the net benefits to the merchant from -1 to 1 would decrease the profits only by about 1%. Thus the sensitivity of the profits to these strategies is very low. The demand and thereby the market share of a payment card itself reacts more sensitively with changes of up to 20%.

A final observation is that despite 9 cards being present in the market, the fraction of cash transactions remains high at about 35%, implying frequent mismatches

[5] It has also been confirmed that the demand for payment cards had stabilized a considerable time before the end of the experiment.

Table 13.3. Random payment card strategies

Card number	Consumer fixed fee	Merchant fixed fee	Consumer net benefits	Merchant net benefits	Marketing costs
1	3.16	3.95	-0.66	-0.88	0.69
2	6.91	7.36	-0.44	-0.41	8.67
3	6.85	7.61	-0.03	0.23	5.85
4	2.87	7.35	-0.13	0.50	14.98
5	8.00	5.63	0.03	-0.34	1.39
6	2.09	4.22	-0.79	-0.62	13.32
7	2.16	2.59	-0.27	0.15	8.99
8	7.02	6.45	0.47	-0.51	12.04
9	2.42	1.77	0.55	0.41	12.72

between the cards subscribed to by consumers and merchants. The reasons for this frequent coordination failure is discussed in (1, 2). Having investigated the demand function for payment cards we can now proceed to evaluate the optimal payment card strategies.

13.4.2 Optimal payment card strategies

With the objective function being the market share of the payment cards and the profits made by card issuers, we can now employ the GPBIL algorithm to optimize the pricing structure of the card issuers. The results of this optimization are discussed in this section.

The results of the optimization using the GPBIL algorithm are presented in table 13.4. From the profit function derived in the previous section we would expect the optimal strategy to consist of a high fixed fee for consumers and negative net benefits, i.e. a transaction fee, for merchants. The negative effect of the fixed fee on the transaction demand would, however, imply that this fee should be limited in size and the fixed fee for merchants should be low. The results confirm these assertions fully.

The negative net benefits to consumers and merchants would make the payment cards less attractive to prospective subscribers and make existing subscribers more likely to cancel their subscription while only having a limited influence on the profits of the issuer. This negative effect is, however, offset by the relatively high marketing effort the issuers make; essentially the revenue generated by the negative net benefits is used for marketing purposes. Hence the negative impact on the payment card switching behavior by applying negative net benefits is offset by marketing activities. We also observe a weak positive relationship between the size of the fixed fee to consumers and the marketing costs, providing further evidence for an offsetting relationship between these costs charged to consumers and marketing efforts.

The high marketing costs by card issuers provide a good example how market participants can get locked into certain strategies by competitive pressures, although they are not beneficial to them and even detrimental to other market participants. Once a card issuer decides to increase his marketing effort, his competitors will have to follow to avoid losing market share. To offset the incurred costs those fees to which market participants react least sensitively, are likely to be increasing, which in our case are the net benefits to consumers and merchants.

In fig. 13.3 we show from a representative sample experiment how the parameters of the nine payment cards evolve over the generations until their values converge as is clearly visible. We clearly see the convergence of the parameters over time. We have also compared the performance of the optimized strategies in a market populated with otherwise random strategies and find that the optimized strategies achieve a significantly higher market share and also outperform the random strategies in term of profits generated. This result provides evidence that the optimization of the strategies has indeed produced strategies that are performing better than randomly generated strategies. Finally, we found that the market share of all competing payment cards are approximately equal, providing evidence for the effectiveness of the learning

Table 13.4. Optimized payment card strategies in 10 experiments. The results denote the converged strategies of all payment cards during the last 100 time steps.

Experiment	Consumer fixed fee	Merchant fixed fee	Consumer net benefits	Merchant net benefits	Marketing costs	Total profits
1	7.57	0.00	-1.00	-1.00	11.11	6,048,995.23
2	5.33	0.00	-1.00	-1.00	7.66	5,275,214.86
3	3.51	0.00	1.00	-1.00	11.81	3,204,527.52
4	6.03	0.00	0.48	-1.00	11.82	4,356,514.63
5	5.46	0.00	-1.00	-1.00	10.49	5,333,885.81
6	6.03	0.00	-1.00	-1.00	13.85	5,562,761.79
7	5.98	0.00	-1.00	-1.00	8.39	5,551,276.47
8	6.48	0.00	-1.00	-1.00	9.97	5,738,453.78
9	5.38	0.00	-1.00	-1.00	10.24	5,299,438.88
10	5.66	0.00	-1.00	-1.00	10.82	5,423,793.36
Mean	5.75	0.00	-0.65	-1.00	10.62	5,179,486.23
Median	5.85	0.00	-1.00	-1.00	10.66	5,378,839.59

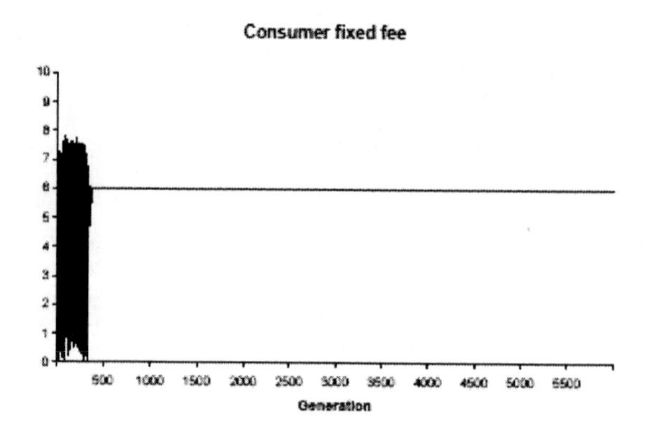

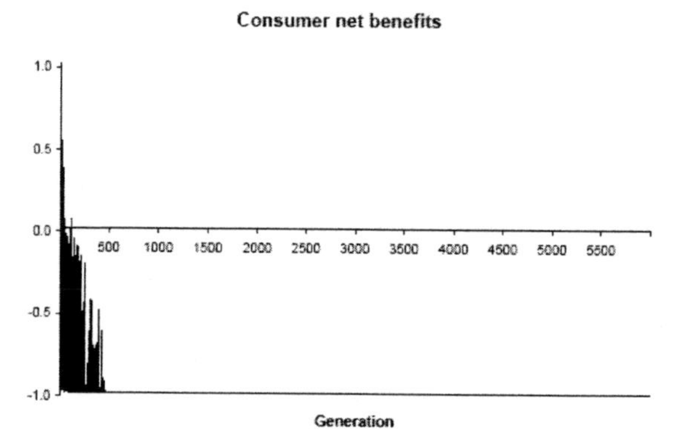

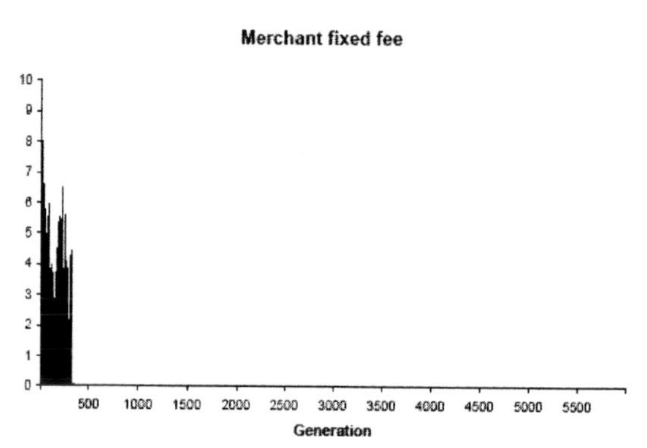

Fig. 13.3. Evolution of strategy parameters in one experiment

algorithm and the convergence of the learning as well as the reaching of the steady state of the computer experiment.

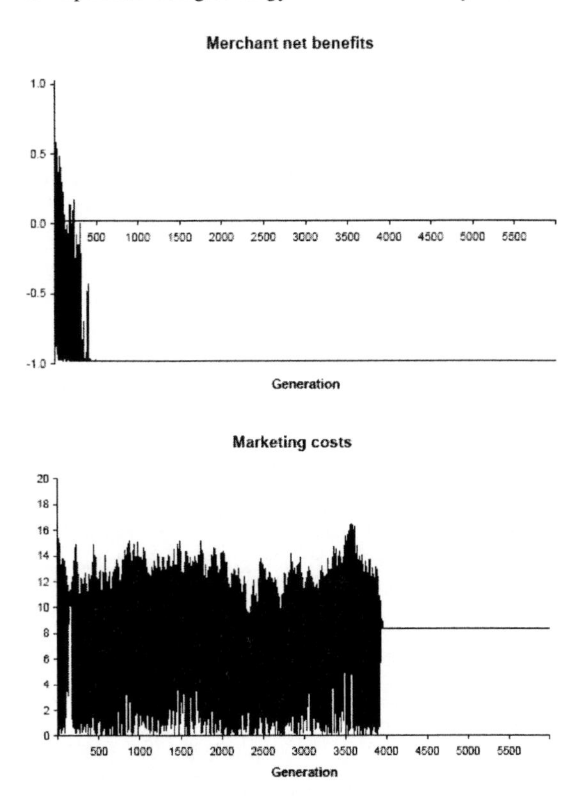

Fig. 13.3. Evolution of strategy parameters in one experiment (ctd.)

13.5 Conclusions

We have developed an artificial payment card market in which consumers and merchants are interacting with each other through payments made for purchases. Based on the usage and acceptance of payment cards, merchants and consumers continuously review their subscriptions to payment cards and card issuers seek to maximize their market share and maximize their profits by setting optimal fees and marketing efforts. Evaluating such a model we were able to derive the demand function for payment cards as well as the profit function of card issuers, observing that most importantly the fixed fees charged by the card issuers drive demand and profits.

The optimized strategies of payment card issuers are characterized by a relative high fixed fee to consumers, no fixed fee to merchants as well as large negative net benefits (i.e. a transaction fee) to consumers and merchants alike and high marketing costs. Such a fee structure with high fixed and transaction fees to consumers can be observed in many markets where substantial annual fees are charged along transaction fees in the form of higher-than-usual interest on purchases or fees on the use of payment cards overseas. Similarly merchants pay a considerable fee for each transaction while not being charged a significant fixed fee. These characteristics are replicated in our model, along with the high marketing costs card issuers often face.

For the first time in the literature we have been able to reproduce realistic properties of the payment card market with our model. While our model can be extended in a wide range of manners, e.g. by using different numbers of competitors, different physical locations of merchants and consumers to name only two possibilities, it provides a first foundation for the analysis of this market which does not limit itself to the interchange fee between different card issuers as commonly done in the literature. It is finally possible to use the model as a basis for the analysis of any proposed regulation of the payment card market, e.g. through the introduction of caps on fees charged by card issuers or limits in benefits granted.

References

[1] Alexandrova-Kabadjova B, Krause A, Tsang E (2006) An agent-based model of interactions in the payment card market. Working Paper 07, Centre for Computational Finance and Economic Agents

[2] Alexandrova-Kabadjova B, Tsang E, Krause A (2006) Market structure and information in payment card markets. Working Paper 06, Centre for Computational Finance and Economic Agents

[3] Baluja S (1994) Population-based incremental learning: A method for integrating genetic search based function optimization and competitive learning. Working paper CMU-CS-94-163, School of Computer Science, Carnegie Mellon University

[4] Chakravorti S (2003) A theory of credit card networks: A survey of the literature. Review of Network Economics 2:50–68

[5] Commission of the European Communities (2007) Communication from the Commission - Sector Inquiry under Art 17 of Regularion 1/2003 in retail banking (Final Report). Tech. rep., Commission of the European Communities

[6] Cruichshank D (2000) Competition in UK Banking. Report to the Chancellor of the Exchequer. http://www.bankreview.org.uk

[7] Evans DS, Schmalensee R (2003) Paying with Plastic: The Digital Revolution in Buying and Borrowing, 2nd edn. MIT Press

[8] Federal Reserve System (2004) The 2004 Federal Reserve Payments Study: Analysis of Noncash Payments Trends in the United States: 2000 - 2003. Tech. rep., Federal Reserve System

[9] Gans J, King S (2002) A Theoretical Analysis of Credit Card Regulation. Working paper 2002-11, Melbourne Business School

[10] Kern M (2006) Parameter adaptation in heuristic search: A population-based approach. PhD thesis, Department of Computer Science, University of Essex

[11] LeBaron B (2006) Agent-based computational finance. In: Tesfatsion L, Judd K (eds) Handbook of Computational Economics, vol 2, North-Holland, chap 9

[12] Rochet JC, Tirole J (2002) Cooperation among competitors: Some economics of payment card associations. The RAND Journal of Economics 33:1–22

[13] Rochet JC, Tirole J (2007) Tying-in two-sided markets and the honour all cards rule. CEPR Discussion Papers 6132, C.E.P.R. Discussion Papers, available at http://ideas.repec.org/p/cpr/ceprdp/6132.html
[14] Schmalensee R (2002) Payment systems and interchange fees. Journal of Industrial Economics 50:103–122
[15] Tesfatsion L (2006) Agent-based computational economics: A constructive approach to economic theory. In: Tesfatsion L, Judd K (eds) Handbook of Computational Economics, vol 2, North-Holland, chap 1
[16] Tesfatsion L, Judd K (eds) (2006) Handbook of Computational Economics, vol 2. North-Holland
[17] Wright J (2003) Optimal card payment systems. European Economic Review 47:587–612
[18] Wright J (2003) Pricing in debit and credit card schemes. Economics Letters 80:305–309

14

Can Trend Followers Survive in the Long-Run? Insights from Agent-Based Modeling

Xue-Zhong He[1]*, Philip Hamill[2], Youwei Li[3]†

[1] School of Finance and Economics, University of Technology, Sydney, Australia. tony.he1@uts.edu.au
[2] School of Business, Retail and Financial Services, University of Ulster, Coleraine, UK. pa.hamill@ulster.ac.uk
[3] School of Management and Economics, Queen's University Belfast, UK. y.li@qub.ac.uk

Summary. This chapter uses a simple stochastic market fraction (MF) asset pricing model to investigate market dominance, profitability, and how traders adopting fundamental analysis or trend following strategies can survive under various market conditions in the long/short-run. This contrasts with the modern theory of finance which relies on the paradigm of utility maximizing representative agents and rational expectations assumptions which some contemporary theorists regard as extreme. This school of thought would predict that trend followers will be driven out of the markets in the long-run. Our analysis shows that in a MF framework this is not necessarily the case and that trend followers can survive in the long-run.

14.1 Introduction

The modern theory of finance relies on the paradigm that asset prices are the outcome of the market interaction of utility maximizing representative agents who are rational when forming expectations about future market outcomes. The assumption that agents rationally impound all relevant information into their trading decisions produces price changes which are random, and consequently exhibit random walk behaviour. The representative agent assumption, as argued by Friedman (14), leads to the conjecture that irrational traders (also called less informed traders or chartists) could profit in the short-run, but are expected to perish in the long-run while rational traders (also called informed traders or fundamentalists) should be the only long-run survivors.

Despite all the evidence presented in academic journals that security prices follow random walks the use of technical trading rules is widespread amongst financial

* Acknowledgement: Xue-Zhong He would like to thank the financial support from the Australian Research Council (ARC) under discovery grant (DP0450526), and the University of Technology, Sydney under a Research Excellence Grant (REG).

† Corresponding author.

X.-Z. He et al.: *Can Trend Followers Survive in the Long-Run? Insights from Agent-Based Modeling*, Studies in Computational Intelligence (SCI) **100**, 253–269 (2008)
www.springerlink.com

market practitioners as evidenced by the growth in hedge funds employing quantitative trading strategies: trend following is one of the most popular trading rules. Also, finance theory provides limited economic explanations for the well documented stylized facts reported in the empirical finance literature (see Pagan (30)). The existence of excess volatility (asset return is more volatile relative to the dividends and underlying cash flows), volatility clustering (high/low asset return fluctuations tend to cluster), either positive or negative skewness and excess kurtosis (compared to normally distributed returns), and long-range dependence (insignificant autocorrelations (ACs) of raw returns and hyperbolic decline of ACs of the absolute and squared returns, see Ding et al. (13)) are difficult to accommodate within the established theoretical structure of market efficiency and rational expectations (see, for example, Shiller (31)). In practice, GARCH methodology has been successful modeling volatility clustering and capturing the short-run dynamics of volatility, but fails to provide an economic explanation.

Agent-Based Modeling is an alternative paradigm which may provide an appropriate theoretical and methodological framework to explain the stylized facts. For those new to this area we refer them to the survey papers by Hommes (20) and LeBaron (24) for the recent developments in this literature. In contrast to the traditional assumptions of investor homogeneity and rational expectations, the latter of which is regarded as an extreme informational assumption (see, for example, Cochrane (11)), agent-based models allow for heterogeneous agents, potentially showing bounded rational behaviour, who have different attitudes to risk and different expectations about the future evolution of prices. This approach has been shown to be able to characterize the dynamics of financial asset returns. The works of Arthur *et al.* (2), Brock and Hommes (3), Chiarella (6), Chiarella and He (7) and (8), Day and Huang (12) LeBaron (21), Levy *et al.* (25), Lux (28), among others, are examples of this approach. Agent-Based Models attempt to explain various types of market behaviour and to replicate the well documented empirical features of actual financial markets.

Heterogeneous agent-based models have had success in explaining market behavior and reproducing stylized facts. However, there are few works explicitly investigating whether irrational traders can survive in the long-run. In this paper a market fractions (MF) model with heterogeneous traders - fundamentalists (rational investors who believe the market price is mean reverting to fundamental price) and trend followers (irrational investors who believe market price will follow the trend generated from historical prices) - participate in a simple stochastic asset-pricing and wealth dynamics framework to investigate market dominance, profitability, and if they survive in the short/long-run. Although the techniques discussed in Arnold (1) may be useful for analyzing the stochastic model the mathematical analysis of nonlinear stochastic dynamical systems is difficult in general. Therefore, this analysis is conducted through Monte Carlo simulation (see Li *et al.* (26) and (27) for a more systematic study on this). The results from our simulation analysis show that, as expected, fundamentalists survive in the long-run and their profitability improves as they become increasingly confident of their forecasts of fundamental value. More interestingly, the key insight from our analysis is that trend followers can survive in the long-run even though they don't engage in information processing to establish the fundamental value of the asset to inform their trading strategy. In the context of our MF model this can be explained by the learning mechanism encapsulated in the trend followers forecasts. Our results also show that trend followers' profitability increases as their market share increases and when fundamentalists become naive traders.

This chapter is organized as follows. Section 14.2 outlines a market fraction model with heterogeneous agents. In this model the market clearing price is set by a market maker who adjusts the market price in response to aggregate excess demand in the market. Next, the expectations and learning mechanisms for the fundamentalists and trend followers is introduced. The latter part of this section develops the full market fraction stochastic model for asset prices and wealth dynamics. In section 14.3 the profitability and survivability of fundamental and trend following strategies is explored under alternative scenarios. Section 14.4 discusses the key insights from the study and identifies logical extensions for future research.

14.2 Heterogeneous Beliefs and Market Fractions

Intuitively, market population fractions among different types of traders play an important role in financial markets. Markets can be driven by certain types of investors at different time periods. This is particularly the case in either a bull or bear market. Empirical evidence from Taylor and Allen (32) suggests that at least 90% of the traders place some weight on technical analysis, such as moving average and trend following rules, over various time horizons. In particular, traders rely more on technical analysis, as opposed to fundamental analysis, at shorter horizons. As the time horizons increases more traders rely on fundamental rather than technical analysis. In addition, there are a proportion of traders who do not change their strategies over all time horizons. This situation is consistent with money-managers following a longer-term value investing strategy. Their time horizon for realizing gains is often years as opposed to days for the technical analyst, see Haugen (16) and Chan and Lakonishok (5). Theoretically, the study by Brock and Hommes (3) shows that when different groups of traders have different expectations about future prices and dividends compete between trading strategies and choose their strategy according to an evolutionary *fitness measure* the corresponding deterministic system can exhibit very complicated, and even, chaotic dynamics. The adaptive switching mechanism proposed by Brock and Hommes (3) is an important element of the adaptive belief model. It is based on both a *fitness function* and a discrete choice probability.

In this chapter we employ the market fraction (MF) model introduced in He and Li (17). It is a simplified version of Brock and Hommes' framework which assumes that the market fraction parameters among heterogeneous agents are fixed. Apart from its mathematical tractability this simplification has a number of distinct advantages. First, it clearly identifies how different market fractions influence the market price. In Brock and Hommes' framework this is difficult due to the amplifying effect of the exponential function used in the discrete choice probability which makes the market fractions very sensitive to price changes and the stated fitness functions. Second, our model doesn't allow agents to switch between trading strategies. This makes it easier to characterize market dominance, profitability, and survivability. Finally, it is important to understand how the trading strategies are linked to aspects of price behaviour. Having market fractions among heterogeneous agents as fixed parameters allows for an explicit examination of how market fractions influence price behaviour.

The MF model considered in the following discussion, introduced in He and Li (17), follows the standard discounted value asset pricing model with heterogeneous agents. The market clearing price is arrived at via a market maker scenario in line with Day and Huang (12) and Chiarella and He (10) rather than the Walrasian auctioneer scenario used in Brock and Hommes (4). We focus on a simple case in which there are three classes of participants in the asset market: two groups of traders, fundamentalists and trend followers, and a market maker.

14.2.1 Market Fraction and Market Maker

Consider an asset pricing model with one risky asset and one risk-free asset. It is assumed that the supply of the risk-free asset is perfectly elastic with a gross return of $R = 1 + r/K$, where r stands for a constant risk-free rate per annum and K stands for the trading frequency. Typically, $K = 1, 12, 52$ and 250 for trading periods of a year, month, week, and a day. To calibrate the stylized facts observed from daily price movements in financial markets K is set equal to 250.

Let P_t be the price (ex-dividend) per share for the risky asset at time t and $\{D_t\}$ be the stochastic dividend process for the risky asset. Then the wealth of a typical investor h at $t+1$ is given by

$$W_{h,t+1} = RW_{h,t} + [P_{t+1} + D_{t+1} - RP_t]z_{h,t}, \tag{14.1}$$

where $W_{h,t}$ and $z_{h,t}$ are the wealth and the number of shares of the risky asset purchased by investor h at t. Let $E_{h,t}$ and $V_{h,t}$ be the *beliefs* of type h traders about the conditional expectation and variance of quantities at $t+1$ based on their information set at time t. Where the excess capital gain on the risky asset at $t+1$ is denoted by R_{t+1}, that is

$$R_{t+1} = P_{t+1} + D_{t+1} - RP_t. \tag{14.2}$$

Then it follows from (14.1) and (14.2) that

$$E_{h,t}(W_{t+1}) = RW_t + E_{h,t}(R_{t+1})z_{h,t}, \qquad V_{h,t}(W_{t+1}) = z_{h,t}^2 V_{h,t}(R_{t+1}), \tag{14.3}$$

where $z_{h,t}$ is the demand by agent h for the risky asset. Assume that trader h has a constant absolute risk aversion (CARA) utility function with the risk aversion coefficient a_h (that is $U_h(W) = -e^{-a_h W}$), the optimal demand $z_{h,t}$ for the risky asset is determined by maximizing the expected utility of wealth, that is

$$z_{h,t} = \frac{E_{h,t}(R_{t+1})}{a_h V_{h,t}(R_{t+1})}. \tag{14.4}$$

Given the heterogeneity and the nature of asymmetric information among traders we consider two trading strategies corresponding to two types of boundedly rational traders: fundamentalists and trend followers. Their beliefs are defined in the following discussion. Assume the market fraction of the fundamentalists and trend followers is n_1 and n_2 with risk aversion coefficient a_1 and a_2, respectively. Let $m = n_1 - n_2 \in [-1, 1]$. Obviously, $m = 1, -1$ correspond to the cases when all the traders are fundamentalists or trend followers. Assume a zero supply of outside shares. Then, using (14.4), the aggregate excess demand per investor ($z_{e,t}$) is given by

$$z_{e,t} \equiv n_1 z_{1,t} + n_2 z_{2,t} = \frac{1+m}{2} \frac{E_{1,t}[R_{t+1}]}{a_1 V_{1,t}[R_{t+1}]} + \frac{1-m}{2} \frac{E_{2,t}[R_{t+1}]}{a_2 V_{2,t}[R_{t+1}]}. \tag{14.5}$$

To complete the model we assume that the market is cleared by a market maker. The role of the market maker is to take a long (when $z_{e,t} < 0$) or short (when $z_{e,t} > 0$) position so as to clear the market. At the end of period t, after the market maker has carried out all transactions, he or she adjusts the price for the next period in the direction of the observed excess demand. Let μ be the speed of price adjustment of the market maker. To capture unexpected market news or noise created by *noise traders* we introduce a noisy demand term $\tilde{\delta}_t$ which is an i.i.d. normally distributed random variable with $\tilde{\delta}_t \sim \mathcal{N}(0, \sigma_\delta^2)$. In this paper, we assume a constant volatility noisy demand and the volatility is related to an average fundamental price level. This noisy demand may also depend on the market price. Theoretically, how the price dynamics are influenced by adding different noisy demand is still a difficult problem. Here, we focus on the constant volatility noisy demand case and use Monte Carlo simulations and statistical analysis to gain some insights into this problem. Based on these assumptions the market price is determined by

$$P_{t+1} = P_t + \mu z_{e,t} + \tilde{\delta}_t. \tag{14.6}$$

Using (14.5), this becomes

$$P_{t+1} = P_t + \frac{\mu}{2} \left[(1+m) \frac{E_{1,t}[R_{t+1}]}{a_1 V_{1,t}[R_{t+1}]} + (1-m) \frac{E_{2,t}[R_{t+1}]}{a_1 V_{2,t}[R_{t+1}]} \right] + \tilde{\delta}_t. \tag{14.7}$$

We use fig. 14.1 to illustrate the general role of heterogeneous expectation for our two groups of agents and how the market cleaning price is arrived at. Let $p_{t,t+1}^{f,e}$ and $p_{t,t+1}^{c,e}$ be the expected price at time $t+1$ for the fundamentalists and chartists conditional upon their information set at time t. The market maker aggregates the demand from agents' heterogeneous expectations to form the expected market price $p_{t,t+1}^{e}$ at time t and adjusts the market price at time $t+1$ accordingly. The important feature of this structure is that the price generating mechanism is driven by expectations feedback. Observed market prices are then used to form expectations for the next period which in turn feeds back into the price generating process.

It should be pointed out that market maker behaviour in this model is highly stylized. For instance, the inventory of the market maker built up as a result of the

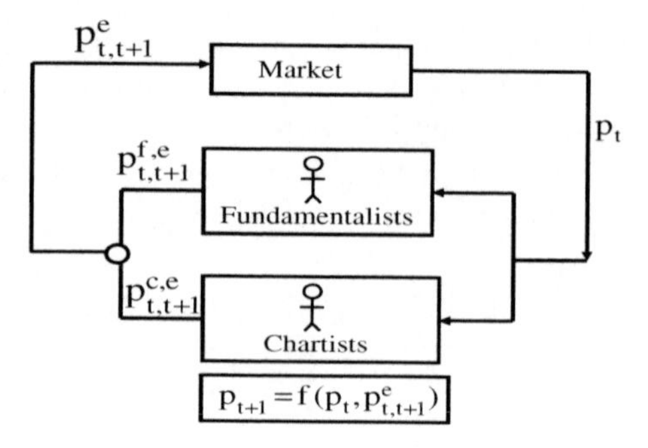

Fig. 14.1. Expectations Feedback

accumulation of various long and short positions is not considered. This could affect his or her behaviour and the market maker price setting role in (14.7) could be a function of the inventory. Allowing μ to be a function of inventory would be one way to model such behaviour. In this chapter it is best thought of as a market friction. One of the aims of our analysis is to understand how this friction affects the market dynamics.

14.2.2 Fundamentalists

We assume that $F_t = \{P_t, P_{t-1}, \cdots, D_t, D_{t-1}, \cdots\}$ is the common information set at time t. Apart from the common information set, the fundamentalists have *superior* information on the fundamental value, P_t^*, of the risky asset and they also realize the existence of non-fundamental traders such as trend followers to be introduced in the following discussion. They believe that the stock price may be driven away from the fundamental value in the short-run, but it will eventually converge to the fundamental value in the long-run. The speed of convergence measures their confidence level in the fundamental value. More precisely, we assume that the fundamental value satisfies a stationary random walk process (as we know that the fundamental value driven by this random walk process can be negative)

$$P_{t+1}^* = P_t^*[1 + \sigma_\varepsilon \tilde{\varepsilon}_t], \qquad \tilde{\varepsilon}_t \sim \mathcal{N}(0,1), \qquad \sigma_\varepsilon \geq 0, \qquad P_0^* = \bar{P} > 0, \qquad (14.8)$$

where $\tilde{\varepsilon}_t$ is independent of the noisy demand process $\tilde{\delta}_t$. We assume the conditional mean and variance of the fundamental traders are

$$E_{1,t}(P_{t+1}) = P_t + \alpha(P_{t+1}^* - P_t), \qquad V_{1,t}(P_{t+1}) = \sigma_1^2, \qquad (14.9)$$

where σ_1^2 stands for a constant variance for the price. Here parameter $\alpha \in [0,1]$ is the speed of price adjustment toward the fundamental value. It measures their level of confidence in fundamental value. Specifically, for $\alpha = 1$ the fundamental traders are fully confident about the fundamental value and adjust their expected price at the next period instantaneously to the fundamental value. When $\alpha = 0$ fundamentalists become naive traders. In general, the fundamental traders believe that markets are efficient and prices converge to their fundamental value. An increase (decrease) in α indicates that the fundamental traders have high (low) confidence in their estimated fundamental value, leading to a quick (slow) adjustment of the expected price towards the fundamental price.

14.2.3 Trend followers

Unlike the fundamentalists, trend followers are technical traders who believe that future price changes can be predicted from various patterns or trends generated from the history of prices. The trend followers are assumed to extrapolate the latest observed price change over prices' long-run sample mean and to adjust their variance estimate accordingly. More precisely, their conditional mean and variance are assumed to satisfy

$$E_{2,t}(P_{t+1}) = P_t + \gamma(P_t - u_t), \qquad V_{2,t}(P_{t+1}) = \sigma_1^2 + b_2 v_t, \tag{14.10}$$

where $\gamma, b_2 \geq 0$ are constants, and u_t and v_t are the sample mean and variance, respectively, which follow a learning process. The parameter γ measures the extrapolation rate. High (low) values of γ correspond to strong (weak) extrapolation from the trend. The coefficient b_2 measures the influence of the sample variance on the conditional variance estimated by the trend followers who believe in a more volatile price movement. Various learning schemes can be used to estimate the sample mean u_t and variance v_t (see Chiarella and He (9), (10) for related studies on heterogeneous learning and asset pricing models with heterogeneous agents whose conditional mean and variance follow various learning processes). In this chapter we assume that

$$u_t = \delta u_{t-1} + (1-\delta)P_t, \tag{14.11}$$

$$v_t = \delta v_{t-1} + \delta(1-\delta)(P_t - u_{t-1})^2, \tag{14.12}$$

where $\delta \in [0,1]$ is a constant. This is the limiting process of a *geometric decay process* where the memory lag length tends to infinity. Basically, a geometric decay probability process $(1-\delta)\{1, \delta, \delta^2, \cdots\}$ is associated with historical prices $\{P_t, P_{t-1}, P_{t-2}, \cdots\}$. The parameter δ measures the geometric decay rate. For $\delta = 0$, the sample mean $u_t = P_t$, which is the latest observed price, while $\delta = 0.1, 0.5, 0.95$ and 0.999 gives a half life of 0.43 day, 1 day, 2.5 weeks and 2.7 years, respectively. The selection of this process is two fold. First, traders tend to put a high weight on the most recent prices and less weight on the more remote prices when they estimate the sample mean and variance. Second, it has the mathematical advantage of analytical tractability.

14.2.4 The Complete Stochastic Model

To simplify our analysis we assume that the dividend process D_t follows a normal distribution $D_t \sim \mathcal{N}(\bar{D}, \sigma_D^2)$. The expected long-run fundamental value $\bar{P} = \bar{D}/(R-1)$ and the unconditional variances of price and dividend over the trading period are related by $\sigma_D^2 = q\sigma_1^2$. In this study, we choose $\sigma_1^2 = \sigma_{\bar{P}}^2/K$ and

$q = r^2$. This can be justified as follows. Let $\sigma_{\bar{P}}$ be the annual volatility of P_t and $\bar{D}_t = rP_t$ be the annual dividend. Then the annual variance of the dividend $\bar{\sigma}_D^2 = r^2\sigma_{\bar{P}}^2$. Therefore $\sigma_D^2 = \bar{\sigma}_D^2/K = r^2\sigma_{\bar{P}}^2/K = r^2\sigma_1^2$. For all numerical simulations in this paper we choose $\bar{P} = \$100, r = 5\%$ p.a. $\sigma = 20\%$ p.a., $K = 250$. Correspondingly, $R = 1 + 0.05/250 = 1.0002, \sigma_1^2 = (100 \times 0.2)^2/250 = 8/5$ and $\sigma_D^2 = 1/250$. Based on assumptions (14.9)-(14.10), the fundamentalists' optimal demand is

$$z_{1,t} = \frac{1}{a_1(1+q)\sigma_1^2}[\alpha(P_{t+1}^* - P_t) - (R-1)(P_t - \bar{P})]. \tag{14.13}$$

In particular, when $P_t^* = \bar{P}$,

$$z_{1,t} = \frac{(\alpha + R - 1)(\bar{P} - P_t)}{a_1(1+q)\sigma_1^2}. \tag{14.14}$$

Similarly, from (14.10), (using $\bar{D} = (R-1)\bar{P}$) the trend followers' optimal demand is

$$z_{2,t} = \frac{\gamma(P_t - u_t) - (R-1)(P_t - \bar{P})}{a_2\sigma_1^2(1+q+bv_t)}, \tag{14.15}$$

where $b = b_2/\sigma_1^2$. Subsisting (14.13) and (14.15) into (14.7), the price dynamics under a market maker is determined by the following 4-dimensional stochastic difference system (**SDS** hereafter)

$$\begin{cases} P_{t+1} = P_t + \dfrac{\mu}{2}\left[\dfrac{1+m}{a_1(1+q)\sigma_1^2}[\alpha(P_{t+1}^* - P_t) - (R-1)(P_t - \bar{P})] \right. \\ \qquad \left. + (1-m)\dfrac{\gamma(P_t - u_t) - (R-1)(P_t - \bar{P})}{a_2\sigma_1^2(1+q+bv_t)}\right] + \tilde{\delta}_t, \\ u_t = \delta u_{t-1} + (1-\delta)P_t, \\ v_t = \delta v_{t-1} + \delta(1-\delta)(P_t - u_{t-1})^2, \\ P_{t+1}^* = P_t^*[1 + \sigma_\varepsilon \tilde{\varepsilon}_t]. \end{cases} \tag{14.16}$$

Using Monte Carlo simulation and statistical analysis, He and Li (17) found that the long-run behaviour and convergence of the market prices and various under and over-reaction autocorrelation patterns of returns can be characterized by the dynamics, including the stability and bifurcations, of the underlying deterministic system. In the relation to the ability of the MF model to characterize the stylized facts, especially the long-range dependence in volatility, He and Li (18) demonstrate that agent heterogeneity, risk-adjusted trend chasing through the geometric learning process, and the interplay of noisy fundamental and demand processes and the underlying deterministic dynamics can be the source of power-law distributed fluctuations. In particular, the noisy demand plays an important role in the generation of insignificant autocorrelations (ACs) on returns, while the significant decaying AC patterns of the absolute returns and squared returns are more influenced by the noisy fundamental process. A statistical analysis based on Monte Carlo simulations is conducted to characterize the decay rate. Realistic estimates of the power-law decay indices and the (FI)GARCH parameters are presented. This analysis provides some insights into the understanding of financial markets. In the following discussion we introduce measures of wealth dynamics and we explore the potential of the MF model to characterize the profitability and survivability of fundamental and trend following strategies.

14.2.5 Wealth Dynamics and Shares

We assume that traders' wealth follows a stochastic process. To be able to measure the wealth dynamics among different trading strategies and to examine the market dominance and price behaviour we introduce two wealth measures. The first measures the absolute level of the wealth share (or proportion) of the representative agent from each type, called the *absolute wealth share* for short, which is defined by

$$w_{1,t} = \frac{W_{1,t}}{W_{1,t}+W_{2,t}}, \qquad w_{2,t} = \frac{W_{2,t}}{W_{1,t}+W_{2,t}}, \tag{14.17}$$

where $W_{1,t}$ and $W_{2,t}$ are the wealth at time t of the representative trader of the fundamentalists and trend followers, respectively. This measure can be used to measure the evolutionary performance or profitability of the two trading strategies: As $w_{1,t}$ ($w_{2,t}$) increases the profitability of the fundamentalists (trend followers) increases. The second measures the overall market wealth share, called the *market wealth share* for short, of the different trading strategies. It is defined as the market fraction weighted average of the absolute wealth proportions,

$$\bar{w}_{1,t} = \frac{(1+m)W_{1,t}}{(1+m)W_{1,t}+(1-m)W_{2,t}}, \quad \bar{w}_{2,t} = \frac{(1-m)W_{2,t}}{(1+m)W_{1,t}+(1-m)W_{2,t}} \tag{14.18}$$

A high market wealth share $\bar{w}_{1,t}$ ($\bar{w}_{2,t}$) indicates market dominance of the fundamentalists (trend followers) with respect to the overall market wealth. Let $V_{1,t} = 1/W_{1,t}$ and $V_{2,t} = 1/W_{2,t}$. Then it follows from (14.1) that

$$V_{1,t+1} = \frac{V_{1,t}}{R+R_{t+1}z_{1,t}V_{1,t}}, \qquad V_{2,t+1} = \frac{V_{2,t}}{R+R_{t+1}z_{1,t}V_{2,t}}.$$

Note that

$$\frac{V_{1,t}}{V_{1,t}+V_{2,t}} = \frac{1/W_{1,t}}{1/W_{1,t}+1/W_{2,t}} = \frac{W_{2,t}}{W_{1,t}+W_{2,t}},$$
$$\frac{V_{2,t}}{V_{1,t}+V_{2,t}} = \frac{1/W_{2,t}}{1/W_{1,t}+1/W_{2,t}} = \frac{W_{1,t}}{W_{1,t}+W_{2,t}}$$

and, therefore, the absolute wealth shares are determined by

$$w_{1,t} = \frac{V_{2,t}}{V_{1,t}+V_{2,t}}, \qquad w_{2,t} = \frac{V_{1,t}}{V_{1,t}+V_{2,t}} \tag{14.19}$$

and the market wealth shares are governed by

$$\bar{w}_{1,t} = \frac{(1+m)V_{2,t}}{(1+m)V_{2,t} + (1-m)V_{1,t}}, \quad \bar{w}_{2,t} = \frac{(1-m)V_{1,t}}{(1+m)V_{2,t} + (1-m)V_{1,t}}. \tag{14.20}$$

For these wealth measures it is difficult to obtain explicit closed form expressions in terms of (stationary) state variables. In this study we use the auxiliary functions $(V_{1,t}, V_{2,t})$ and numerical simulations to study the wealth dynamics of the fundamentalists and trend followers and the market impact of the two different trading strategies.

14.3 Wealth Accumulation, Profitability and Survivability

Friedman (14) argued that irrational traders (such as the trend followers in our model) may do better than rational traders (such as the fundamentalists) in the short-run, but over the long-run they will be driven out of the market and rational traders will be the only long-run survivors. We now justify Friedman's hypothesis by analyzing the wealth dynamics of our heterogeneous market fraction model in which traders' beliefs are time invariant. Consequently, we examine profitability and survivability of both types of trading strategies. The dynamics of the speed of price adjustment and the market fraction are considered in the following discussion.

14.3.1 Dynamics of the Price Adjustment Speed α

In our model, the market price is related to the fundamental price through the activity from the fundamentalists. Therefore, how the market price reflects the fundamental price depends on the reaction of the fundamentalists to the fundamental price. It is believed that chartists may perform better when the market price is far away from the fundamental price, while the fundamentalists may do better when the market price reflect the fundamental price. Hence it is interesting to analyze the wealth dynamics when the fundamentalists behave differently. In the following we examine the wealth dynamics of the model when the confidence level of the fundamentalists on the fundamental price, measured by parameter α, changes. To assess the impact of α we choose parameters set to

$$\gamma = 2.1, \ \delta = 0.85, \ \mu = 0.43, \ m = 0, \ w_{1,0} = 0.5, \ \alpha = 1, 0.5, 0.1, 0 \tag{14.21}$$

by fixing market fraction m and varying α. For each set of parameters, we run one simulation over 20,000 time periods in order to see possible limiting behaviours.

Fig. 14.2 demonstrates the absolute wealth share accumulations for the fundamentalists with $\alpha = 1, 0.5, 0.1, 0$ and keeping all the other conditions the same. This figure shows that (i) trend followers survive in the long-run for $\alpha = 1, 0.5$ and 0.1 in the sense that their absolute wealth share does not vanish, although they accumulate less wealth shares over the time period; (ii) the trend followers do better than the fundamentalists when $\alpha = 0$; (iii) the profitability of the fundamentalists improves

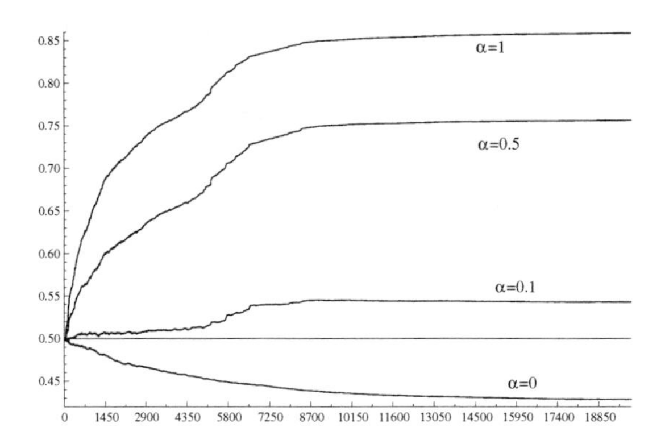

Fig. 14.2. Time series of the absolute wealth accumulation of the fundamentalists $w_{1,t}$ with $\alpha = 1, 0.5, 0.1$ and 0.

as α increases (i.e. as they become more confident in their estimated fundamental value). For $\alpha = 0$, the absolute wealth share of the fundamentalists is dropped from 50% to about 43%, while for $\alpha = 0.1, 0.5$ and 1, it is increased from 50% to about 55%, 76% and 86%, respectively. These results are further confirmed when we run Monte Carlo simulations. We ran 1,000 independent simulations and discard the first 1,000 time periods to wash out the possible initial noise effect. The results are given in fig. 14.3. For four values of α we plot the average market price (left column), return (middle column), and the fundamentalists absolute wealth share accumulation (right column). The initial wealth share for both types of traders are equal $w_{1,0} = 0.5$. Because of $m = 0$, both the absolute and market wealth shares are the same.

14.3.2 Dynamics of the Market Fraction m

Intuitively it seems that the market price is partially determined by the market dominance of different players in the market. Therefore we would expect that the market fraction, measured by parameter m, will influence the market price and the consequent performance of fundamentalists and chartists.

Given that both α and m have a similar impact on the local stability of the deterministic system (see He and Li (17)), we can demonstrate that they play a similar role in terms of wealth accumulation. Again, by running one simulation over 20,000 time periods, fig. 14.4 shows the absolute wealth share accumulations for the fundamentalists for three different values of $m = -0.95, 0$ and 0.5 with $\alpha = 0.5, \gamma = 2, \mu = 0.5, \delta = 0.85, w_{1,0} = 0.5$. In this case, the fundamentalists form their conditional expectation by taking an average of the latest market price and fundamental price. In all four cases, (i) the fundamentalists accumulate more wealth share than the trend followers in the long-run (an increase from 50% to about 70-75%), however, the trend followers survive in the long-run and they can even accumulate more wealth share in the short-run when they dominate the market (this is the case when $m = -0.95$, which corresponds to 97.5% of trend followers and 2.5%

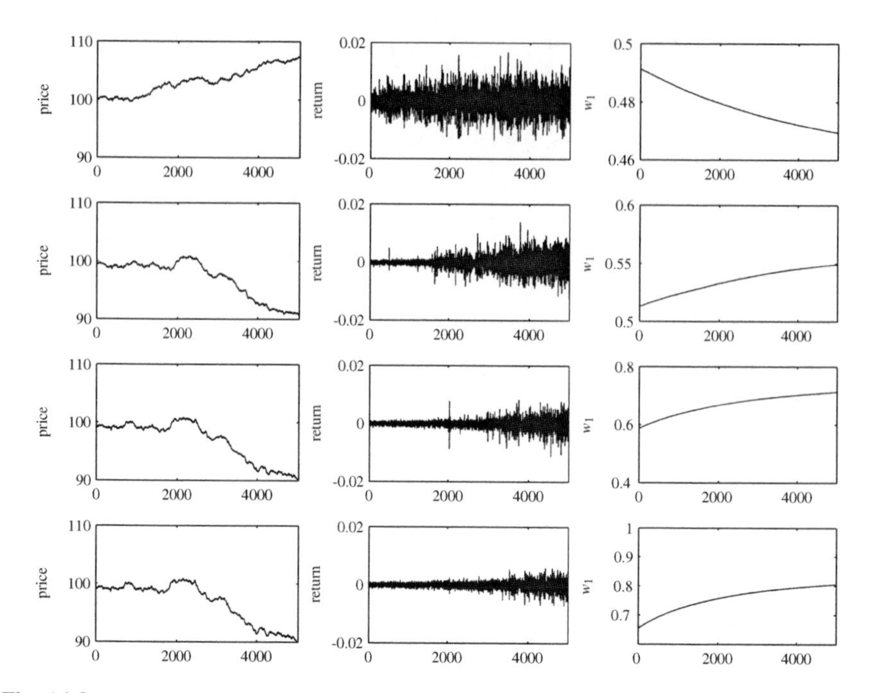

Fig. 14.3. Average Monte Carlo time series of market prices, returns, absolute wealth share and market wealth share of the fundamentalists with $\alpha = 0, 0.1, 0.5$ and 1.

of the fundamentalists); (ii) the profitability of the fundamentalists improves as m increases (i.e. as the market fraction of the fundamentalists increases). Essentially, we have shown that both α and m have a similar effect on profitability and survivability for fundamentalists and trend followers. Comparison of fig.14.2 and fig. 14.4 indicates that the parameter α affects wealth accumulation more than the parameter m does.

When the fundamentalists are naive traders (i.e. $\alpha = 0$ and $E_{1,t}(P_{t+1}) = P_t$) fundamental price doesn't influence how they form their conditional expectation. We choose

$$\alpha = 0,\ \gamma = 1,\ \mu = 0.4,\ \delta = 0.85,\ w_{1,0} = 0.5,\ m = -1, -0.5, 0, 0.5, 1. \quad (14.22)$$

Again, for each set of parameters we run one simulation over 20,000 time periods such that the corresponding limiting behaviour becomes clear. Fig. 14.5 illustrates the absolute wealth share accumulations of the fundamentalists with different market fraction $m = -1, -0.5, 0, 0.5, 1$, and keeping all the other conditions the same. They converge to different constant levels for different values of m in the long-run. Note that, unlike the market price, the absolute wealth shares are independent from the market fraction m and they are calculated for the given market price series. In particular, when $m = \pm 1$ the market price is affected only by one type of traders, but the absolute wealth share accumulations can still be calculated based on the market

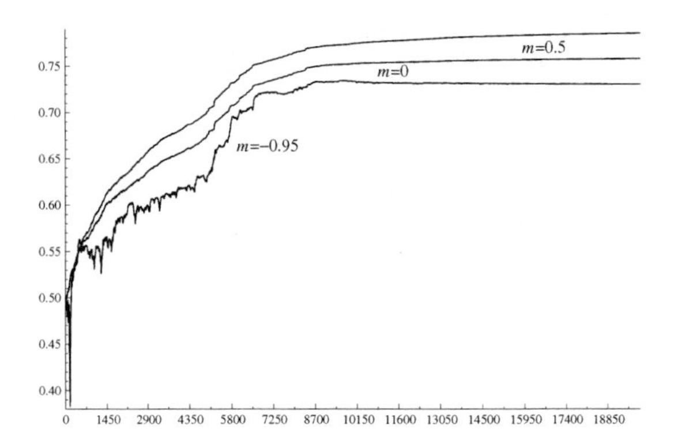

Fig. 14.4. Time series of the absolute wealth accumulation of the fundamentalists $w_{1,t}$ with $m = -0.95, 0, 0.5$ and $\alpha = 0.5$, $\gamma = 2$, $\mu = 0.5$, $\delta = 0.85$, $w_{1,0} = 0.5$.

price. Fig. 14.5 shows that, overall, no one does significantly better by accumulating significant higher absolute wealth share than the others. For $m = 1$ trend followers don't affect the market price. In this case, the long-run absolute wealth share accumulation of the fundamental trading strategy stays just above the average level, indicating that the trend followers will survive in long-run, although they have no impact on the market price and accumulate less absolute wealth share. For $m \neq 1$ trend followers do slightly better by accumulating a higher absolute wealth share. When $m = -1$ fundamentalists don't affect the market price. Under this scenario the trend followers accumulate more absolute wealth share in long-run. Overall, the profitability of the fundamentalists improves as m increases (i.e. as their market population share increases). These results are further confirmed when we run Monte Carlo simulations. The results in fig. 14.6 include the average market price, return, and absolute wealth share accumulation for the fundamentalists. The initial wealth share for both types of traders are equal $w_{1,0} = 0.5$. For different values of m, the market wealth shares are different. It is also interesting to see that the average market price increases, rather than decreases in the first case, stochastically. Given the naive expectation of the fundamentalists this may be due to the trend chasing activity of the trend followers.

The above analysis leads to the following implications for profitability and survivability:

- Although the trend followers have no information on the fundamental value they survive in the long-run and can even out-perform fundamentalists in the short-run. This may be due to the learning mechanism they are engaged in.
- Fundamentalists' profitability increases as they become more confident in their estimates of fundamental value or they dominate the market.
- When the fundamentalists become naive traders, trend followers do better and they accumulate a higher wealth share. In addition, their profitability improves as their market population share increases.

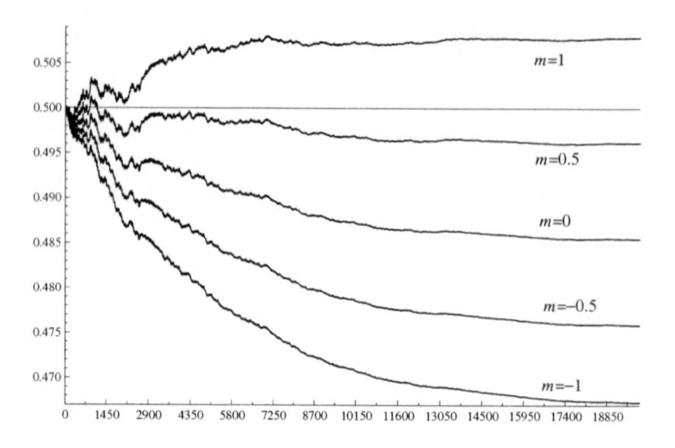

Fig. 14.5. Time series of the absolute wealth accumulation of the fundamentalists $w_{1,t}$ with $m = -1, -0.5, 0, 0.5, 1$ and $\alpha = 0$, $\gamma = 1$, $\mu = 0.4$, $\delta = 0.85$, $w_{1,0} = 0.5$.

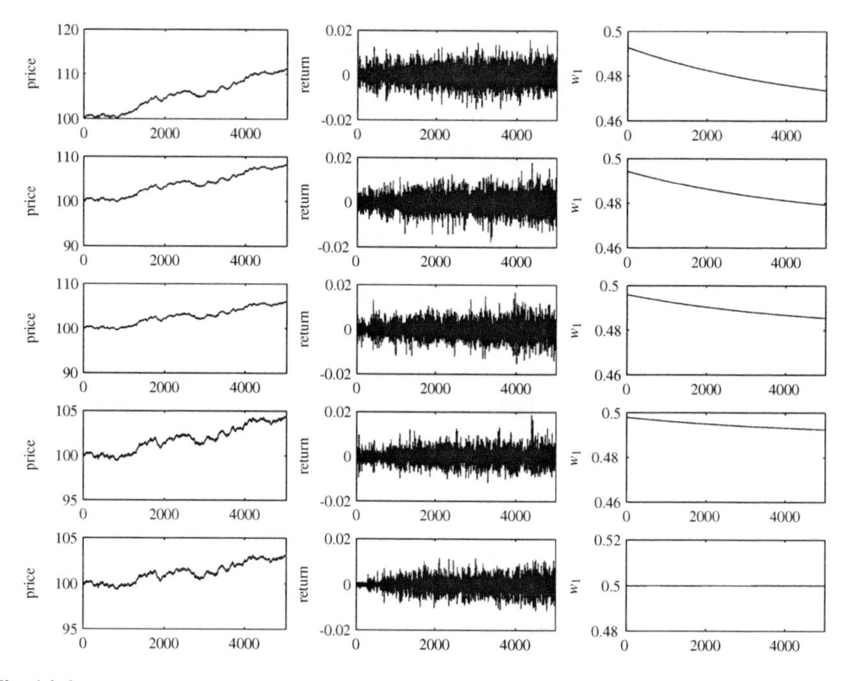

Fig. 14.6. Average Monte Carlo time series of market prices, returns and absolute wealth share of the fundamentalists with $\alpha = 0$, $\gamma = 1$, $\mu = 0.4$, $\delta = 0.85$, $w_{1,0} = 0.5$, and $m = -0.5$ (top row), 0 (second row), 0.5 (third row), 1 (4-th row), -1 (the last row).

The wealth share measures used in this paper compare the relative performance between two trading strategists. Survivability of the chartists is measured by their positive wealth share generated from their trading strategy. Overall, we have shown the short/long-run profitability for both the fundamental and trend following trading strategies, and the long-run survivability for the trend following strategy. This result partially verifies Friedman's argument that the chartists may do better in the short-run, but the market will be dominated by the fundamentalists in the long-run. However, in contrast to Friedman's argument, chartists do survive in the long-run due to the learning mechanism.

14.4 Conclusion

In this study a market fractions model with heterogeneous traders in a simple asset-pricing and wealth dynamics framework is employed to investigate market dominance, profitability, and whether investors who follow fundamental and trend following strategies survive in the short/long-run. Two measures of wealth are introduced to assess the connection between market dominance and wealth dynamics.

The conclusions drawn from the statistical analysis, based on Monte Carlo simulations, show that when the market is dominated by fundamentalists both their absolute and market wealth proportions increase significantly in the long-run. On the other hand, when trend followers dominate the market and the deviation between fundamental price and the trending price series is small, their wealth proportions fails to increase significantly. When the deviation is large trend followers' proportion of wealth increase.

The level of confidence fundamentalists have in the convergence of market price to fundamental value impacts upon trend followers wealth proportions. When they are less confident of convergence the market price can be driven away by trend followers and fundamentalists' wealth proportions decrease. Though, the trend followers' commensurate increase in wealth is relatively small. As fundamentalists become increasingly confident in convergence trend followers' wealth proportions can be reduced dramatically over the long-run. The key finding from our analysis which differs from what would be predicted by Friedman is that trend followers do survive in the long-run in all scenarios, while in the short-run they can out-perform fundamentalists which is consistent with the traditional view.

A potentially fruitful line for future research would be to explore the impact of information costs. In practice, information costs can be a non-trivial component of fundamental analysis. Integrating a range of plausible costs structures into a MF model could elucidate their affect on fundamentalists wealth share. In contrast, chartists' reliance on costless information and learning may explain their survival in the long-run.

References

[1] Arnold, L (1998) Random Dynamical Systems. Springer-Verlag, Berlin

[2] Arthur W, Holland J, LeBaron B, Palmer R, Tayler P (1997) Asset Pricing Under Endogeneous Expectations in an Artifical Stock Market. Economic Notes 26(2):297–330.

[3] Brock W, Hommes C (1997) A Rational Route to Randomness. Econometrica 65:1059–1095
[4] Brock W, Hommes C (1998) Heterogeneous Beliefs and Routes to Chaos in a Simple Asset Pricing Model. Journal of Economic Dynamics and Control 22:1235–1274
[5] Chan, L, Lakonishok J (2004) Value and Growth Investing: Review and Update. Financial Analysts Journal 60:71–86
[6] Chiarella C (1992) The Dynamics of Speculative Behaviour. Annals of Operations Research 37:101–123
[7] Chiarella C, He X (2001) Asset Pricing and Wealth Dynamics under Heterogeneous Expectations. Quantitative Finance 1:509–526
[8] Chiarella C, He X (2002) Heterogeneous Beliefs, Risk and Learning in a Simple Asset Pricing Model. Computational Economics 19:95–132
[9] Chiarella C, He X (2003a) Dynamics of Beliefs and Learning under α_l-processes - Heterogeneous Case. Journal of Economic Dynamics & Control 27:503–531
[10] Chiarella C, He X (2003b) Heterogeneous Beliefs, Risk and Learning in a Simple Asset Pricing Model with a Market Maker. Macroeconomic Dynamics 7:503–536
[11] Cochrane J (2001) Asset Pricing. Princeton University Press, Princeton
[12] Day R, Huang W (1990) Bulls, Bears and Market Sheep. Journal of Economic Behaviour and Organization 14:299–329
[13] Ding Z, Granger C, Engle R (1993) A Long Memory Property of Stock market Returns and a New Model. Journal of Empirical Finance 1:83–106
[14] Friedman M (1953) The Case for Flexible Exchange Rate. In: Essays in positive economics. Chicago, University of Chicago Press
[15] Gaunersdorfer A (2000) Endogenous Fluctuations in a Simple Asset Pricing Model with Heterogeneous Agents. Journal of Economic Dynamics and Control 24:799–831
[16] Haugen R (2003) The New Finance: Overreaction, Complexity and Uniqueness. Prentice Hall, Upper Saddle River, NJ
[17] He X, Li Y (2007a) Heterogeneity, Convergence, and Autocorrelations. Quantitative Finance in press
[18] He X, Li Y (2007b) Power-law behaviour, Heterogeneity, and Trend Chasing. Journal of Economic Dynamics and Control 31:3396–3426
[19] Hommes C (2002) Modeling the Stylized Facts in Finance Through Simple Nonlinear Adaptive Systems, Proceedings of National Academy of Science of the United States of America, 99, 7221–7228.
[20] Hommes C (2006) Heterogeneous Agent Models in Economics and Finance, Handbook of Computational Economics. Volume 2, Edited by K.L. Judd and L. Tesfatsion, Elsevier Science.
[21] LeBaron B (2000) Agent-based Computational Finance: Suggested Readings and Early Research. Journal of Economic Dynamics & Control 24:679–702
[22] LeBaron B (2001) A Builder's Guide to Agent-based Financial Markets. Quantitative Finance 1(2):254–261

[23] LeBaron B (2002) Calibrating an Agent-based Financial Market to Macroeconomic Time Series. Technical report, Brandeis University, Waltham, MA

[24] LeBaron B (2006) Agent-based Computational Finance. In Judd K. and Tesfatsion L. (ed) Handbook of Computational Economics Volume 2. Elsevier Science.

[25] Levy M, Levy H, Solomon S (2000) Microscopic Simulation of Financial Markets. Academic Press, New York.

[26] Li Y, Donkers B, and Melenberg B (2006a) Econometric Analysis of Microscopic Simulation Models. CentER Discussion Papers 2006-99, Tilburg University. Available at: `http://ssrn.com/abstract=939518`.

[27] Li Y, Donkers B, and Melenberg B (2006b) The Nonparametric and Semiparametric Analysis of Microscopic Simulation Models. CentER Discussion Papers 2006-95, Tilburg University. Available at: `http://ssrn.com/abstract=939510`.

[28] Lux T (1998) The Social-economic Dynamics of Speculative Markets: Interacting Agents, Chaos, and the Fat Tails of Return Distributions Journal of Economic Behaviour & Organization 33:143–165

[29] Lux T, Marchesi M (1999) Scaling and Criticality in a Stochastic Multi-agent Model of a Financial Markets. Nature 397(11):498–500

[30] Pagan A (1996) The Econometrics of Financial Markets. Journal of Empirical Finance 3:15–102

[31] Shiller R (2003) From Efficient Markets Theory to Behavioural Finance. Journal of Economic Perspectives 17(1):83–104

[32] Taylor M, Allen H (1992) The Use of Technical Analysis in the Foreign Exchange Market. Journal of International Money and Finance 11:304–314

15

Co-Evolutionary Multi-Agent System for Portfolio Optimization

Rafał Dreżewski, Leszek Siwik

Department of Computer Science
AGH University of Science and Technology, Kraków, Poland.
drezew@agh.edu.pl, siwik@agh.edu.pl

Summary. Co-evolutionary techniques for evolutionary algorithms can enhance the adaptive capabilities of evolutionary algorithms and help maintain population diversity. In this chapter the concept and a formal model of an agent-based realization of a predator-prey co-evolutionary algorithm is presented. The resulting system is applied to the problem of effective portfolio building and is compared to classical multi-objective evolutionary algorithms.

15.1 Introduction

Evolutionary Algorithms (EAs) are global search and optimization techniques based on analogies to the Darwinian model of natural evolution (3). EAs have demonstrated efficiency and robustness as global optimization techniques. However, in the case of some problems (for example, multi-modal optimization, multi-objective optimization, and dynamic problems) EAs can show a negative tendency to lose population diversity. Typically, both experiments and formal analysis show that for multi-modal problem landscapes a simple EA will locate a single solution (27). If we are interested in localizing multiple solutions (like in the case of so-called "multi-modal optimization problems"), special techniques should be used. *Niching and speciation methods* for EAs (27) are aimed at forming and maintaining subpopulations (species) throughout the search process, thereby allowing the uncovering of all or most of the basins of attraction of local minima. The problem of loss of population diversity also limits the adaptive capabilities of EAs in dynamic environments.

In evolutionary biology the process of co-evolution is defined as the prolonged mutual interaction between two (or more) species. Examples of co-evolutionary interactions include competition for limited resources, predator-prey interaction, host-parasite interaction, mutualism and commensalism etc. Also, sexual selection results from the co-evolution of female mate choice and male displayed traits, where females evolve to reduce the direct costs associated with mating, and males evolve to attract females to mating (*sexual conflict*) (15). It is acknowledged that co-evolution is responsible for bio-diversity, and may lead to speciation (the formation of new species).

R. Dreżewski and L. Siwik: *Co-Evolutionary Multi-Agent System for Portfolio Optimization*, Studies in Computational Intelligence (SCI) **100**, 271–299 (2008)
www.springerlink.com

In *co-evolutionary algorithms* which are, generally speaking, evolutionary algorithms with co-evolutionary mechanisms embedded, the fitness of each individual depends not only on the quality of its solution to a given problem (like in the case of EAs) but also (or solely) on other individuals' fitness. Such techniques are applicable in the case of problems for which the fitness function formulation is difficult or impossible (like game strategies), or where there is a need to improve the adaptive capabilities of EA, or where there is a need to maintain useful population diversity.

Because many financial and economic decision and optimization problems are multi-modal (there exist many comparable solutions) and / or multi-objective (multiple, possibly conflicting, objectives) different techniques for maintaining population diversity in EAs may be found useful and applicable. In the case of such problems, an intelligent computer system can provide alternative solutions to the decision maker, allowing him to make a final decision based on his experience. In order to do so, evolutionary algorithms must maintain a high level of population diversity—otherwise it simply will not be able to provide many different solutions to the given problem.

Besides the positive effect of maintaining population diversity, co-evolutionary algorithms also provide us with useful analogies between co-evolution, financial markets, and generally speaking market-oriented economic systems. These include for example "arms races" between capitalist enterprises and financial institutions (comparable to predator-prey or host-parasite interactions). Such "arms races" help avoid economic stagnation. These "Red Queen effects" ("It takes all the running you can do, to keep in the same place.") can be observed in market and economic processes. Capitalist enterprises need to continually innovate, merely to "keep in the same place".

Co-evolutionary mechanisms can also be found useful when we are interested in socio-economic modeling and simulations, for example simulation of antagonistic and non-antagonistic interactions between different classes and groups in society (generally speaking problems of social stratification).

In the case of multi-objective optimization problems, which are the main subject of this chapter, the loss of population diversity may mean that the population locates in areas far away from the Pareto frontier or that individuals are located only in selected areas of Pareto frontier. In the case of multi-objective problems with many local Pareto frontiers (defined by Deb in (7)) the loss of population diversity may result in locating only local Pareto frontier instead of the global one.

The notion of an "agent" is now very well established in the area of social science (psychology, sociology, and economy), artificial intelligence, and computer modeling and simulation. According to J. Ferber (13) an agent can be defined as a physical or virtual entity which can act within an environment, can communicate with other agents, tries to realize some goals or optimize its fitness function, possesses some resources, may observe the environment (but only in a restricted way), possesses restricted knowledge about the environment, has some abilities and may offer some services to other agents, may reproduce, acts in the way that leads to the realization of its own goals taking into account the possessed resources, abilities, and knowledge acquired during the observation of the environment and communication with other agents.

A multi-agent system is composed of the following elements (13): an environment, a set of objects situated within the system which can be observed, created, destroyed and modified by agents (which are active entities), a set of agents, a set of relations between objects (including agents), a set of operations which allow agents to observe, create, destroy, "consume", and modify objects, and finally the operators which represent the operations performed by agents and the reaction of the environment. The above features of multi-agent systems makes them an ideal tool for social and economic simulation as they include all the tools necessary for modeling and simulation of different kinds of societies, social structures, modes of production, competing or co-operating enterprises, social mechanisms of conflict and co-operation, and so on.

Evolutionary multi-agent systems (EMAS) are multi-agent systems, in which the population of agents evolves (agents can die, reproduce and compete for limited resources). The model of *co-evolutionary multi-agent system (CoEMAS)* (8) introduces additionally the notions of species, sexes, and interactions between them. CoEMAS allows modeling and simulation of different co-evolutionary interactions, which can serve as the basis for constructing the techniques of maintaining population diversity and improving adaptive capabilities of such systems. CoEMAS systems with sexual selection and host-parasite mechanisms have already been applied with promising results to multi-objective optimization problems (9, 10).

Co-evolutionary multi-agent systems have of course all the advantages and mechanisms of multi-agent systems, which can be used in artificial life modeling and simulations (especially in the area of psychology, sociology and economy). Additionally, we can utilize the evolutionary optimization of agents and co-evolutionary interactions between them. This is a very promising area for future interdisciplinary research including for example, psychological, social and economic simulations which can embed emergent phenomena in society and economy, the problems of social stratification, the role of conflict in the society, antagonistic and non-antagonistic conflicts between classes and groups, the effects of particular economic policy, the role of the state and institutions in economy and society, the role of ideology, its role in the reproduction of relations of production, social power, and stratification, etc.

In the following sections an introduction to multi-objective optimization problems is presented. Then, we concentrate on previous research on techniques for maintaining population diversity in multi-objective evolutionary algorithms. Next, the co-evolutionary multi-agent system with population diversity maintaining technique based on predator-prey interactions is formally described. The presented system is applied to problem of effective portfolio building. Results from the experiments with the CoEMAS system are then compared to other classical evolutionary techniques' results.

15.2 Multi-Objective Optimization

The most natural process of decision making for human beings consists in analyzing many—often contradictory—factors and searching for a compromise among them.

Such decision processes are known as *multi-criteria decision making (MCDM)*. Obviously, human beings are equipped with natural abilities for making multi-criteria decisions. While these natural gifts may be sufficient in everyday life they are not sufficient in more complex technical, business or scientific decision environments.

In such cases a *decision maker*, has to be equipped with appropriate mathematical and computing techniques to make a proper decision. The most common, *MCDM* process is based on an appropriately defined *multi-objective optimization problem (MOOP)*. Following (7)—the *multi-objective optimization problem* in its general form is defined as follows:

$$MOOP \equiv \begin{cases} Minimize/Maximize \;\; f_m(\bar{x}), \quad m = 1,2\ldots,M \\ Subject\ to \qquad g_j(\bar{x}) \geq 0, \;\; j = 1,2\ldots,J \\ \qquad\qquad\qquad h_k(\bar{x}) = 0, \;\; k = 1,2\ldots,K \\ \qquad\qquad x_i^{(L)} \leq x_i \leq x_i^{(U)}, \;\; i = 1,2\ldots,N \end{cases}$$

The set of constraints—both constraint functions (equalities $h_k(\bar{x})$, inequalities $g_j(\bar{x})$) and decision variable bounds (lower bounds $x_i^{(L)}$ and upper bounds $x_i^{(U)}$) — define all possible (feasible) decision alternatives ($\mathscr{D}$).

Because there are many criteria—to indicate which solution is better than another—a specialized ordering relation has to be introduced. To avoid problems with converting minimization to maximization problems (and vice versa of course) an operator $\lhd$ can be defined. Then, notation $\bar{x}_1 \lhd \bar{x}_2$ indicates that solution $\bar{x}_1$ is better than solution $\bar{x}_2$ for particular objective. Now, the crucial concept of Pareto optimality i.e. the so-called dominance relation, can be defined. It is said that solution $\bar{x}_A$ dominates solution $\bar{x}_B$ ($\bar{x}_A \prec \bar{x}_B$) if and only if:

$$\bar{x}_A \prec \bar{x}_B \Leftrightarrow \begin{cases} f_j(\bar{x}_A) \not\rhd f_j(\bar{x}_B) \;\; for \;\; j = 1,2\ldots,M \\ \exists i \in \{1,2,\ldots,M\}: \;\; f_i(\bar{x}_A) \lhd f_i(\bar{x}_B) \end{cases}$$

A solution in the Pareto sense of the multi-objective optimization problem means determining all the non-dominated alternatives from the set $\mathscr{D}$. The Pareto-optimal set consists of globally optimal solutions. However there may also exist locally optimal solutions, which constitute locally non-dominated set (*local Pareto-optimal set*) (7). The set $\mathscr{P}_{local} \subseteq D$ is local Pareto-optimal set if (41):

$$\begin{aligned} &\forall \mathbf{x}^a \in \mathscr{P}_{local} : \nexists \mathbf{x}^b \in D \text{ such that} \\ &\qquad \mathbf{x}^b \succeq \mathbf{x}^a \wedge \left\|\mathbf{x}^b - \mathbf{x}^a\right\| < \varepsilon \wedge \left\|F(\mathbf{x}^b) - F(\mathbf{x}^a)\right\| < \delta \end{aligned}$$

where $\|\cdot\|$ is a distance metric and $\varepsilon > 0$, $\delta > 0$. The set $\mathscr{P} \subseteq D$ is global Pareto-optimal set if (41):

$$\forall \mathbf{x}^a \in \mathscr{P} : \nexists \mathbf{x}^b \in D \text{ such that } \mathbf{x}^b \succeq \mathbf{x}^a \tag{15.1}$$

These locally or globally non-dominated solutions create (in the criteria space) local ($\mathscr{P}\mathscr{F}_{local}$) or global ($\mathscr{P}\mathscr{F}$) Pareto frontiers that can be defined as follows:

$$\mathscr{P}\mathscr{F}_{local} = \left\{\mathbf{y} = F(\mathbf{x}) \in \mathbb{R}^M \;\mid\; \mathbf{x} \in \mathscr{P}_{local}\right\} \tag{15.2a}$$

$$\mathscr{PF} = \{\mathbf{y} = F(\mathbf{x}) \in \mathrm{I\!R}^M \mid \mathbf{x} \in \mathscr{P}\} \tag{15.2b}$$

Multi-objective problems with one global and many local Pareto frontiers are called *multi-modal multi-objective problems* (7).

During the last twenty years of research on evolutionary multi-objective algorithms (EMOAs) many techniques have been proposed. Generally, all of these techniques and algorithms can be classified as elitist, which give the best individuals in the current population the opportunity to be directly carried over to the next generation, or non-elitist ones (7).

15.3 Selected Issues of Maintaining Population Diversity in Evolutionary Multi-Objective Algorithms

In order to maintain useful population diversity and introduce speciation (processes of forming species—subpopulations—located in different areas of solution space) special techniques, like niching mechanisms and co-evolutionary models, are used.

Niching techniques are primarily applied in problems of multi-modal optimization, but they are also used in evolutionary multi-objective algorithms. Such techniques promote useful population diversity and make possible the creation of species located within the basins of attraction of local minima or in different parts of the Pareto frontier. Various niching techniques have been proposed. All these techniques promote niche formation via the modification of the mechanism for selecting individuals for new generation (*crowding model* (26)), the modification of the parent selection mechanism (*fitness sharing technique* (16) or *sexual selection mechanism* (33)), or restricted application of selection and/or recombination mechanisms (by *grouping* individuals into subpopulations (20) or by introducing the environment with some topography, in which the individuals are located (1, 5)).

The fitness-sharing technique was used in Hajela and Lin, which illustrated the use of a weighting method in a genetic algorithm for multi-objective optimization (17). The weights were encoded in genotype and fitness sharing was used in objective space in order to introduce the diversity of the weights. Fitness sharing in the objective space was also used by Fonseca and Fleming in their multi-objective genetic algorithm using a Pareto-based ranking procedure (14). In the niched Pareto genetic algorithm (NPGA) (18) fitness sharing mechanism is used in objective space during the tournament selection in order to decide which individual wins (when the mechanism based on domination relation fails to choose the winner). In the non-dominated sorting genetic algorithm (NSGA) (37) the fitness sharing is performed in decision space, within each set of non-dominated individuals separately, in order to maintain high population diversity. In the strength Pareto evolutionary algorithm (SPEA) (41) a special type of fitness sharing is used in order to maintain diversity. The fitness sharing in SPEA forms niches, not on the basis of distance but, on the basis of Pareto dominance.

As noted above, co-evolutionary techniques for EAs are applicable in cases where the fitness function formulation is difficult (or even impossible). Co-evolutionary algorithms are also applicable in cases when we want to maintain population diversity, introduce speciation, open-ended evolution, "arms races", and improve adaptive capabilities of EAs—especially in dynamic environments. As the result of ongoing research many co-evolutionary models and techniques have been proposed. Generally, each belongs to one of two classes: competitive (30) or co-operative (32). In competitive co-evolution based systems two (or more) individuals compete in a game and their "competitive fitness functions" are calculated based on their relative performance in that game (6). In co-operative co-evolutionary algorithms a problem is decomposed into sub-problems and each sub-problem is then solved by different subpopulation (32). Each individual from the given subpopulation is evaluated within a group of randomly chosen individuals coming from different sub-populations. Its fitness value depends on how well the group solved the problem and on how well the individual assisted in the solution.

Laumanns, Rudolph and Schwefel (22) proposed co-evolutionary algorithm with a predator-prey model and a spatial graph-like structure for multi-objective optimization. Deb introduced a modified algorithm in which predators eliminated prey not only on the basis of one criterion but on the basis of the weighted sum of all criteria (7). Li proposed other modifications to this algorithm (23). The main difference was that both predators and prey were allowed to migrate within the graph. The model of cooperative co-evolution was also applied to multi-objective optimization (19).

Sexual selection resulting from female-male co-evolution is considered to be one of the ecological mechanisms responsible for biodiversity and sympatric speciation (15, 39). All the work on sexual selection mechanism for multi-objective evolutionary algorithms focuses on using this mechanism for maintaining population diversity, so that individuals are evenly distributed over the Pareto frontier. Allenson proposed a genetic algorithm with sexual selection for multi-objective optimization (2). In his technique the number of sexes was the same as the number of criteria of the given problem and individuals of the given sex were evaluated only according to one criterion (associated with their sex). Sex of the child was determined randomly and it replaced the worst individual from its sex. Allenson also introduced sexual selection mechanism. For each individual the partner for reproduction was selected on the basis of individual's preferences coded within its genotype. Lis and Eiben proposed a multi-sexual genetic algorithm (MSGA) for multi-objective optimization (25). They also used one sex for each criterion. If a recombination operator was used during the reproduction (this was decided randomly) then partners for reproduction were chosen from each sex separately with the use of ranking mechanism and the offspring was created with the use of special multi-parent crossover operator. The sex of generated offspring was the same as the sex of the parent that provided most of genes. After the population of next generation was created the group of Pareto-optimal individuals was selected and this group was merged with the group of Pareto-optimal individuals from previous generations. During this phase, dominated individuals were removed from the set of Pareto-optimal individuals. Bonissone and Subbu (4) continued work on Lis and Eiben's algorithm. They proposed additional mechanisms for determining the sex of offspring: random and based on phenotype (child had the sex associated with the criterion for which it had the best fitness).

Co-evolution of species and sexes are biological mechanisms which contribute to biodiversity and sympatric speciation. However these mechanisms are not widely used as a way of maintaining useful genetic diversity in evolutionary algorithms. It seems that co-evolution and sexual selection can be used as a basis for constructing niching and speciation mechanisms (which promote the formation of species located within basins of attraction of different local optima or in different areas of Pareto frontier) but this is still an open issue and the subject of ongoing research.

15.4 Co-Evolutionary Multi-Agent System with Population Diversity Maintaining Mechanism

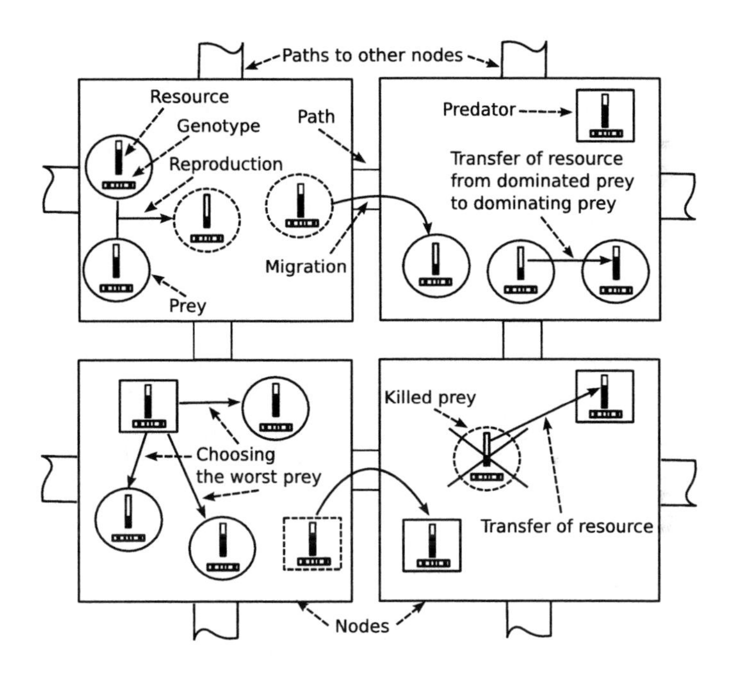

Fig. 15.1. CoEMAS with predator-prey mechanism

The system presented in this section is based on the CoEMAS model—a general model of co-evolution in a multi-agent system (8). The most important component of the population diversity—maintaining mechanism are predator-prey co-evolutionary interactions (see fig. 15.1). The spatial structure of EMAS systems also plays the role of diversity maintaining mechanism but it is rather the mechanism of secondary importance. The first prototypes of the CoEMAS with predator-prey interactions were presented in (11, 12). In the following sections, the system used in experiments is described with the use of ideas, notions, and relations introduced in the general model for co-evolution in a multi-agent system.

15.4.1 CoEMAS

The co-evolutionary multi-agent system with predator-prey interactions (*CoEMAS*) is defined as follows (8):

$$CoEMAS = \langle E, S, \Gamma, \Omega \rangle \qquad (15.3)$$

where E is the environment of the *CoEMAS* system, S is the set of species ($s \in S$) that exist and co-evolve in *CoEMAS*, Γ is the set of resource types (the amount of type γ resource which is possessed by the given element of the system will be denoted by r^{γ}), Ω is the set of information types (the information of type ω, which can be used or possessed by the given element of the system is denoted by i^{ω}). Two information types ($\Omega = \{\omega_1, \omega_2\}$) and one resource type ($\Gamma = \{\gamma\}$) are used. Information of type ω_1 denotes nodes to which agent can migrate. Information of type ω_2 denotes the prey that are located within the particular node in time t.

The selection mechanism is based on the closed circulation of resource within the system. The overall amount of resources is constant. Resources can be possessed by the agents, and transferred from dominated prey to dominating prey, and from prey to predators during killing prey. The environment E is defined in the following way:

$$E = \left\langle T^E, \Gamma^E = \emptyset, \Omega^E = \Omega \right\rangle \qquad (15.4)$$

where T^E is the topography of the environment E, Γ^E is the set of resource types that exist within the environment, and Ω^E is the set of information types that exist within the environment. The topography of the environment $T^E = \langle H, l \rangle$, where H is a directed graph with the cost function c defined ($H = \langle V, B, c \rangle$, V is the set of vertices, B is the set of arches). In the case of the presented system, every node is connected with its four neighbors, which results in the torus-like environment. The $l : A \rightarrow V$ (A is the set of agents) function makes it possible to locate particular agent in the environment space. Vertex v is given by:

$$v = \left\langle A^v, \Gamma^v = \Gamma^E, \Omega^v = \Omega^E \right\rangle \qquad (15.5)$$

A^v is the set of agents that are located within the vertice v. There are two types of information in the vertice. The first one includes all vertices that are connected with the vertice v:

$$i^{\omega_1, v} = \{u : u \in V \wedge \langle v, u \rangle \in B\} \qquad (15.6)$$

The second one includes all agents of species *prey* that are located within the vertice v:

$$i^{\omega_2, v} = \{a^{prey} : a^{prey} \in A^v\} \qquad (15.7)$$

15.4.2 Species

The set of species $S = \{prey, pred\}$. The prey species (*prey*) is defined as follows:

$$prey = \langle A^{prey}, SX^{prey} = \{sx\}, Z^{prey}, C^{prey} \rangle \qquad (15.8)$$

where SX^{prey} is the set of sexes which exist within the *prey* species, Z^{prey} is the set of actions that agents of species *prey* can perform, and C^{prey} is the set of relations of *prey* species with other species that exist in the *CoEMAS*. The set of actions Z^{prey} is defined as follows:

$$Z^{prey} = \{die, get, give, accept, seek, clone, rec, mut, migr\} \tag{15.9}$$

where:

- *die* is the action of death (prey dies when it is out of resources);
- *get* action gets some resource from another a^{prey} agent located within the same node, which is dominated by the agent that performs *get* action or is too close to it in the criteria space;
- *give* action gives some resource to another agent (which performs *get* action);
- *accept* action accepts partner for reproduction when the amount of resource possessed by the prey agent is above the given level;
- *seek* action seeks for another prey agent that is dominated by the prey performing this action or is too close to it in criteria space. This action is also used in order to find the partner for reproduction when the amount of resource is above the given level and agent can reproduce;
- *clone* is the action of producing offspring (parents give some of their resources to the offspring during this action);
- *rec* is the recombination operator (intermediate recombination is used (3));
- *mut* is the mutation operator (mutation with self-adaptation is used (3));
- The *migr* is the action of migrating from one node to another. During this action an agent loses some of its resource.

The set of relations of *prey* species with other species that exist within the system is defined as follows:

$$C^{prey} = \left\{ \xrightarrow{prey,get-}, \xrightarrow{pred,give+} \right\} \tag{15.10a}$$

The first relation models intra species competition for limited resources ("-" denotes that as a result of performing *get* action the fitness of another prey is decreased):

$$\xrightarrow{prey,get-} = \{\langle prey, prey \rangle\} \tag{15.10b}$$

The second one models predator-prey interactions ("+" denotes that when prey gives all its resources to the predator, the predator fitness is increased):

$$\xrightarrow{pred,give+} = \{\langle prey, pred \rangle\} \tag{15.10c}$$

The predator species (*pred*) is defined as follows:

$$pred = \left\langle A^{pred}, SX^{pred} = \{sx\}, Z^{pred}, C^{pred} \right\rangle \tag{15.11}$$

All the symbols used have analogical meaning as in the case of *prey* species—see eq. (15.8). The set of actions Z^{pred} is defined as follows:

$$Z^{pred} = \{seek, get, migr\} \tag{15.12}$$

where:

- The *seek* action allows finding the "worst" (according to the criteria associated with the given predator) prey located within the same node as the predator;
- *get* action gets all resources from the chosen prey,
- *migr* action allows predator to migrate between nodes of the graph H—this results in losing some of the resources.

The set of relations of *pred* species with other species that exist within the system are defined as follows:

$$C^{pred} = \left\{ \xrightarrow{prey,get-} \right\} \tag{15.13a}$$

This relation models predator-prey interactions:

$$\xrightarrow{prey,get-} = \{\langle pred, prey \rangle\} \tag{15.13b}$$

As a result of performing *get* action and taking all resources from selected prey, it dies.

15.4.3 Prey Agents

Agent a of species *prey* ($a \equiv a^{prey}$) is defined as follows:

$$a = \langle gn^a, Z^a = Z^{prey}, \Gamma^a = \Gamma, \Omega^a = \Omega, PR^a \rangle \tag{15.14}$$

The genotype of agent a consists of two vectors (chromosomes): $\mathbf{x}$ of real-coded decision parameters values and σ of standard deviation values, which are used during mutation with self-adaptation. $Z^a = Z^{prey}$ (see eq. (15.9)) is the set of actions which agent a can perform. Γ^a is the set of resource types used by the agent, and Ω^a is the set of information types.

The partially ordered set of profiles includes resource profile (pr_1), reproduction profile (pr_2), interaction profile (pr_3), and migration profile (pr_4):

$$PR^a = \{pr_1, pr_2, pr_3, pr_4\} \tag{15.15a}$$

$$pr_1 \trianglelefteq pr_2 \trianglelefteq pr_3 \trianglelefteq pr_4 \tag{15.15b}$$

Each profile pr is defined as follows:

$$pr = \langle \Gamma^{pr}, \Omega^{pr}, M^{pr}, ST^{pr}, GL^{pr} \rangle \tag{15.16}$$

where Γ^{pr} is the set of resource types used in the pr profile ($\Gamma^{pr} \subseteq \Gamma^a$). Ω^{pr} is the set of information types ($\Omega^{pr} \subseteq \Omega^a$). M^{pr} is the set of informations (the model) which represent the agent's knowledge about the environment and other agents. ST^{pr} is the partially ordered set ($ST^{pr} \equiv \langle ST^{pr}, \preccurlyeq \rangle$) of strategies which agent can use in order to realize the active goal of the given profile. The relation $\preccurlyeq$ is defined as follows:

$$\preccurlyeq = \left\{ \langle st_i, st_j \rangle \in ST^{pr} \times ST^{pr} : \text{ strategy } st_i \text{ has equal or higher priority than strategy } st_j \right\} \tag{15.17}$$

The single strategy $st \in ST^{pr}$ is composed of actions, which performing (in the given order) leads to the realization of a *pr* profile's active goal:

$$st = \langle z_1, z_2, \ldots, z_k \rangle, \quad st \in ST^{pr}, \quad z_i \in Z^a \tag{15.18}$$

GL^{pr} is the partially ordered ($GL^{pr} \equiv \langle GL^{pr}, \preccurlyeq \rangle$) set of goals. The relation $\preccurlyeq$ is defined in the following way:

$$\preccurlyeq = \left\{ \langle gl_i, gl_j \rangle \in GL^{pr} \times GL^{pr} : \text{ the goal } gl_i \text{ has equal or higher priority, than the goal } gl_j \right\} \tag{15.19}$$

Now we can define the $\trianglelefteq$ relation (see eq. (15.15)):

$$\trianglelefteq = \left\{ \langle pr_i, pr_j \rangle \in PR^a \times PR^a : \text{the realization of active goals of the profile } pr_i \text{ has the equal or higher priority than the realization of the active goals of profile } pr_j \right\} \tag{15.20}$$

By "active goal" (denoted by gl^*) we mean the goal gl which should be realized in the given time step.

The Process of Realizing Goals and Choosing the Strategies

The defined above partially ordered sets of profiles (PR^a), goals (GL^{pr}) and strategies (ST^{pr}) are used by agent for selecting the goal and strategy for its realization. The whole process of decision making is realized in the following way:

1) Agent a activates the profile with highest priority ($pr_i \in PR^a$), which has the active goal $gl_j^* \in GL^{pr_i}$.
2) If there is more than one active goal in the set GL^{pr_i} then the goal which has the highest priority is chosen for realization (let us assume that this goal is gl_j^*).
3) Next, such strategy for the realization of the goal gl_j^* is chosen from the set ST^{pr_i} that it has the highest priority, it is possible to realize it in the given time, and it does not contradict with the goals of profiles with the lower priority than profile pr_i (let us assume that this strategy is $st_k \in ST^{pr_i}$).
4) If the realization of the chosen strategy is accomplished with success then the gl_j becomes a non-active goal.
5) Next, again activities from 1) are realized.

The Profiles

The processes of realizing goals and choosing the strategies by prey agent are illustrated in fig. 15.2. The goal of the pr_1 (resource) profile is to keep the amount of resources above the minimal level or to die. In order to realize this goal an agent can use

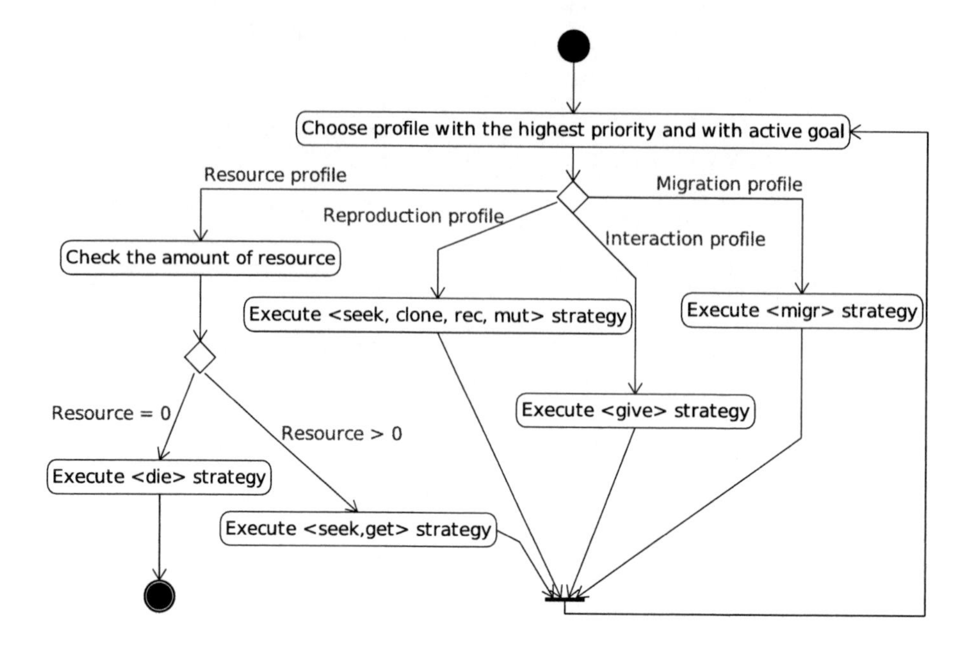

Fig. 15.2. The process of realizing goals and choosing the strategies by a prey agent

the following strategies: $\langle die \rangle$, $\langle seek, get \rangle$. This profile uses the model $M^{pr_1} = \{i^{\omega_2}\}$ (see eq. (15.7)).

The only goal of the pr_2 (reproduction) profile is to reproduce. In order to realize this goal an agent can use the strategy of reproduction: $\langle seek, clone, rec, mut \rangle$. The model is defined in the following way: $M^{pr_2} = \{i^{\omega_2}\}$.

The goal of the pr_3 (interaction) profile is to interact with predators with the use of strategy $\langle give \rangle$.

The goal of the pr_4 (migration) profile is to migrate within the environment. In order to realize this goal the migration strategy is used: $\langle migr \rangle$. The model used is defined as follows: $M^{pr_4} = \{i^{\omega_1}\}$ (see eq. (15.6).) As a result of migrating, the prey loses some resource.

15.4.4 Predator Agents

An agent a of species *pred* is defined analogically to *prey* agent (see eq. (15.14)). There exist two main differences. The genotype of a predator agent consists of information about the criterion associated with the given agent. The set of profiles consists only of two profiles, a resource profile (pr_1), and a migration profile (pr_2): $PR^a = \{pr_1, pr_2\}$, where $pr_1 \trianglelefteq pr_2$.

The processes of realizing goals and choosing the strategies by predator agent are illustrated in fig. 15.3. The goal of the pr_1 (resource) profile is to keep the amount of resource above the minimal level with the use of strategy $\langle seek, get \rangle$. The model

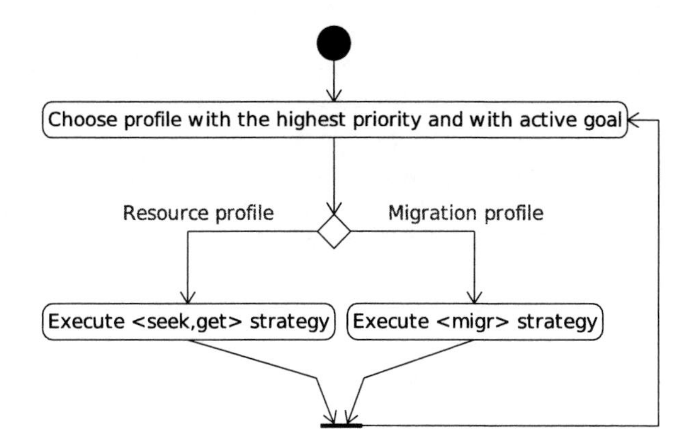

Fig. 15.3. The process of realizing goals and choosing the strategies by a predator agent

used within this profile is defined as follows: $M^{pr_1} = \{i^{\omega_2}\}$. The goal of pr_2 (migration) profile is to migrate within the environment. In order to realize this goal, the migration strategy $\langle migr \rangle$) is used. The model of the environment is defined in the following way: $M^{pr_2} = \{i^{\omega_1}\}$. The realization of the migration strategy results in losing some of the resource possessed by the agent.

15.5 Building an Effective Investment Portfolio

The developed co-evolutionary agent-based system has been tested in (11, 12) using well known benchmark problems such as: the Kursawe problem (21), Laumanns problem (22), and—recently also—the set of Zitzler test problems ZDT1—ZDT6 (41) where solving each next problem algorithm which is being tested has to deal with the more and more difficult and challenging characteristics starting from continuous and convex Pareto frontier, through concave or disconnected problems until multi-objective multi-modal problem (discussion about consequences of concavity, discontinuity or multimodality of the Pareto frontier can be found in (7)).

When analyzing the behavior and characteristics of co-evolutionary computation techniques in general, and agent-based co-evolutionary techniques in particular (especially such approaches as predator-prey, or host-parasite approaches)—it is natural that one of the first associations to such techniques (and obviously one of possible applications of such computational techniques) are financial and investments markets in particular. Entrepreneurs, SMEs, corporations—all of them all the time have to be better, more innovative, cheaper, more effective etc. than the others. That is why, the free market is so dynamic, all the time some enterprises introduce some organizational, financial or technological innovations and the rest of market-game participants has to respond to such changes introducing another innovations, products etc—so, we are witnesses to a continuing arms race. The range of dependencies that can be seen on the market can be pretty wide—from cooperation, through competition until antagonism. As can be read in (31)—such a situation is not the best one for all the market players (the situation when all participants of market game are the "winners" is not possible—always some of them have to lose). There is no doubt however, that (only) thanks to such strong relationships, influences and interactions—the common organizational, technological and economical development and progress are possible—and in that way, extremely desirable phenomenon called "invisible hand of market" by Adam Smith is realized. Of course, the most desirable situation is the perfect competition—but even the most developed markets only bring nearer and nearer to such a situation—mainly because of conditions (third condition in particular) required by "perfect competition". Mentioned three conditions of perfect competition are:

1. There are many buyers and many sellers in particular branch.
2. There are mainly small enterprises in the market.
3. The buyers and the sellers possess the full and perfect knowledge about the market (uncertainty and information asymmetry do not take place).

Fulfilling especially the third condition is very difficult and if so, it is no wonder that both, competitive situation as well as possible interactions and relationships among market-players can vary in a (mentioned above) wide range. It is obvious however, that in a Darwinian world—all activities of each participant of the market game are conformed to one overriding goal—to survive and to gain more and more wealth. From the interactions with another enterprise's point of view it can be realized by: eliminating from the market as many weak rivals as possible and taking over their customers, products, delivery channels etc. (so by being "predator"), by sucking out of another (stronger) enterprise's customers, technologies, products etc. (so by being "parasite"), by supplementing partners' portfolio with additional products, technologies, customers etc.—and vice versa (so by living in symbiosis) etc. etc. It is seen clearly, that one of the most important activity of all market-game participants is co-existence with co-development—and from the computational intelligence point of view we would say—co-evolution. Because (generally speaking of course and under additional conditions) participants of the market game are autonomous entities (from the computational intelligence point of view we would say—agents), they are distributed, they act asynchronously, and they interact with another entities to achieve common goal—prosperity and wealth—in natural way applying co-evolutionary multi agent systems seems to be the perfect approach for modeling such phenomenons and environments. This is the first motivation of our experiments. But why "building effective portfolio". Well, we are working and perceiving co-evolutionary multi agent systems not only as modeling techniques but also as computational techniques. When we finished preliminary tests with benchmark problems—we wanted to run such systems against real—because of above stated motivation market-oriented—problems. Additionally, our goal was running one of proposed approaches against challenging, combinatorial, well defined and well-known multi-objective optimization problem where arm race interactions can be observed to test our predator-prey co-evolutionary multi-agent system. Building effective portfolio seems to be the perfect candidate test problem fulfilling all above mentioned requirements.

We know now why building an effective portfolio has been selected as a test problem. Unfortunately, the next problem arises. How should this problem be formally defined. Practically, there are some well known models describing the construction of an effective portfolio i.e. Modern Portfolio Theory (MPT), the one-factor Sharpe model, CAPM—Capital Asset Pricing Model, APT—Arbitrage Pricing Theory, Post Modern Portfolio Theory (PMPT) and so on. The starting point for modern considerations about building efficient portfolio is the Nobel prize winner Harry Markowitz' Modern Portfolio Theory (MPT) (1952) (28, 29), or its extension proposed in 1958 by James Tobin (38)—consisting in introducing risk-free assets to the model. Those research resulted in defining for the first time formal foundations of *risk—rate of return* investing decision making and defining so-called Capital Market Line (CML) with the following equation:

$$R = R_f + (\frac{R_M - R_f}{S_M}) * S \tag{15.21}$$

where:

R - rate of return;
S - standard deviation;
R_M - rate of return of market portfolio;
S_M - standard deviation of market portfolio.

It turned out, after introducing to the model the risk-free assets that effective portfolio(s) belong(s) to the segment of the above defined line. Markowitz' portfolio analysis (and its expansion by J. Tobin) makes some strong and important assumptions. The most significant are:

- The goal of investor is to maximize his wealth;
- Investors are characterized by risk aversion;
- Investing horizon is the same for all investors;
- A suitable measure of risk level is the standard deviation of rates of return from the "average" rate of return of the market portfolio;
- Investors make a decision on the basis of only rates of return and standard deviation;
- No taxes and transaction costs are assumed.

The above theory lays the foundations of modern capital investments. The Capital Asset Pricing Model (CAPM) was proposed by J. Traynor (40), J. Lintner (24), J. Mossin and formalized by W. Sharpe (36)—and it was based of course on previous work of Markowitz and his MPT theory. This time, in this model, not only the Capital Market Line but also the so-called Security Market Line (SML) is crucial. The SML is defined as follows:

$$R_i = R_f + \beta_i * (R_M - R_f) \tag{15.22}$$

where $R_M - R_f$ - is the so-called premium for risk. CAPM is the most popular effective-portfolio building model. One may ask why this very model was not used during our tests. Well, mainly because of its complexity and shortcomings. On the basis of the critique of CAPM (Roll's Critique)—Arbitrage Pricing Theory (APT) was proposed by Stephen A. Ross in mid-1970s (35). APT can be described using the following equation:

$$R_i = a_i + b_{i1}F_1 + b_{i2}F_2 + \cdots + b_{im}Fm + e_i \quad (15.23)$$

APT assumes that rates of return depends on m factors. Coefficient b_{ij} indicates how sensitive the rate of return on the R_ith asset is to changes in F_{ij} factor. The APT model makes several assumptions:

- The number of F factors used in the model can not be higher than the number of assets and—more importantly
- In the market we have the perfect competition.

In the 1990s, Post Modern Portfolio Theory was proposed. The notion of PMPT was used for the first time probably by B.M. Rom and K.W. Ferguson in 1993 (34). The PMPT model is based on three main assumptions and observations:

1. The risk measure in MPT is symmetrical—i.e. returns above average or target rates of returns are as risky as returns below this value—whereas from investor's point of view—really risky are returns below the target (minimum or average) value, and the return above those values are perceived as a risk premium. It was observed and stated already by Markowitz, confirmed by Sharpe and another researchers—but mainly because of computational difficulties PMT was based on symmetrical measure.
2. A much better measure of risk (downside risk in this case) is continuous formula rather than its discrete version.
3. A much better index of rate of return is the Sortino ratio rather than the Sharpe ratio.

Taking all the pros and cons into consideration—because it was the first attempt at applying the proposed algorithm to building an effective portfolio—we decided to use the one-factor Sharpe model during our experiments. This model is discussed below. The meaning of symbols used in the definitions below, are as follows:

p - the number of assets in the portfolio;
n - the number of periods taken into consideration (the number of rates of return taken to the model);
α_i, β_i - coefficients of the equations;
ω_i - percentage participation of i-th asset in the portfolio;
e_i - random component of the equation;
R_{it} - the rate of return in the period t;
R_{mt} - the rate of return of market index in period t;
R_m - the rate of return of market index;
R_i - the rate of return of the i-th asset;
R_p - the rate of return of the portfolio;
s_i^2 - the variance of the i-th asset;

${s_{e_i}}^2$ - the variance of the random index of the i-th asset;
${s_{e_p}}^2$ - the variance of the portfolio;
$\overline{R_i}$ - arithmetic mean of rate of return of the i-th asset;
$\overline{R_m}$ - arithmetic mean of rate of return of market index;

The algorithm (based on the one-factor Sharpe model) of computing the expected risk level and, generally speaking, income expectation related to the portfolio of p assets is as follows:

1. Compute the arithmetic means on the basis of rate of returns;
2. Compute the value of α coefficient:

$$\alpha_i = \overline{R_i} - \beta_i \overline{R_m} \quad (15.24)$$

3. Compute the value of β coefficient:

$$\beta_i = \frac{\sum_{t=1}^{n}(R_{it} - \overline{R_i})(R_{mt} - \overline{R_m})}{\sum_{t=1}^{n}(R_{mt} - \overline{R_m})^2} \quad (15.25)$$

4. Compute the expected rate of return of asset i:

$$R_i = \alpha_i + \beta_i R_m + e_i \quad (15.26)$$

5. Compute the variance of random index:

$${s_{e_i}}^2 = \frac{\sum_{t=1}^{n}(R_{it} - \alpha_i - \beta_i R_m)^2}{n-1} \quad (15.27)$$

6. Compute the variance of market index:

$${s_m}^2 = \frac{\sum_{t=1}^{n}(R_{mt} - \overline{R_m})^2}{n-1} \quad (15.28)$$

7. Compute the risk level of the investing portfolio:

$$\beta_p = \sum_{i=1}^{p}(\omega_i \beta_i) \quad (15.29)$$

$${s_{e_p}}^2 = \sum_{i=1}^{p}(\omega_i^2 {s_{e_i}}^2) \quad (15.30)$$

$$risk = \beta_p^2 {s_m}^2 + {s_{e_p}}^2 \quad (15.31)$$

8. Compute the portfolio rate of return:

$$R_p = \sum_{i=1}^{p}(\omega_i R_i) \quad (15.32)$$

The goal of the optimization is to maximize the portfolio rate of return and minimize the portfolio risk level. The task consists in determining values of decision variables $\omega_1 \ldots \omega_p$ forming the vector

$$\Omega = [\omega_1, \ldots, \omega_p]^T \tag{15.33}$$

where $0\% \leq \omega_i \leq 100\%$ and $\sum_{i=1}^{p} \omega_i = 100\%$ and $i = 1 \ldots p$ and which is the subject of minimization with respect of two criteria:

$$F = [R_p(\Omega) * (-1), risk(\Omega)]^T \tag{15.34}$$

Model Pareto frontiers for two cases (portfolios consisting of three and seventeen stocks set), which are the subject of analysis in the following section, are presented in fig. 15.4.

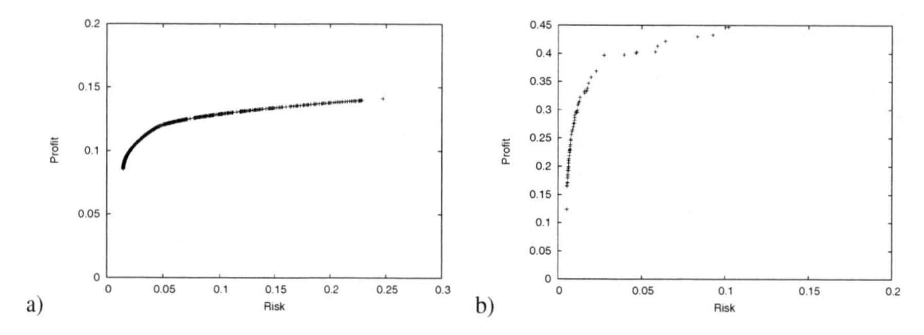

Fig. 15.4. *Building of effective portfolio*: visualization of the model Pareto frontier obtained using utter review method for a) three and b) seventeen stocks set

15.6 Results of Experiments

In this section the results of the experiments are presented. The results obtained by the proposed system are compared with the results obtained by a "classical" (i.e. non agent-based) predator-prey evolutionary strategy (PPES) (22) and another "classical" evolutionary algorithm for multi-objective optimization: the niched pareto genetic algorithm (NPGA) (41). In order to more deeply analyze the results obtained by compared algorithms—values of HV and HVR metrics (their definitions can be found in (7)) are also presented. In the case of optimizing an investing portfolio, each individual in the prey population is represented as a p-dimensional vector. Each dimension represents the percentage participation of the i-th ($i \in 1 \ldots p$) share in the whole portfolio. In this section a summary of two single experiments will be presented.

In the experiments, Warsaw Stock Exchange quotations from 1/1/2003 until 31/12/2005 were used. Simultaneously, the portfolio consists of the following three (experiment I) or seventeen (experiment II) stocks quoted on the Warsaw Stock Exchange: in experiment I: RAFAKO, PONARFEH, PKOBP, in experiment II: KREDYTB, COMPLAND, BETACOM, GRAJEWO, KRUK, COMARCH, ATM, HANDLOWY, BZWBK, HYDROBUD, BORYSZEW, ARKSTEEL, BRE, KGHM, GANT, PROKOM, BPHPBK. The WIG20 is used as the market index proxy.

In fig. 15.5 and fig. 15.6 there are presented Pareto frontiers obtained using CoEMAS, NPGA and PPES algorithms after 1, 300, 500, 700, 900 and 1000 steps in experiment I. As one may notice in this case, the CoEMAS-based frontier is more numerous (especially initially) than NPGA-based and as numerous as the PPES-based one. Unfortunately, the diversity of population in CoEMAS approach is visibly worse than that of the NPGA or PPES-based frontiers. What is more, with time the tendency of CoEMAS-based solver for focusing solutions around small part of the whole Pareto frontier is more and more distinct.

A similar situation can be observed in fig. 15.7 and fig. 15.8 presenting Pareto frontiers obtained by CoEMAS, NPGA and PPES—but this time the portfolio that is being optimized consists of 17 shares. Also this time CoEMAS-based frontier is quite numerous and quite close to the model Pareto frontier but the tendency for focusing solutions around only selected part(s) of the whole frontier is very distinct.

In section 13.1 of this chapter, it was mentioned that the CoEMAS system has been tested using such non-combinatorial test problems as the Kursawe problem, Laumanns problem and the set of Zitzler problems. In these benchmark tests, CoEMAS was definitely the better alternative than NPGA or PPES and the question appears why in the case of building an effective portfolio the situation is the different one. Well, the explanation is as follows. With time, the population of agents consists mainly of mutually non-dominated agents and the situation that during the meetings agent dominates the opponent is more and more unlikely. If so, also gathering additional units of resources is more and more unlikely. Because agents pays in each step with resource for its life—with time the level of its energy falls below the death level and in the consequence it has to be removed from the system. The solution of such a situation is introducing to the system mechanisms similar to the elitism—where elitist agents for instance can migrate to the special island and can not be removed from the system as long as they are non-dominated. As it can be observed in this study, mentioned phenomenon is much more dangerous during solving combinatorial problems, since meeting dominated agents is more unlikely (as simulation time passes) than in the case of continuous problems like Kursawe, Laumanns or Zitzler problems.

In this chapter we decided to present not only Pareto frontiers but also portfolio composition. It is of course impossible in the course of this chapter to present the consecutive portfolios proposed by all non-dominated solutions—that is why we decided to choose average non-dominated solution in first step and then to follow during consecutive steps solutions proposed by this very solution (or its descendant(s)). Such hypothetical non-dominated average portfolios for experiments I and II are presented in fig. 15.9 and in fig. 15.10 respectively (in fig. 15.10 shares are presented from left to right in the order in which they were mentioned above). Generally, it can be said that during experiment I—the average solution proposed by CoEMAS system is a kind of balanced portfolio (percentage share of all three stocks are quite similar, but the percentage participation in the whole portfolio of PONAR is the lowest one and finally PKOBP became the most important "ingredient" of the analyzed portfolio), whereas during experiment II there are more important stocks (with given assumptions and parameters of course)—i.e. HANDLOWY, HYDROBUD, ARKSTEEL.

15.7 Conclusions and Future Work

Co-evolutionary techniques for evolutionary algorithms are applicable in the case of problems for which it is difficult or impossible to formulate an explicit fitness function, where there is need for maintaining useful population diversity, for forming species located in the basins of attraction of different local optima, or when introducing open-ended evolution and "arms races". Such techniques are also widely used in artificial life simulations. Although co-evolutionary algorithms have been the subject of intensive research, their application to multi-modal and multi-objective optimization is still an open problem and many research questions remain unanswered.

In this chapter, the agent-based realization of a predator-prey model within the more general framework of a *co-evolutionary multi-agent system* has been presented. The system was tested against a hard, real-life, multi-objective problem (effective portfolio building) and then compared to two classical multi-objective evolutionary algorithms: PPES and NPGA. CoEMAS was able to form more numerous frontier, however a negative tendency to lose population diversity during the experiment was observed. PPES and NPGA were able to form better-dispersed Pareto frontiers. When the portfolio composition is considered, the average solution proposed by the CoEMAS system was a balanced portfolio when it was composed of three stocks and portfolio with dominating elements when it was composed of seventeen stocks. The results of experiments with effective portfolio building problem show that more research is needed on co-evolutionary mechanisms for maintaining population diversity used in CoEMAS, especially when we want to stably maintain diversity of solutions. It seems that the proposed predator-prey mechanism for evolutionary multi-agent systems may be very useful in the case of hard dynamic and multi-modal multi-objective problems (as defined by Deb (7)).

Future work will include more detailed analysis of the proposed co-evolutionary mechanisms, especially focused on problems of stable maintaining population diversity. The most important part of this research will be the introduction of the elitism mechanism for decentralized agent-based evolutionary computation. Also the comparison of CoEMAS to other classical multi-objective evolutionary algorithms with the use of hard multi-modal multi-objective test problems, and the application of other co-evolutionary mechanisms like symbiosis (co-operative coevolution) are included in future plans. Another, and very important, area of research on co-evolutionary multi-agent systems will be modeling and simulation of socio-economic mechanisms and emergent phenomena.

References

[1] Adamidis P (1998) Parallel evolutionary algorithms: A review. In: Proceedings of the 4th Hellenic-European Conference on Computer Mathematics and its Applications (HERCMA 1998), Athens, Greece

[2] Allenson R (1992) Genetic algorithms with gender for multi-function optimisation. Tech. Rep. EPCC-SS92-01, Edinburgh Parallel Computing Centre, Edinburgh, Scotland

[3] Bäck T, Fogel D, Michalewicz Z (eds) (1997) Handbook of Evolutionary Computation. IOP Publishing and Oxford University Press

[4] Bonissone S, Subbu R (2003) Exploring the pareto frontier using multi-sexual evolutionary algorithms: An application to a flexible manufacturing problem. Tech. Rep. 2003GRC083, GE Global Research

[5] Cantú-Paz E (1998) A survey of parallel genetic algorithms. Calculateurs Paralleles, Reseaux et Systems Repartis 10(2):141–171

[6] Darwen PJ, Yao X (1995) On evolving robust strategies for iterated prisoner's dilemma. In: Yao X (ed) Process in Evolutionary Computation, AI'93 and AI'94 Workshops on Evolutionary Computation, Selected Papers, Springer-Verlag, LNCS, vol 956

[7] Deb K (2001) Multi-Objective Optimization using Evolutionary Algorithms. John Wiley & Sons

[8] Dreżewski R (2003) A model of co-evolution in multi-agent system. In: Mařík V, Müller J, Pěchouček M (eds) Multi-Agent Systems and Applications III, Springer-Verlag, Berlin, Heidelberg, LNCS, vol 2691, pp 314–323, `http://galaxy.uci.agh.edu.pl/~drezew/publications/drezewski2003model.pdf`

[9] Dreżewski R, Siwik L (2006) Co-evolutionary multi-agent system with sexual selection mechanism for multi-objective optimization. In: Proceedings of the IEEE World Congress on Computational Intelligence (WCCI 2006), IEEE

[10] Dreżewski R, Siwik L (2006) Multi-objective optimization using co-evolutionary multi-agent system with host-parasite mechanism. In: Alexandrov VN, van Albada GD, Sloot PMA, Dongarra J (eds) Computational Science — ICCS 2006, Springer-Verlag, Berlin, Heidelberg, LNCS, vol 3993, pp 871–878

[11] Dreżewski R, Siwik L (2007) Co-evolutionary multi-agent system with predator-prey mechanism for multi-objective optimization. In: Beliczynski B, Dzielinski A, Iwanowski M, Ribeiro B (eds) Adaptive and Natural Computing Algorithms, Springer-Verlag, LNCS, vol 4431, pp 67–76

[12] Dreżewski R, Siwik L (2007) Multi-objective optimization technique based on co-evolutionary interactions in multi-agent system. In: Giacobini M (ed) Applications of Evolutionary Computing, Springer-Verlag, LNCS, vol 4448, pp 179–188

[13] Ferber J (1999) Multi-Agent Systems: An Introduction to Distributed Artificial Intelligence. Addison-Wesley

[14] Fonseca C, Fleming P (1993) Genetic algorithms for multiobjective optimization: Formulation, discussion and generalization. In: Genetic Algorithms:

Proceedings of the Fifth International Conference, Morgan Kaufmann, pp 416–423, citeseer.ist.psu.edu/fonseca93genetic.html

[15] Gavrilets S (2003) Models of speciation: what have we learned in 40 years? Evolution 57(10):2197–2215

[16] Goldberg DE, Richardson J (1987) Genetic algorithms with sharing for multimodal function optimization. In: Grefenstette JJ (ed) Proceedings of the 2nd International Conference on Genetic Algorithms, Lawrence Erlbaum Associates, pp 41–49

[17] Hajela P, Lin C (1992) Genetic search strategies in multicriterion optimal design. In: Structural optimization 4, pp 99–107

[18] Horn J, Nafpliotis N, Goldberg DE (1994) A niched pareto genetic algorithm for multiobjective optimization. In: Proceedings of the First IEEE Conference on Evolutionary Computation, IEEE Service Center, Piscataway, New Jersey, pp 82–87

[19] Iorio A, Li X (2004) A cooperative coevolutionary multiobjective algorithm using non-dominated sorting. In: Deb K, Poli R, Banzhaf W, Beyer HG, Burke EK, Darwen PJ, Dasgupta D, Floreano D, Foster JA, Harman M, Holland O, Lanzi PL, Spector L, Tettamanzi A, Thierens D, Tyrrell AM (eds) Genetic and Evolutionary Computation - GECCO 2004, Springer-Verlag, LNCS, vol 3102-3103, pp 537–548

[20] Jelasity M, Dombi J (1998) GAS, a concept of modeling species in genetic algorithms. Artificial Intelligence 99:1–19

[21] Kursawe F (1991) A variant of evolution strategies for vector optimization. In: Schwefel H, Manner R (eds) Parallel Problem Solving from Nature. 1st Workshop, PPSN I, Springer-Verlag, Berlin, Germany, vol 496, pp 193–197, citeseer.ist.psu.edu/kursawe91variant.html

[22] Laumanns M, Rudolph G, Schwefel HP (1998) A spatial predator-prey approach to multi-objective optimization: A preliminary study. In: Eiben AE, Bäck T, Schoenauer M, Schwefel HP (eds) Parallel Problem Solving from Nature — PPSN V, Springer-Verlag, LNCS, vol 1498

[23] Li X (2003) A real-coded predator-prey genetic algorithm for multiobjective optimization. In: Fonseca CM, Fleming PJ, Zitzler E, Deb K, Thiele L (eds) Evolutionary Multi-Criterion Optimization, Second International Conference (EMO 2003), Proceedings, Springer-Verlag, LNCS, vol 2632

[24] Lintner J (1965) The valuation of risk assets and the selection of risky investments in stock portfolios and capital budgets. Review of Economics and Statistics 47:13–37

[25] Lis J, Eiben AE (1996) A multi-sexual genetic algorithm for multiobjective optimization. In: Fukuda T, Furuhashi T (eds) Proceedings of the Third IEEE Conference on Evolutionary Computation, IEEE Press, Piscataway NJ, pp 59–64

[26] Mahfoud SW (1992) Crowding and preselection revisited. In: Männer R, Manderick B (eds) Parallel Problem Solving from Nature — PPSN-II, Elsevier, Amsterdam, pp 27–36, illiGAL report No. 92004

[27] Mahfoud SW (1995) Niching methods for genetic algorithms. PhD thesis, University of Illinois at Urbana-Champaign, Urbana, IL, USA, `citeseer.nj.nec.com/mahfoud95niching.html`
[28] Markowitz H (1952) Portfolio selection. Journal of Finance 7(1):77–91
[29] Markowitz H (1999) The early history of portfolio theory: 1600-1960. Financial Analysts Journal 55(4):5–16
[30] Paredis J (1995) Coevolutionary computation. Artificial Life 2(4):355–375
[31] Paterson R (2002) Compendium of Banking Terms in Polish and English. Foundation of accountancy development in Poland, Warsaw
[32] Potter MA, De Jong KA (2000) Cooperative coevolution: An architecture for evolving coadapted subcomponents. Evolutionary Computation 8(1):1–29
[33] Ratford M, Tuson AL, Thompson H (1997) An investigation of sexual selection as a mechanism for obtaining multiple distinct solutions. Tech. Rep. 879, Department of Artificial Intelligence, University of Edinburgh
[34] Rom B, Ferguson K (1993) Post-modern portfolio theory comes of age. The Journal of Investing Winter
[35] Ross S (1976) The arbitrage theory of capital asset pricing. Journal of Economic Theory 13(3)
[36] Sharpe WF (1964) Capital asset prices: A theory of market equilibrium under conditions of risk. Journal of Finance 19(3):425–442
[37] Srinivas N, Deb K (1994) Multiobjective optimization using nondominated sorting in genetic algorithms. Evolutionary Computation 2(3):221–248
[38] Tobin J (1958) Liquidity preference as behavior towards risk. The Review of Economic Studies 25:65–86
[39] Todd PM, Miller GF (1997) Biodiversity through sexual selection. In: Ch G Langton, et al (ed) Artificial Life V: Proceedings of the Fifth Int. Workshop on the Synthesis and Simulation of Living Systems, Bradford Books, pp 289–299
[40] Treynor J (1961) Towards a theory of market value of risky assets. unpublished manuscript
[41] Zitzler E (1999) Evolutionary algorithms for multiobjective optimization: methods and applications. PhD thesis, Swiss Federal Institute of Technology, Zurich

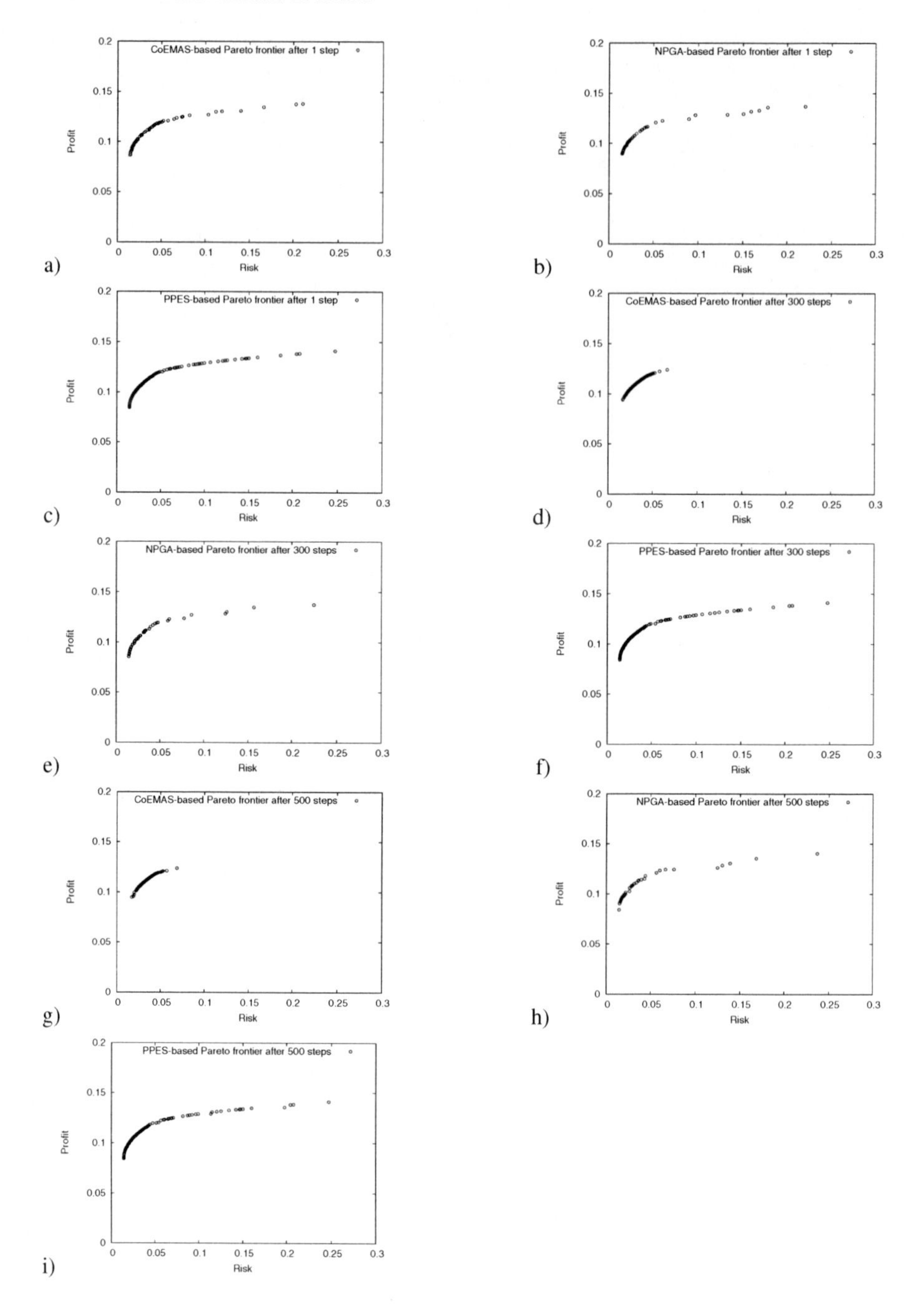

Fig. 15.5. Pareto frontier approximations after 1 (a,b,c), 300 (d,e,f) and 500 (g,h,i) steps obtained by CoEMAS, PPES, and NPGA for building effective portfolio consisting of 3 stocks

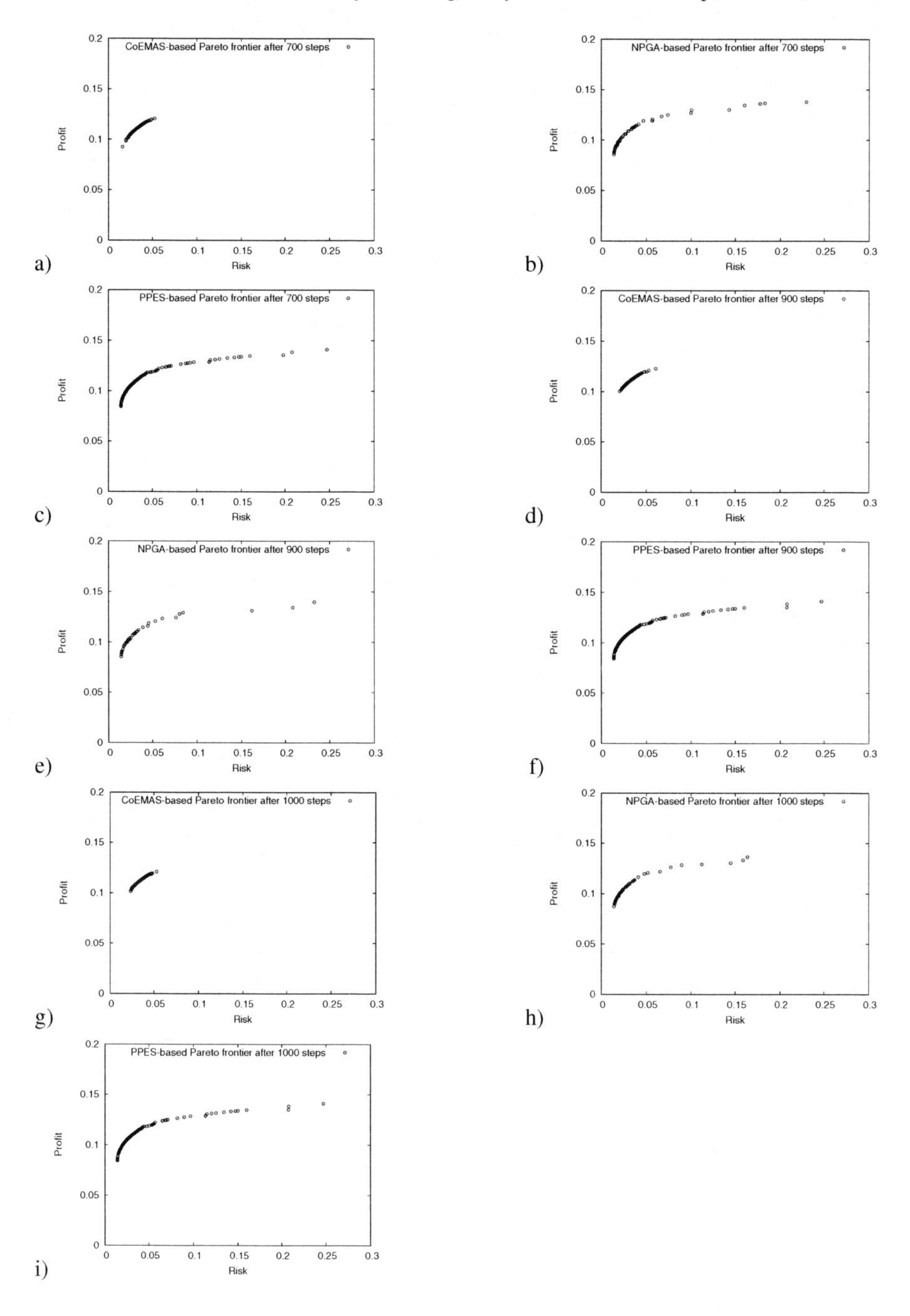

Fig. 15.6. Pareto frontier approximations after 700 (a,b,c), 900 (d,e,f), 1000 (g,h,i) steps obtained by CoEMAS, PPES, and NPGA for building effective portfolio consisting of 3 stocks

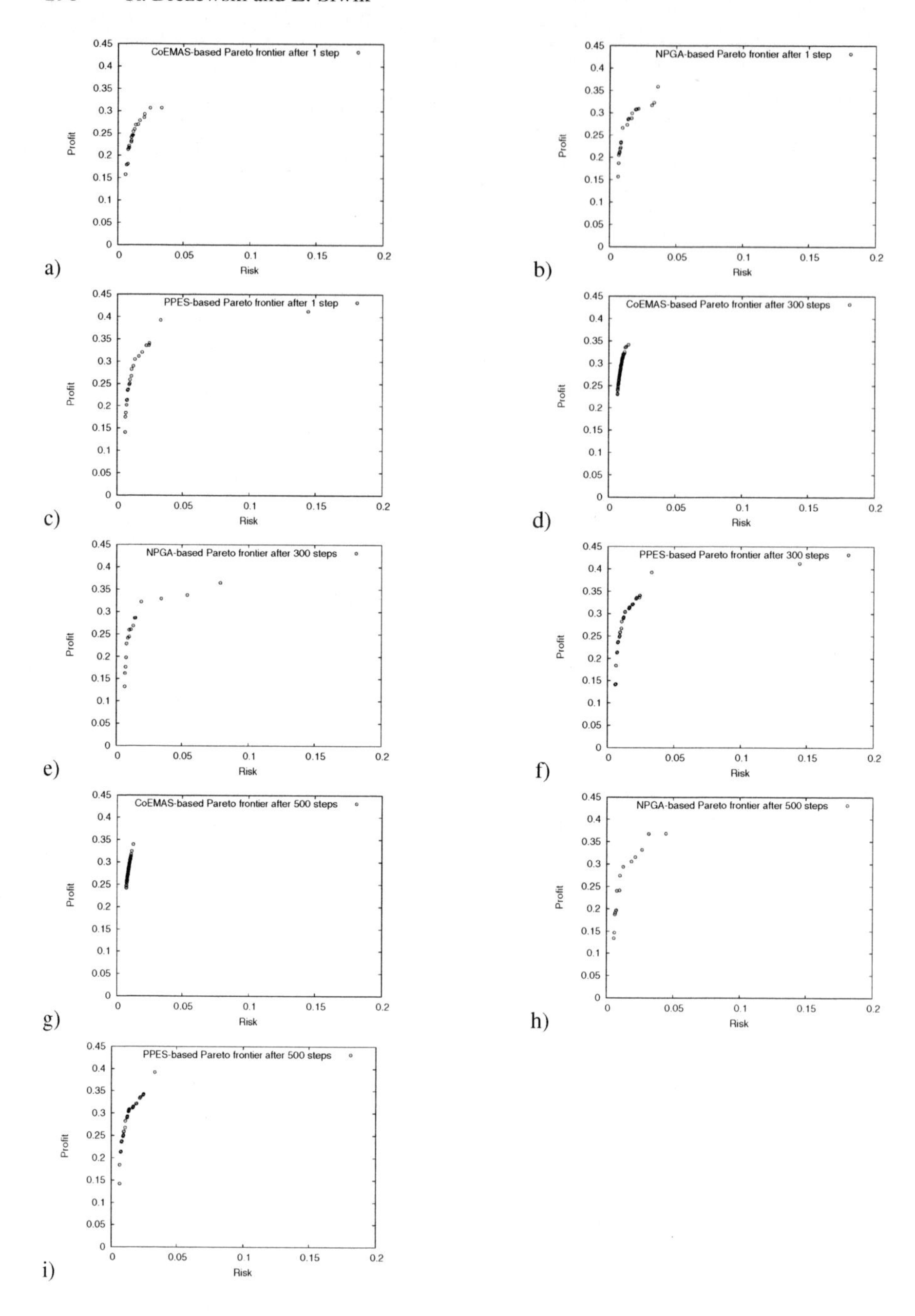

Fig. 15.7. Pareto frontier approximations after 1 (a,b,c), 300 (d,e,f) and 500 (g,h,i) steps obtained by CoEMAS, PPES, and NPGA for building effective portfolio consisting of 17 stocks

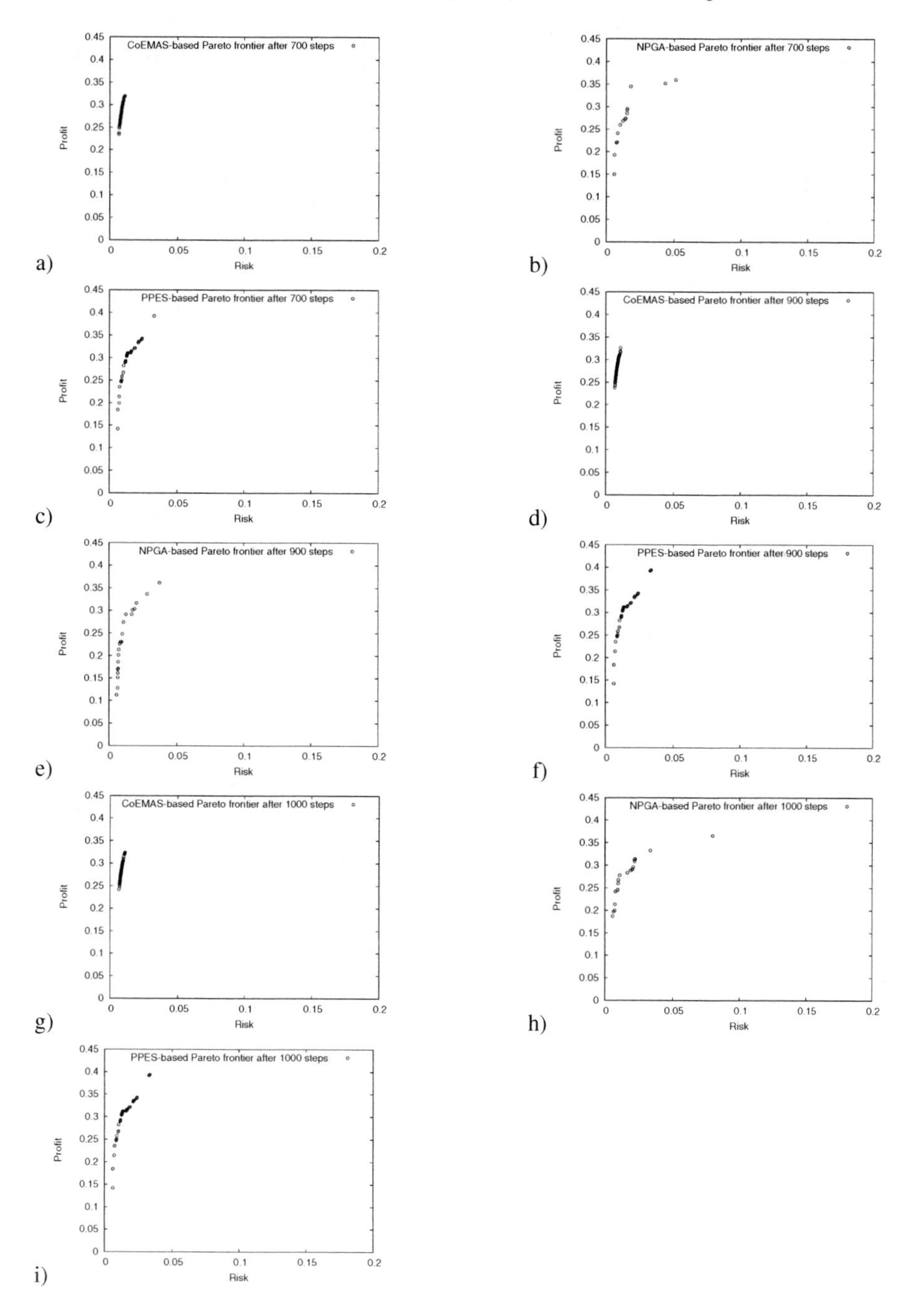

Fig. 15.8. Pareto frontier approximations after 700 (a,b,c), 900 (d,e,f), 1000 (g,h,i) steps obtained by CoEMAS, PPES, and NPGA for building effective portfolio consisting of 17 stocks

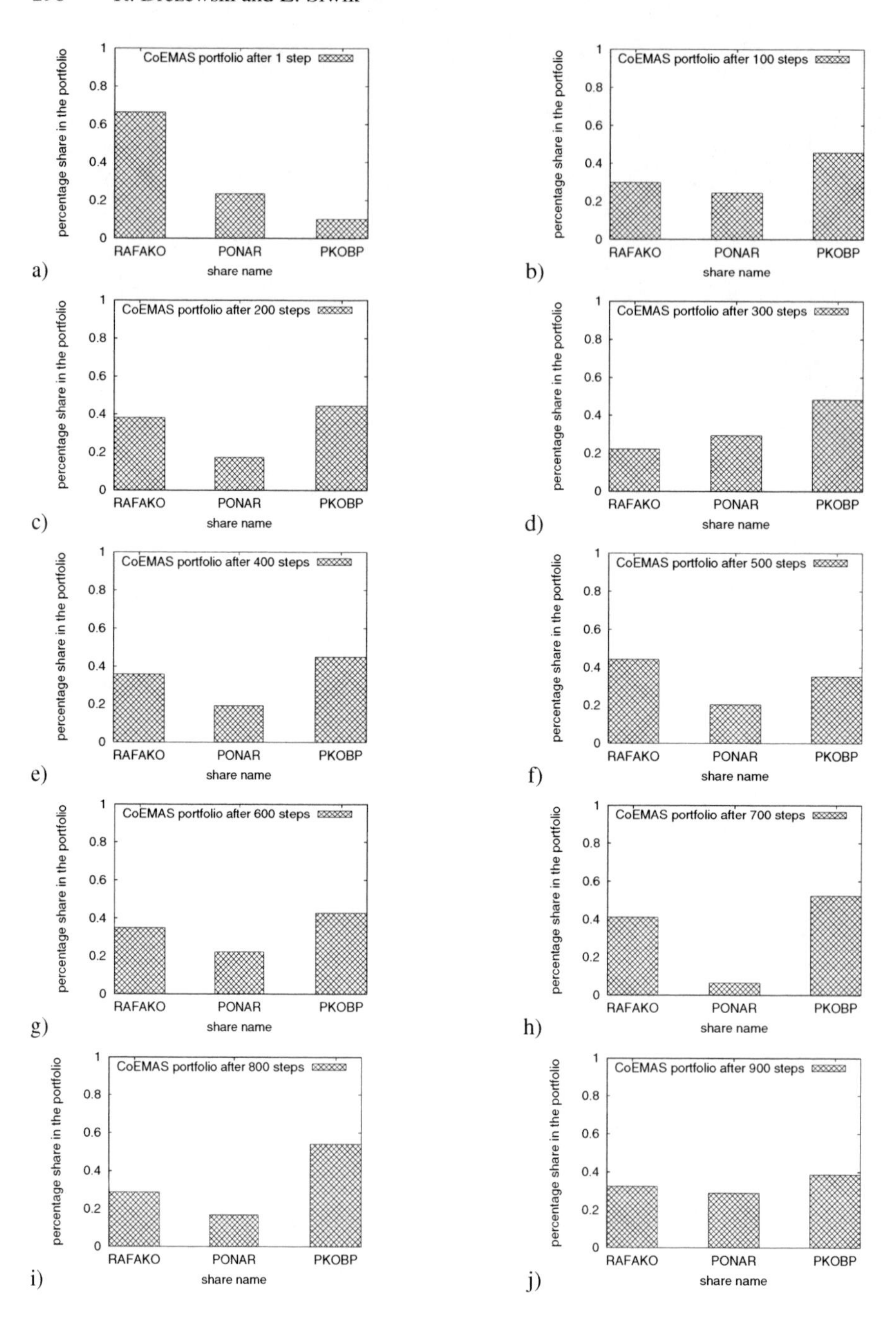

Fig. 15.9. Effective portfolio consisting of three stocks proposed by CoEMAS in consecutive steps

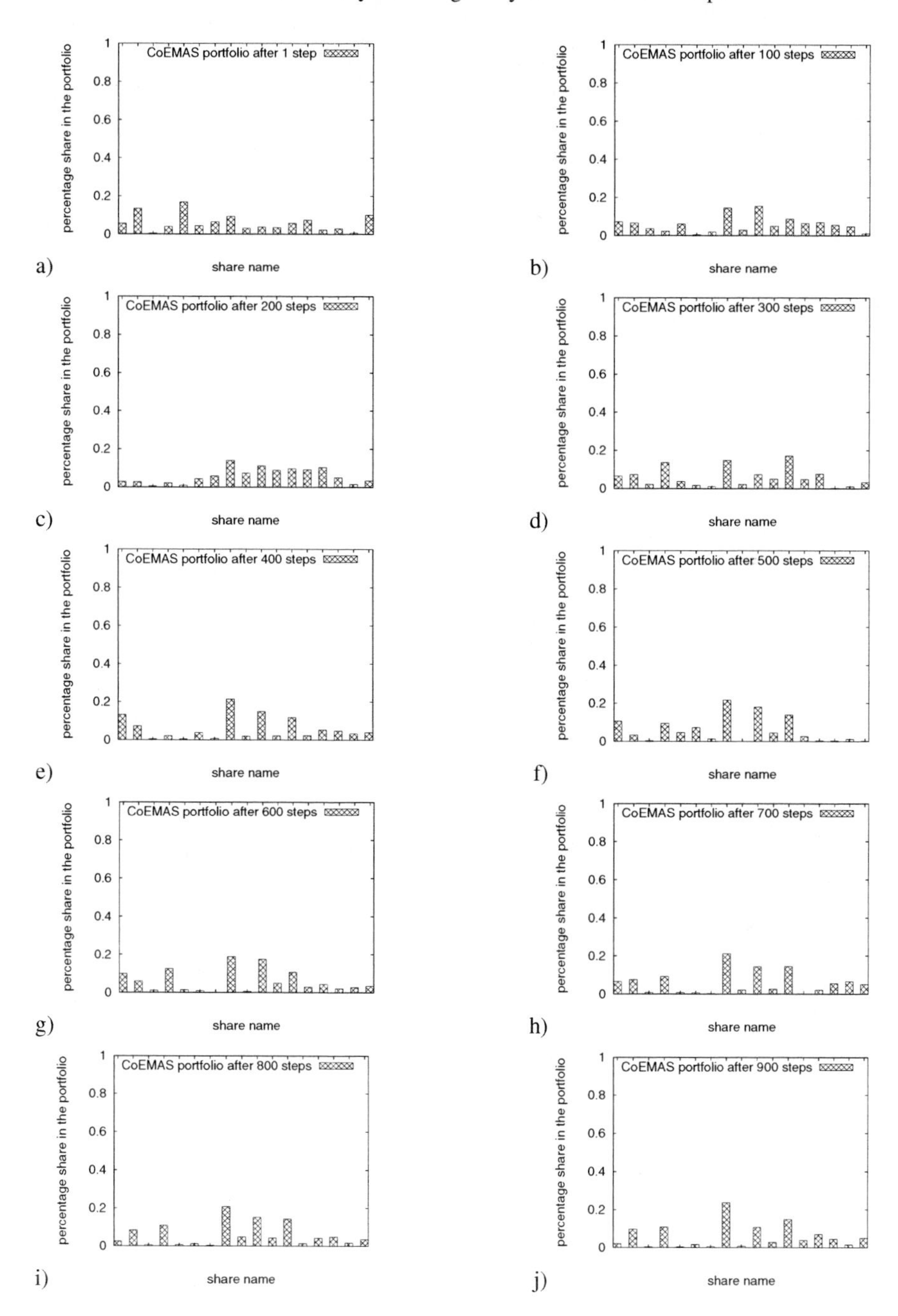

Fig. 15.10. Effective portfolio consisting of seventeen stocks proposed by CoEMAS in consecutive steps

Index